Introduction to

LIVESTOCK AND COMPANION ANIMALS

Second Edition

Jasper S. Lee
Agricultural Educator
Demorest, Georgia

Jim Hutter
Southwest Missouri State University
Springfield, Missouri

Rick Rudd
University of Florida
Gainesville, Florida

Lyle Westrom
University of Minnesota
Crookston, Minnesota

Austin M. Bull
North Carolina A&T University
Greensboro, North Carolina

Chris Embry Mohr
Coordinator, Partners for Agricultural Literacy
Bloomington, Illinois

Jody Pollock
Michigan State University
East Lansing, Michigan

Introduction to
LIVESTOCK AND COMPANION ANIMALS

IST

AgriScience and Technology Series

Jasper S. Lee, Ph.D.
Series Editor

Interstate Publishers, Inc.
Danville, Illinois

Introduction to

LIVESTOCK AND COMPANION ANIMALS

Second Edition

Printed in the United States of America

Library of Congress Catalog Card No. 99-71698

ISBN 0-8134-3177-8

2 3 4 5 6 7 8 9 10   05 04 03 02 01 00

Order from

Interstate Publishers, Inc.

510 North Vermilion Street
P.O. Box 50
Danville, IL 61834-0050

Phone: (800) 843-4774
Fax: (217) 446-9706
Email: info-ipp@IPPINC.com

PREFACE

The animal industry is changing. The changes are radical and have had major impacts on education in animal care and production. A traditional book is no longer appropriate. You will see that this book is different. It has a modern, colorful approach to the animal industry.

Introduction to Livestock and Companion Animals has been revised to serve as an outstanding introductory textbook in animal science. It is about far more than cattle, swine, and chickens—though these are included. It integrates a multitude of perspectives on animal education. Basics of biology are included as the foundation for animal education. Practical connections are addressed in each chapter. Biotechnology has been expanded throughout. Competencies from several states in biological science and animal courses are covered in the book. The basic science areas have been expanded so that science credit can be offered when this book is used.

The content on companion animals has been expanded considerably in the Second Edition. This was done by integrating more content on companion animals in the basic science chapters as well as adding chapters on dogs, cats, and birds, rodents, and reptiles. Animal well-being is stressed throughout.

Traditional areas of animal agriculture are continued in this book. A broad spectrum of animal science is covered. Beef, swine, sheep, dairy, horse, and poultry production are presented. These are followed by chapters on aquatic animals; draft animals; dogs; cats; birds, rodents, and reptiles; and service, laboratory, and exotic animals. Chapters on entrepreneurship and careers conclude the book. The content is designed so that students with limited backgrounds in agriculture can be successful learners.

The book has a student-friendly approach. The chapter design and content are based on research with students and teachers. A clear and consistent approach is used. Line art and photographs clarify and enhance important concepts. The writing level is appropriate for the intended audience. Type face and size have been selected to enhance interest and readability. The content is presented in a systematic manner.

Introduction to Livestock and Companion Animals has so much to offer! Some of the major features implemented in this textbook include:

- full-color design
- hundreds of photographs and drawings
- science-based content
- modern animal subjects
- high appeal to users
- carefully written copy

The book is comfortably a part of the AgriScience and Technology (*IST*) Series produced by Interstate Publishers, Inc.

Take one quick look through this book. You will readily see that it is the introductory animal education book for you!

Jasper S. Lee
Series Editor

ACKNOWLEDGEMENTS

The authors of *Introduction to Livestock and Companion Animals* are indebted to many individuals for help in preparing the book. Their assistance made it possible for the authors to realize the goal of a revised and redirected book on animal production. Some are acknowledged here; others elsewhere in the book.

Associations and businesses acknowledged are:

- Allflex USA, Inc., Dallas/Fort Worth Airport, Texas
- Elite Genetics, Waukon, Iowa
- Harris Farms, Coalinga, California
- National Hereford Hog Record Association, Flandreau, South Dakota
- Wagner Farms, Newton, Mississippi
- Embrex Incorporated, Research Triangle Park, North Carolina
- Gone Fishin' Exotic Pets, Starkville, Mississippi
- Washington Department of Fisheries Hatchery, Wilapa, Washington
- National Swine Registry, West Lafayette, Indiana
- Paso Fino Horse Association, Inc., Plant City, Florida
- Maddox Dairy, Fresno, California
- BarLink Ranch, Madras, Oregon
- College of Veterinary Medicine, Texas A&M University
- Progressive Farmer Magazine, Birmingham, Alabama
- American Breeders Service, DeForest, Wisconsin
- Texas Department of Agriculture, Austin, Texas
- Ankole-Watusi International Registry, Spring Hill, Kansas

- American Rambouillet Sheep Breeders Association, San Angelo, Texas
- The American Donkey and Mule Society, Denton, Texas
- Agriculture Department and Bull Test Station, Hinds Community College, Raymond, Mississippi
- Red Bluff Ranch, Thermopolis, Wyoming
- Luck-E-G Nubians, Blanchard, Idaho
- Boston Market Restaurants, North Carolina
- Gold Kist, Research Triangle Park, North Carolina
- Lucky Three Ranch, Inc., Loveland, Colorado

High schools recognized for their assistance are:

- Park High School, Livingston, Montana
- Alvirne High School, Hudson, New Hampshire
- South Bend High School, South Bend, Washington
- South Panola High School, Batesville, Mississippi
- Brevard High School, Brevard, North Carolina
- Chino Valley High School, Chino Valley, Arizona
- Newton Vocational Center, Newton, Mississippi
- Franklin County High School, Carnesville, Georgia
- Paxton-Buckley-Loda High School, Paxton, Illinois
- Clovis High School, Clovis, California

Special acknowledgment goes to manuscript reviewers Robert C. Albin, Texas Tech University; Richard Joerger, Utah State University; and Tim Chamblee, Mississippi State University.

The assistance of students, staff, and friends of Piedmont College, Demorest, Georgia, is acknowledged. Their help with illustrations and design enhanced the appeal of book.

Each author has several individuals who provided considerable assistance, and these are hereby acknowledged.

The staff of Interstate Publishers, Inc., is thanked for their diligent efforts in producing the book. President Vernie Thomas is acknowledged for his enthusiasm and commitment to agricultural education. Marketing Vice President Dan Pentony is acknowledged for his support in all phases of preparing the book. Kim Romine is acknowledged for her work in the creative design of the book. Jane Weller and Mary Carter are thanked for editorial and graphics assistance.

CONTENTS

Preface v

Acknowledgments vi

Part One—Animals and People

Chapter 1   The Importance of Animals   1

Part Two—Principles of Animal Science

Chapter 2   Animals as Organisms   25

Chapter 3   Animal Nutrition and Feeding   46

Chapter 4   Animal Health   81

Chapter 5   Animal Reproduction 111

Chapter 6   Animal Biotechnology 139

Part Three—Food Animal Technology

Chapter 7   Beef Cattle Production 165

Chapter 8   Swine Production 203

Chapter 9   Sheep and Goat Production 235

Chapter 10   Dairy Production 265

Chapter 11   Poultry Production 301

Chapter 12   Aquatic Animals 333

Part Four—Pleasure and Draft Animal Technology

Chapter 13   Horses 367

Chapter 14   Draft Animals 393

Chapter 15   Dogs 415

Chapter 16   Cats 439

Chapter 17   Birds, Rodents, and Reptiles 457

Chapter 18   Ornamental Fish 485

Chapter 19   Wildlife Animals 511

Part Five—Service, Laboratory, and Exotic Animal Technology

Chapter 20   Service and Safety Animals 527

Chapter 21   Scientific and Laboratory Animals 543

Chapter 22   Exotic Animals 559

Part Six—The Animal Industry

Chapter 23   Entrepreneurship in the Animal Industry 575

Chapter 24   Careers in the Animal Industry 589

Appendixes 609

Glossary 613

Bibliography 625

Index 627

THE IMPORTANCE OF ANIMALS

Animals make the lives of people more enjoyable. They have done so for hundreds of years and continue to be important today. Just think of the ways animals are important to you!

Animals and their products affect everyone. In fact, animal products are likely close by as you are reading this book! Look around the room. Is anyone wearing leather shoes or wool clothing? What is on the lunch menu for today? How many students in your class have a pet at home? All of these uses make animals important to us.

Animal care and production form a large industry. The industry varies from a cattle ranch in the West, to the pet store in a big city, to the veterinarian in a suburb. You may keep animals or want to have a career in the animal industry. Learning about animals is exciting!

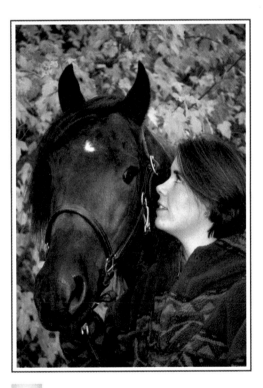

1-1. People often develop close bonds with animals. (Courtesy, Paso Fino Horse Association, Inc.)

1

OBJECTIVES

This chapter provides basic information on the animal industry. It has the following objectives:

1 Explain the meaning of the animal industry

2 Identify ways animals help people

3 Identify trends in the animal industry

4 Trace the domestication of animals

5 Describe the role of science in animal production

6 List the names of common animals based on sex classification and age

7 Explain the meaning of animal well-being

8 Identify environmental concerns associated with animal production

TERMS

animal domestication
animal industry
animal marketing
animal processing
animal production
animal selection
animal services
animal supplies
animal well-being
botany
by-product
castration
companion animal
dairy cattle
environment
euthanasia

factory farm
health
life science
livestock
meat animal
mohair
neutering
nutrition
poultry
ration
reproduction
science
spaying
veterinary medicine
zoology

THE ANIMAL INDUSTRY

The *animal industry* is all of the activities involved in raising animals and meeting the needs of people for animals and animal products. It includes many important steps that make modern production possible. The animal industry assures people of quality products. It uses practices that provide proper care of animals. The animal industry is made of three major areas: production, supplies and services, and marketing and processing.

ANIMAL PRODUCTION

Animal production is raising animals for food, companionship, and other uses. How animals are produced has changed. Livestock are raised on farms or ranches, often on a large scale. Other animals may be raised in homes, aquaria, pens, and other small-scale facilities. A good example is the use of kennel facilities to raise dogs.

Livestock

Livestock are mammals produced on farms and ranches for food and other purposes. They are often given special care to assure good health and growth. Common livestock include cattle, swine (hogs), sheep, horses, and goats. Animals now being produced more widely include llamas, bison, and elk.

1-2. Beef cattle convert pasture grasses to meat for human food.

1-3. A well-trained companion animal, such as this boxer, provides its owner with relaxation and personal satisfaction.

Companion Animals

A *companion animal* is an animal that is used to provide humans with fun and friendship through close association. Companion animals are often called pets. Common companion animals include dogs, cats, fish, rabbits, hamsters, gerbils, ferrets, and snakes. Some companion animals also provide service such as dogs for people who are visually impaired.

Poultry

Poultry is a group of fowl (birds) that are raised for use as food and other products such as feathers. Both meat and eggs are produced by poultry. Special kinds of feed, facilities, and care are used to keep poultry healthy and growing. Common kinds of poultry are chickens, turkeys, ducks, and geese. New species that are being farmed include quail, ostrich, and emu.

 1-4. White Leg-horn chickens are bred for egg production. (Courtesy, Agricultural Research Service, USDA)

Other Animals

Many kinds of animals are produced. Food fish, laboratory animals, and exotic animals are included. These animals help people in many ways. Many of the animals are covered in other chapters of this book.

1-5. Bison once roamed much of the Western United States. Today, most are in protected herds to promote their well-being.

SUPPLIES AND SERVICES

Keeping and producing animals requires supplies and services. With livestock, some are produced on the farm or ranch, such as hay for feed. Most supplies come from off the farm.

Animal supplies are the inputs needed in animal production. They are typically obtained from an agribusiness that provides animal supplies. These include feed mills and factories that make the equipment needed with animals. Dealers make them available in the local area. Common supplies include manufactured feed, cages and feeding equipment, animal medicines, fencing material, and hauling equipment.

Animal services include the help that producers need to raise animals efficiently. Veterinarians help assure good animal health. Farriers are people skilled in putting shoes on horses. People skilled in shearing sheep (clipping off the wool) provide the service to wool producers. People in animal services may go from one ranch or kennel to another providing help.

1-6. Companion animal owners often seek the services of animal care facilities.

MARKETING AND PROCESSING

When animals or their products are ready for the consumer, they are marketed. *Animal marketing* is the process of moving animals and their products from the producer to the consumer. It includes many functions. With companion animals, it may include getting animals into stores for customers to buy. With food animals, it involves preparing the animal products for consumption.

Preparing animal products usually involves processing. *Animal processing* is preparing animals or their products for the consumer. What is done varies with the product. Pork is an example. The swine are slaughtered and butchered into the desired pieces of meat. These cuts are delivered to your local grocery store. The store may place the meat in attractive packages for displaying in a refrigerated case. Milk and egg products go through a similar marketing process.

All animal products are carefully packaged and stored to assure good quality. Meat, eggs, and milk are inspected by trained government officials and industry personnel. Products that are not wholesome are rejected and are not used for food.

1-7. Lamb carcasses are being chilled in a processing plant.

WAYS ANIMALS HELP PEOPLE

Most people are well aware of common animal products. The hamburger, turkey sandwich, and fish filet are foods that we enjoy eating. Animals are used for more than food. They fill important roles as pets, in sports, and to do work.

In processing, very little of an animal is wasted. Many parts of an animal's carcass that were formerly wasted are now used. Many medicines are made from certain animal organs and tissues.

These materials are known as by-products. A *by-product* is a product made from the parts of animals that are not used for food, such as bone meal for fertilizer.

FOOD

Animals provide many foods that people enjoy. These foods are high in nutrients and help us live healthy lives. Without good food, people are not healthy and do not grow as they should.

Foods from animals primarily include meat, milk, and eggs. Some animals give us more than one kind of food product, such as chickens that provide both meat and eggs.

We get meat from animals we raise and from wild animals. A *meat animal* is an animal raised for its meat. The animals that we raise for food include cattle, fish, turkeys, chickens, swine, and sheep. Horses are used as meat animals in some countries.

Wild animals used for food include deer, rabbit, quail, and fish. Wild animals are not included as livestock. They are known as game or wildlife. Game is wildlife hunted for food or other uses. The meat from animals is known by different names, such as beef from cattle and pork from swine.

Milk is primarily from cattle. Cattle specially grown to produce milk are known as *dairy cattle*. Goats and a few other animals are sometimes milked.

1-8. Grilling-out hamburgers is a popular summertime activity.

Table 1-1. Common Food Products From Animals

Animal	Food Item Name*
Meat	
Cattle	
Younger than 3 months	Veal
Older animals	Beef
Swine (hogs)	Pork
Sheep	
Young (less than 1 year)	Lamb
Older (older than 1 year)	Mutton
Goat	Goat mutton
Chicken	
Young (less than 12 weeks)	Broiler
Neutered young male	Capon
Old hen	Hen
Turkey	Turkey
Fish	Fish
Milk	
Dairy cattle	Milk
Goats	Goat's milk
Eggs	
Chickens	Eggs
Fish	Caviar

*Variety meat is the general name for food made from organs and glands of different meat animals. Potted meat animals and vienna sausage are examples.

Eggs are primarily from chickens. A few other species may produce eggs for human food, including guineas and ducks. In some instances, people enjoy fish eggs, known as caviar.

CLOTHING

Clothing is made from many different animal by-products. In addition, some animals are raised specifically for products to make clothing. These include mink for their fur and certain breeds of sheep raised primarily for wool.

1-9. Clothing is made from fibers and animal skins.

Clothing may be made from animal skin (hide) or hair that grows on the skin. Bones, antlers, and other animal parts may be used. The skin of animals is sometimes known as leather. Cattle, sheep, kangaroo, and snakes are a few animals whose skin is used in making belts, shoes, jackets, and other items. The hair or fleece from several different animals is used to make clothing, such as wool from sheep and fur from mink. A special quality clothing is made from **mohair**, a product of angora goats. Feathers from geese may be used in stuffing pillows.

The most valuable animal product used in clothing is the pearl. Often used in jewelry, the pearl is formed by an oyster when sand or another foreign body gets inside its shell. Some oysters are grown to produce pearls, known as cultured pearls.

COMPANIONSHIP AND PLEASURE

Companionship and pleasure are important to people. A companion animal is an animal that provides benefits that help people

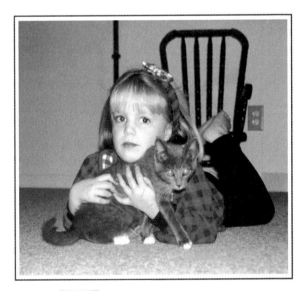

1-10. People of all ages enjoy pets.

enjoy life. Examples include dogs, cats, and ornamental fish. Horses and ponies are examples of animals kept as pets that are too big to stay inside a home. Some animals are used for sporting events, such as horse racing and goat roping. Many different kinds of pleasure animals are found in North America. Dog and horse racing, rodeoing, and similar events are recreational events that involve animals.

SERVICE

Some animals help people. Service animals are animals that assist people in living and work. They are used in many ways and may be given special training. Dogs may be used to lead visually impaired people, herd sheep, or guard property. Some animals, such as rats and monkeys, are carefully used in laboratories for finding new medicines to cure human diseases. Dogs in police canine units are helpful against crime. Horses and mules are used as draft animals. They are strong and have the ability to pull heavy loads.

TRENDS IN THE ANIMAL INDUSTRY

Having the animals and animal products we want involves a number of different activities. How we produce animals has changed greatly. Animal producers use supplies and services to help animals grow. Once food animals and their products have been produced, they must be made available to the consumer.

LARGER PRODUCERS

The nature of animal production has become larger and more specialized. There are fewer producers today in North America. A half century ago, many people raised animals for food purposes. Many homes had a few chickens and a cow or so for milk. Nearly every farm was diversified and kept a few animals in addition to crops. Times have changed!

Today, livestock are raised on specialized farms or ranches that are often quite large. New approaches are being used to grow them. Only one kind of animal may be grown. It is becoming unusual to find a large-scale farm that has more than one kind of animal.

FACTORY FARMS

Factory farms are emerging. A *factory farm* is a large-scale operation that efficiently produces animals in carefully controlled environments. The animals are often confined in houses specially designed to provide for their well-being.

Hog production has rapidly moved to a factory farm approach. One producer may have a large operation that produces far more hogs than several smaller farms.

Many corporations grow for a specific market. You will likely see some of these products in your local grocery store, such as a brand label on poultry.

SOME SMALL PRODUCERS REMAIN

Many small producers are still involved with animals. Service animals, such as draft horses, are often raised on small farms. Companion animals are raised by pet growers or individuals who raise them for a hobby. Animals used in laboratories are often carefully raised under close supervision so that quality standards are met.

DEMAND FOR QUALITY

Consumers are increasingly concerned about the quality of animal products. Greater emphasis is on providing products that are wholesome. Consumers are also concerned about the conditions under which animals are raised. Some consumers are willing to pay more for animals that have not been produced in factory-type situations.

1-11. This modern hog facility at the University of Tennessee–Martin bears little resemblance to pig pens of the past.

DOMESTICATION OF ANIMALS

Animals have not always been tame. Some are still wild today. Raising animals began hundreds of years ago. People found that it was easier to tend them than to hunt them in the wild. People also found that they could do a better job of raising the animals if they understood their needs.

Animal domestication is taking animals from nature and raising them in a controlled environment. It involves taming wild animals and helping them adapt to being raised by humans.

Many years ago, humans were hunters and gatherers. They hunted for meat and eggs and gathered food from wild plants. Humans then began to have gardens near their home and raise animals for convenience.

Domestication is important in the history of how human societies developed. Domestication began thousands of years ago with goats, donkeys, and a few other animals. People are striving to domesticate new species today,

Connection

ANIMAL TALK

These two horses are communicating. What do you think they are saying?

Animals do not "talk" but many of them communicate. They share information with members of their species. The communication may involve both sounds and visual signals.

Many animals can make and detect vocal sounds. Horses, for example, snort, neigh, nicker, squeal, make mating calls, and "talk" with foals. All of these are part of the daily life of a horse.

Animals signal with ears, tails, eyes, nostrils, and other body parts. How the parts are moved will quickly tell an experienced animal handler what is on the mind of some animals, such as one that is about to kick or bite. Signs also tell about the health condition of an animal.

Animals may not "talk" but they can communicate!

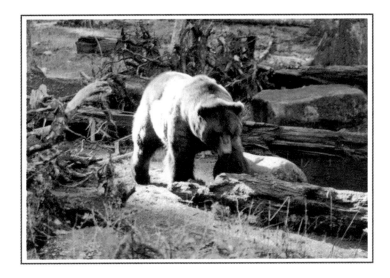

1-12. Some animals have not been domesticated, such as this grizzly bear in a Washington wildlife park. (The grizzly is listed as a threatened species by the U.S. Fish and Wildlife Service.)

such as catfish, alligators, and elk. The dog was one of the first companion animals to be domesticated.

SCIENCE AND TECHNOLOGY IN LIVESTOCK AND POULTRY PRODUCTION

Successfully producing animals requires education. The animal industry depends on people educated in many areas to produce the animals we need. Science and technology play significant roles in education about animal production.

SCIENCE AREAS

Science is knowledge about the world we live in. It deals with facts and relationships among the facts. People who have an education in science are better animal producers. Animal science is the special application of basic science in the production of animals.

Areas of life science are very important in successful animal production. *Life science* is about living things, and includes *zoology*, which is the study of animals. The study includes the structure and functions of animals and their parts. It also includes studying how to keep them healthy by meeting their needs.

Another important life science that is indirectly related to animals is botany. *Botany* is the study of plants. Since most animals eat plants or plant products, knowing how to produce plants is essential.

Technology is the use of science in the work and lives of people. Producers of animals use technology about animals in their work. Every time a disease is treated or an animal fed, some form of technology is used. The technology could be in the design of equipment used, the way the treatment is given, or in other ways.

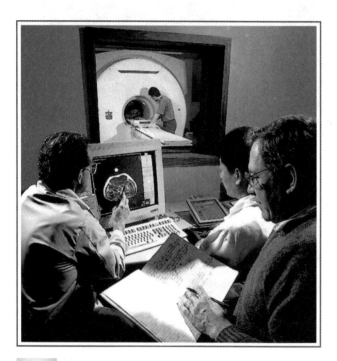

1-13. Technology improves animal production practices. (Courtesy, Agricultural Research Service, USDA)

AGRISCIENCE

Agriscience is the use of science in producing animals and plants. The plants include flowers, grain crops, and forests. Some of the plants are used as food for animals. Many areas of science are involved. Producers need education in basic areas of science. Thousands of high schools have classes in agriscience to help people learn about animal production.

ANIMAL SCIENCE AND TECHNOLOGY

Science and technology are important in the production of animals. Information from many areas is used in the animal industry. The most widely used information areas are in animal selection, nutrition, animal health, reproduction, and environment.

Animal Selection

Animal selection is choosing animals to achieve desired goals. Animal production begins with selection. Animals and humans share many similarities. Animals are individuals, each with its own personality, body type, and

1-14. Hogs have been improved so that pork is sometimes referred to as "white meat."

rate of growth. Carefully select the animals to raise. Choose those that fill a need in some unique way. After all, you want a demand to exist for the animal you raise!

Breeding animals are carefully selected for their traits. This is because scientists know that the offspring of animals resemble their parents. If parents have a certain trait, their offspring are highly like to have the trait.

Animal Nutrition

A major role of food animal production is to convert feed into human food. Some animals, such as cattle and sheep, can use bulky feed materials for growth and maintenance of their bodies. Humans have no use for grass

1-15. Dorset rams being conditioned for breeding season. This requires good nutrition, positive health care, and attention to reproductive potential.

and hay as a raw material, but can consume the "hay" when it is converted to beef by a steer.

Nutrition is the study of the kind of food an animal needs and how it is used by the animal. Much research has been done to learn the nutritional needs of animals. Several areas of science are involved in helping animal producers provide for the needs of animals.

There are many considerations that must be taken into account when preparing a ration for an animal. A *ration* is an animal's diet. It is what the animal consumes each day. The importance of having a proper ration is illustrated here with a lamb. Lambs require certain nutrients in order to grow. Their digestive systems can only handle certain kinds of feed. A lamb will not grow well on a diet of only hay.

Animal Health

Health is the condition in which an animal is free of disease and all parts of the animal are functioning properly. Many different diseases can attack animals if not prevented. Animal producers can take steps to keep animals healthy.

Veterinary medicine is the area that deals with the health and well-being of animals. Veterinarians are people trained in methods of preventing and treating disease and other animal health problems. They often have assistants and technicians who can provide valuable help.

Producers know that it is important to prevent disease. Once an animal gets a disease, it is often very difficult to successfully provide a treatment. The best approach is to keep animals healthy and avoid disease problems. Regular vaccination can be used to prevent some diseases. Having a good place for animals to live helps to keep them healthy. Owners of companion animals should also keep their animals healthy through preventative medicine and care.

Animal Reproduction

Animals must reproduce themselves. *Reproduction* is the process by which offspring are produced. People who raise animals need to understand the reproduction process. Large numbers of offspring are often desired.

Efficiently having new animals is essential for most animal producers. For example, a hog producer wants piglets to be born at a certain time and in good health condition. The producer also wants the females to bare as many pigs as practical. This requires careful attention to the reproductive processes in hogs.

People who have companion animals and other animals may want to limit reproduction. They may want to keep animals from producing offspring. This can be achieved with knowledge of the reproduction process.

Animal Environment

A good environment is essential for an animal to survive and grow properly. *Environment* is the surroundings of an animal. Some animals require shelter; others can survive in outside conditions without shelter. Producers need to know the environment required and manage animals within the environment.

A good example of the importance of environment is with chicken production. Small, newly-hatched chicks must be in a protected house. The house must have controlled temperature and humidity. Equipment in the house must provide food and water. The chicks must also be protected from predators and other hazards.

ANIMAL CLASSIFICATION

Animals are classified and named by specie, age, and sexual condition. This approach is important in correctly identifying animals. The common name of an animal is the most popular way of identification. For example, we refer to canines as "dogs."

The age of an animal gives another distinction. Mature animals and immature animals have different names. For example, a puppy is a young dog.

The sexual condition of an animal is important. Animals are either male or female. Some animals are altered sexually, known as *neutering*. Neutering is done to keep an animal from reproducing or to get other desired traits. For example, neutering may increase the rate of growth and quality of meat a beef animal produces. In males, neutering involves removing the testicles, which is known as *castration*. With females, neutering is known as spaying. *Spaying* is removing the ovaries or cutting the fallopian tubes so eggs cannot enter the uterus where they may be fertilized and develop as an embryo and fetus.

Tables 1-2 and 1-3 summarize important terms for animals based on their species, age, and sexual condition.

Table 1-2. Classification of Livestock Based on Age and Sexual Condition

Category	Cattle	Goat	Sheep	Hog	Chicken	Turkey
Name	Bovine	Caprine	Ovine	Porcine	Galine	Meleagris
Mature Male	Bull	Buck	Ram	Boar	Rooster	Tom
Mature Female	Cow	Doe	Ewe	Sow	Hen	Hen
Young Male	Bull	Buck Kid	Ram Lamb	Shoat	Cockerel	Tom Poult
Young Female	Heifer	Doeling	Ewe Lamb	Gilt	Pullet	Hen Poult
Altered Male	Steer	Wether	Wether	Barrow	Capon	—
Altered Female	Spayed	Spayed	Spayed	Spayed	—	—
Newborn	Calf	Kid	Lamb	Pig	Chick	Poult
Group	Herd	Herd	Flock	Drove	Flock	Flock

Table 1-3. Classification of Selected Companion Animals Based on Age and Sexual Condition

Category	Dog	Cat	Horse	Guinea Pig	Rabbit
Name	Canine	Feline	Equine	Cavy	Lagomorph
Mature Male	Stud	Tom	Stallion	Male Cavy	Buck
Mature Female	Bitch	Queen	Mare	Female Cavy	Doe
Young Male	Intact	—	Colt	—	—
Young Female	Bitch	—	Filly	—	—
Altered Male	Neuter	Gib	Gelding	—	—
Altered Female	Spay	Spay	Spayed	—	—
Newborn	Puppy	Kitten	Foal	Kit	Kit
Group	Pack	Bevy	Herd	Group	Group

ANIMAL WELL-BEING

Animal well-being is caring for an animal so that its needs are met and it does not suffer. The needs of animals vary depending on their species. Producers need to study the natural conditions under which animals grow best and use that information in caring for their needs.

Animals should not be confined in places or ways that threaten their well-being. Some people refer to this as humane treatment of animals. This applies to producers of animals for food as well as to people who have companion and other animals. For example, leaving a dog in a parked car on a hot day is cruel and may lead to disease or death.

1-16. This young dog has been tied to a small tree. How does the environment provide for the well being of the puppy? What would you do to improve the situation?

Most producers take steps that assure that an animal has a far better environment than would be present if living in the wild. Using vaccinations to prevent disease is definitely a positive step in animal well-being. Providing proper feed and water is also another positive practice in animal well-being. Facilities that provide protection from a harsh environment are also important. Steps are taken to prevent suffering such as using anesthetic prior to surgery. Animals should never be whipped and beaten. Never leave an animal tied up longer than necessary. Protect animals from bad weather.

Animals are sometimes injured or diseased. They may be suffering from pain or disability associated with the injury or disease. Sometimes animals get struck by cars when they get in a road. People face a dilemma about what to do. They know that there is no hope for the badly injured animal. The animal should likely be euthanatized. *Euthanasia* is killing an animal that is suffering. Euthanasia provides an easy and painless death for an animal. Euthanasia is sometimes controversial. Only trained, qualified individuals should practice euthanasia.

ENVIRONMENTAL CONCERNS WITH ANIMAL PRODUCTION

We must be aware of the potential problems animals can cause in our environment. This is true if raised as companions, meat, fur, fun, or other reasons.

All animals create waste. Some occasionally die or become diseased. This creates a need to dispose of the dead animal. Animals produce body wastes that require disposal. Factory-type farms create large amounts of waste. The wastes can contaminate water and cause health problems. Animals can cause soil erosion, make an area aesthetically displeasing, and create unpleasant odors.

Permits are needed for some types of animal production. Other areas have zoning laws that restrict animal production. Landlords may restrict keeping companion animals in rental property. Leash laws and permit regulations should always be followed.

DISPOSING OF ANIMAL WASTE

When animals defecate and urinate, the waste products must be disposed of properly. Wastes allowed to remain in an area can attract insects and become a source of disease. Animal housing must be cleaned regularly to prevent disease. Animal waste that is allowed to drain into the ground or into a water source can contaminate surface and/or ground water and cause health problems for wildlife and eventually people.

Even the waste produced by small animals must be disposed of properly. Although the animal is small, the waste produced is proportionate with their size. Some small animals can be trained to use litter boxes so that waste management is easier. Small animals need to have their cages cleaned regularly to keep them healthy!

Waste from animal facilities should be disposed of without contaminating soil or water. Many animal producers use animal waste as fertilizer for

1-17. A compost bin is used on this poultry farm to dispose of dead birds and other wastes.

crops. Small animal waste can be composted. However you dispose of your animals' waste you must have a good plan to do it properly!

OTHER ENVIRONMENTAL CONCERNS

Large animals can create erosion and cause an area to become unsightly. For example, hogs on pasture may root their snouts in the ground looking for food. They can tear up an area of turf with little effort. When left unattended, these areas often become mud pits and are not very pleasing to look at. This also causes soil erosion that can pollute surface water and cause the loss of topsoil. Large animals that wade in water or cross creeks loosen soil and cause it to erode. This degrades water.

1-18. Range land of North America is ideal for sheep production. (Courtesy, National Lamb and Wool Grower)

REVIEWING

MAIN IDEAS

The animal industry is a large and diverse field. It includes the production of animals as well as supplies and services and product marketing and processing. Animal domestication has had a big impact on how animals affect the lives of people. We depend on animal products and services every day.

Science and technology are an integral part of modern animal production. Selecting and producing animals that will meet consumer needs involves using science. Providing animals with a balanced diet and adequate care to ensure animal health uses science and technology.

Animal production is the foundation of the livestock and poultry industry. Farm animal production is moving away from the small operator to commercial operations capable of raising large numbers of livestock. Others involved in animal production include pet shops, laboratory technicians, and private individuals.

Animals are classified by their specie, common name, sexual condition, and age.

Animals are used in many ways. Some are used primarily for their meat and food products. Others are used for pleasure, companionship, service, and power. Overall, animals make a huge contribution to the quality of human life.

Animals do pose concerns for the environment. The waste they produce must be properly disposed of. Animals should be managed so they do not cause soil erosion or cause aesthetic damage to property.

QUESTIONS

Answer the following questions using correct spelling and complete sentences.

1. What is the animal industry?

2. What are the major areas of the animal industry?

3. What are four ways animals are used? Explain each.

4. What is animal domestication?

5. What are the trends in the animal industry? Briefly explain each.

6. How is science used in the animal industry?

7. What are the five most important areas of the animal industry that involve science and technology?

8. What is a factory farm?

9. What is a neutered animal? How does neutering vary for males and females?

10. What is animal well-being?

EVALUATING

Match the term with the correct definition. Write the letter by the term in the blank provided.

a. poultry	e. ration	i. botany
b. livestock	f. castration	j. health
c. euthanasia	g. mohair	
d. nutrition	h. zoology	

_____ 1. Mammals produced on farms and ranches for food and other purposes.

_____ 2. The condition of an animal in being free of disease.

_____ 3. Fowl raised for food and other products.

_____ 4. The daily diet of an animal.

_____ 5. Providing painless and easy death of a diseased animal.

_____ 6. The study of the food needs of an animal.

_____ 7. The study of plants.

_____ 8. The study of animals.

_____ 9. Removing the testicles of a male.

_____ 10. The hair of an Angora goat.

EXPLORING

1. Take a field trip to a farm, kennel, or other facility that produces animals. Determine the kinds of animals that are produced. Learn the major practices followed to produce the animals. What is involved in raising the animals? Prepare a written report on your findings. Give an oral report in class.

2. Investigate how animal production has changed in the last 20 or so years. Interview an animal producer who has been in business for a long time. Ask how the production practices have changed. Discuss probable reasons for farmers raising

fewer livestock species and the shrinking numbers of animal producers. Prepare a written report on your findings. Give an oral report in class.

3. Investigate how animals and animal products are marketed in your local community. Tour a livestock auction, egg processing plant, pet store, or other facility. Interview the manager about the nature of the activities involved in marketing the animals. Summarize your findings in a written or oral report.

4. Assess your interests in animals. Indicate how you feel about the items below. Check the response that best represents your feelings about an activity with animals. Discuss your responses with your agriculture teacher. This should help you better understand your interest in animals. Areas that you are unsure about and don't like should be explored. You may find that you really like the activity!

	like	unsure	don't like
feed an animal	___	___	___
groom an animal	___	___	___
care for a sick animal	___	___	___
give a shot to prevent disease	___	___	___
haul an animal	___	___	___
care for a baby animal	___	___	___
train an animal	___	___	___
care for where animals live	___	___	___
exhibit an animal at a show	___	___	___
use science to study animals	___	___	___

2

ANIMALS AS ORGANISMS

More than one million different kinds of animals are on the Earth. Many of these animals or their products are sources of food and clothing. You probably use many of these products in your daily living!

Animals are very complex creatures. By studying the functions and needs of animals, people are better able to raise them. Research is studying new areas of animal physiology, genetics, and environment. The research should help find better ways of raising animals.

The most common animals have the same basic systems and needs. Each has cells, tissues, organs, and organ systems. Slight variations among these animals distinguish them from each other. Begin your study of animals by learning about the major external and internal parts and their functions. That will help you excel in caring for animals.

2-1. Performing surgery on a horse requires a thorough knowledge of horse anatomy and consideration of its well-being. (Courtesy, Veterinary Medical Center, Texas A&M University)

OBJECTIVES

This chapter covers the basic biology of animals. It has the following objectives:

1 Classify animals by taxonomy

2 Explain animal names and list examples

3 Describe the major animal groups

4 Identify the life processes of animals

5 Explain the structural basis of animals

6 Describe the anatomy and physiology of animals

7 Explain the major organ systems of animals

TERMS

anatomy
cell
cell division
cell specialization
circulatory system
classification system
common name
digestive system
excretory system
homeostasis
integumentary system
invertebrate
Kingdom Animalia
lymphatic system
mammal
meiosis
mitosis

muscular system
nervous system
nervous tissue
organ
organism
organ system
physiology
protoplasm
reproductive system
respiratory system
scientific name
skeletal system
taxonomy
tissue
vertebrae
vertebrate

ANIMAL TAXONOMY

Animals are organisms. An ***organism*** is any living thing. Organisms carry out processes needed to remain alive and continue their species.

Taxonomy is the science of classifying organisms. Most classifications of animals are due to physical characteristics. A ***classification system*** is used to distinguish different animals from each other. The hierarchy of classification is used to show the relationships and differences among animals. Table 2-1 shows the hierarchy of classification.

Table 2-1. Hierarchy of Classification	
Category	**Description**
Species	Reproductively isolated populations with physical similarities
Genus	Contains related species
Family	Contains related genera
Order	Contains related family
Class	Contains related orders
Phylum	Contains related classes
Kingdom	Contains related phyla

Moving from bottom to top—more characteristics in common.

Moving from top to bottom—common characteristics are more distinctive.

KINGDOM ANIMALIA

As organisms, animals are in the ***Kingdom Animalia***. Though all animals are in this Kingdom, they also have unique differences. Some have fur, while others have scales or feathers. Some fly, while others walk or swim.

Organisms in the Kingdom Animalia share three traits:

1. Animals are made up of cells. (Cells are the "building blocks" of living organisms.)

2. Animals can move about on their own. (Moving about is known as locomotion. A few animals move about very little, such as the oyster.)

3. Animals get their food from other sources. (Animals take in food, transform and store it, and release it as energy.)

STRUCTURES AND FUNCTIONS

Within the Kingdom Animalia, animals are grouped many ways. This grouping is based on the structures and functions of the body.

Animals are classified based on the presence of vertebrae. **Vertebrae** are bones that form a segmented spinal column. This segmented column is commonly known as a backbone. The presence of vertebrae is associated with the kind of skeletal system of an animal.

Animals are classified into two groups based on vertebrae: vertebrates and invertebrates. A **vertebrate** is an animal with a backbone. This is as opposed to an **invertebrate**, which is an animal without a backbone. Examples of invertebrates include shrimp, crawfish, honeybees, spiders, mites, earthworms, and snails.

Vertebrates belong to the Phylum Chordata and to the Subphylum Verbrata. They have many common characteristics. They have vertebrae (bones and cartilage), which surround the nerve cord. Bones make up the internal skeleton that provides the body's framework. They have a skull that protects the brain. Vertebrates have an axial skeleton made of the backbone and skull. Paired limbs (arms and legs) are attached to the axial skeleton. Finally, muscles provide movement by being attached to the skeleton.

2-2. A horse is an example of a vertebrate (in the Phylum Chordata and the Subphylum Verbrata).

SCIENTIFIC NAMES

Animals have common and scientific names. A ***common name*** is the name used in everyday conversation. A problem with common names is that they tend to vary from one place to another.

A ***scientific name*** is based on the taxonomy of an animal species. Every animal has a two-part scientific name. The name is made of its genus and species. Scientific names are written with the two words in italics or underlined. Only the first letter of the genus is capitalized. Table 2-2 shows the common and scientific names of selected animals.

Table 2-2. Common and Scientific Names of Selected Animals	
Common Name	**Scientific Name**
Hog (domestic)	*Sus scrofa*
Cattle	*Bos taurus* (no hump, European origin) *Bos indicus* (hump, Indian origin)
Sheep	*Ovis aries*
Spanish goat	*Capra pyrenaica*
Horse	*Equus caballus*
Cat (domestic)	*Felis domesticus*
Common hamster	*Cricetus cricetus*
Golden hamster	*Cricetus auratus*
Chicken	*Gallus domesticus*
Turkey	*Meleagris gallopavo*
Channel catfish	*Ictalurus punctatus*
Rainbow trout	*Oncorhynchus mykiss*
Llama	*Lama glama*

MAJOR ANIMAL GROUPS

Animals are in several classes. The three main classes are birds (Aves), bony fish (Osteichthyes), and mammals (Mammalia). Snakes, shellfish, turtles, and alligators are examples in other classes.

BIRDS

Some 9,000 species of birds have been identified. Birds belong to the class Aves. Some birds can live in the air, others on land, and others largely in the water. Most birds are wild; a few have been domesticated. Chickens, turkeys, and ducks are often raised. People are eating more chicken and turkey because they are lower in fat and cholesterol. Wild ducks, geese, pheasant, and quail are popular game birds among sport hunters.

Birds are covered with feathers. They are lightweight because of their hollow bones. Their wings help to lift them and make them better flyers. Birds have two legs and internal organ systems similar to most other animals.

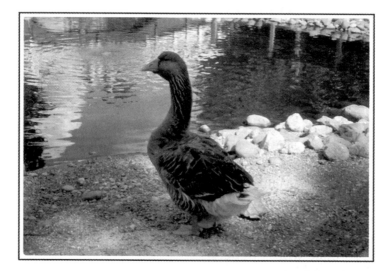

2-3. Birds, such as this Toulouse goose, belong to the Class Aves and are very different from mammals and fish.

FISH

Fish are aquatic animals. They live in water. There are three classes of fish, but the bony fish (Osteichthyes) are the ones commonly raised for food. Scientists have identified about 25,000 species of fish.

Many fish are caught from rivers and oceans, but aquafarming is increasing in popularity. Catfish are the most commonly raised fish. Trout, tilapia, salmon, and baitfish are also raised on aquafarms. Many people raise several species of ornamental fish at home or at their business for pets or decorations.

Fish are covered with skin. Some also have scales. The head, tail, and trunk form the three major parts of the body. The mouth, eyes, and gills are found on the head, while the trunk contains the internal organs. Strong muscles in the tail help the fish to move about.

2-4. Fish, such as these koi, are quite different from aves and mammals.

Fish are physiologically similar to other animals. However, fish filter oxygen from the water as it passes over their gills. Some lay eggs that are fertilized outside the body; others give birth to live young. The body temperature is regulated by the temperature of the water.

Mammals

Mammals belong to the class Mammalia. A ***mammal*** is a species of animal that has mammary glands. Female mammals have mammary glands that produce milk for the newborn young. Male and female mammals mate to reproduce. The female will carry the developing embryo in her uterus.

Mammals have hair, a well-developed brain, a lower jawbone with teeth, and a heart with four chambers. Unlike fish, mammals internally regulate their body temperature. Examples of mammals include cattle, goats, hogs, horses, and sheep.

Male mammals have rudimentary or undeveloped mammary glands. Female mammary glands develop into mature organ systems during puberty.

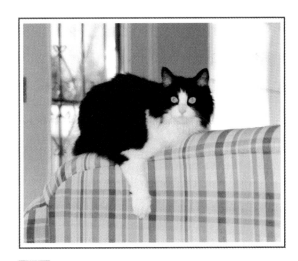

2-5. As a mammal, a cat is quite a bit different from aves and bony fish. (Cats prey on aves and fish!)

Many of the animals used for food, companionship, and in other ways are mammals. Examples include cattle, hogs, sheep, dogs, cats, hamsters, and rats.

LIFE PROCESSES

Animals carry out certain processes in order to remain alive. When these processes stop, the animal is no longer living. It has died.

Life is based on chemical processes that occur in the protoplasm. ***Protoplasm*** is the mixture of water and proteins in cells. The nature of the chemical processes determines the kind of animal and life functions that it is performing. As complex organisms, animals carry out a number of life processes that support the living condition in protoplasm.

The life processes are:

- Getting and using food—Food provides the nutrients for an animal to carry out life processes. Without proper food, an animal will not grow well and may become diseased and die.

- Movement—Movement includes processes that occur internally as well as locomotion. The ability to move is important for most animals to get their food. It is also important in escaping danger and raising young.

- Circulation—Circulation is moving blood, nutrients, oxygen, and wastes throughout the body.

- Respiration—Respiration is the process of providing oxygen and nutrients to cells. This releases the energy in the nutrients and produces some waste materials.

- Growth and repair—These two processes are important in young and mature animals. Animals grow by increasing the number and size of cells. Repair is replacing cells that have been damaged or are worn and no longer function.

- Secretion—Secretion is the production of liquid substances containing hormones and other materials. It is important in eating and digesting food as well as other life processes.

- Sensation—Sensation is the ability of an animal to respond to its environment. It receives information, such as pain, and responds to provide comfort.

- Reproduction—Reproduction is the process by which new members of a species are created. It is not essential for the life of an organism but is essential for a species to perpetuate itself. Animal producers are concerned about promoting and controlling reproductive processes.

THE STRUCTURAL BASIS OF ANIMALS

The structural basis of animals begins with cells. A *cell* is the basic structure of life. Cells have a definite internal organization and relationship. Most body cells can be seen only with a microscope. Some cells are large such as a chicken egg.

Cells vary in size and shape. Some are quite long, such as a cell in a muscle in the leg of an animal. Other cells are nearly square. All animal cells have membranes that surround the internal parts and create a definite structure. The cell membrane regulates what enters and leaves a cell. Cells have a nucleus that controls cell activity. The nucleus contains chromosomes and genetic material. The nucleus is surrounded by cytoplasm within the cell membrane. The protoplasm is all of the liquid substance inside the membrane.

2-6. Major parts of a typical animal cell.

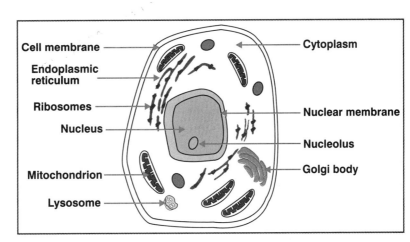

Cell membrane
Endoplasmic reticulum
Ribosomes
Nucleus
Mitochondrion
Lysosome

Cytoplasm
Nuclear membrane
Nucleolus
Golgi body

CELL DIVISION

Cell division is the process of a cell splitting into two cells. The process involves several steps. The new cells contain identical genetic material.

Two types of cell division occur: mitosis and meiosis. *Mitosis* is cell division for growth and repair. The process involves several phases. Division by one cell results in two cells. Animal growth depends on rapid cell division. Good nutrition is needed to promote cell division.

Meiosis is cell division for sexual reproduction. With meiosis, division results in four cells known as daughter cells. Each daughter cell has half the number of chromosomes of the parent cell. The process of fertilization provides the other half of the chromosomes.

CELL SPECIALIZATION

Cell specialization is difference in cells that allow them to perform unique activities. Organisms have several different kinds of cells. These different cells form groups that work together to perform functions. For example, the cells of eyes are specialized to allow vision to occur.

Tissue

Groups of cells form tissues. A *tissue* is a cluster of cells that are alike in structure and activity. Tissues perform specific functions, such as those that form the skin or bones. Tissues may group together to form organs.

There are four types of animal tissue:

■ Epithelial Tissue—Epithelial tissue covers body surfaces and lines body cavities. Its purpose as skin is to protect the body. Epithelial tissue contains cilia that are hairlike extensions. These cilia move materials. For example, cilia in the lungs move dirt and other impurities out of the body. The main organ system of the epithelial tissue is skin.

■ Connective Tissue—Connective tissue holds and supports body parts. It binds body parts together providing support and protection. It also fills spaces, stores fat, and forms blood cells. Connective tissue helps to form muscular and skeletal cells.

■ Muscular Tissue—Muscular tissue creates movement of body parts. It is also known as contractile tissue. These tissues not only aid in movement of the

2-7. A raccoon— a wild animal—is comprised of many cells, tissues, organs, and organ systems.

entire animal, but also allow the respiratory and digestive systems to function.

■ Nervous Tissue—*Nervous tissue* responds to stimuli and transmits nerve impulses. Nervous tissue contains neurons that conduct impulses away from the cell. Environmental stimuli cause reactions to occur.

Organs

An *organ* is a group of tissues with a similar function. The tissues work together so that the organ can carry out its purpose. Livers and lungs are examples of organs. Some organs work with other organs to make life possible.

Organ Systems

An *organ system* is a group of organs working together to carry out a specific activity. These are the major systems of the body. An example is the digestive system. The mouth, stomach, and intestines form the digestive system. (Each system is covered in more detail later in the chapter.)

ANATOMY AND PHYSIOLOGY OF ANIMALS

Anatomy and physiology are important in knowing how organisms live and go about life processes. *Anatomy* is the study of the form, shape, and appearance of animals. *Physiology* is the study of the functions of the cells, tissues, organs, and systems of an organism. Knowing anatomy is essential to understanding physiology.

All animals have organ systems that maintain homeostasis. *Homeostasis* is the relative constancy of the internal environment. An example of the homeostasis is the work of the circulatory system. The circulatory system carries nutrients from the digestive system. It also carries oxygen from the lungs to the cells from the respiratory system via the blood. The circulatory rids the body of metabolic waste from the cells into the excretory system. The nervous system coordinates the functions of the other systems.

Understanding life processes is important. Knowing the workings of the inner body involves anatomy and physiology of livestock. Internal and external features are closely related. Figures 2-8 through 2-14 illustrate the external parts of several animals.

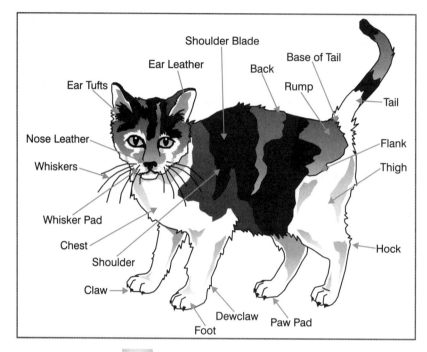

2-8. Major external parts of a cat.

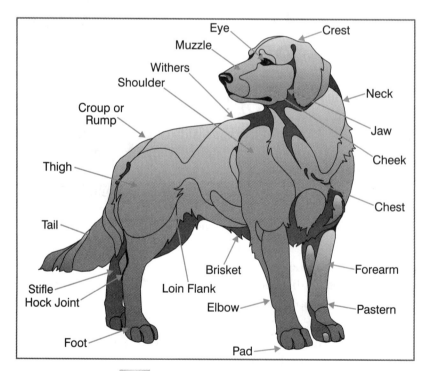

2-9. Major external parts of a dog.

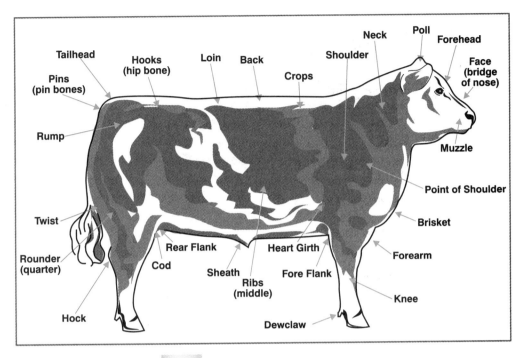

2-10. Major external parts of a steer.

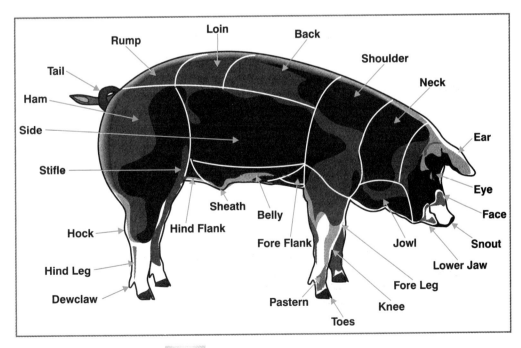

2-11. Major external parts of a hog.

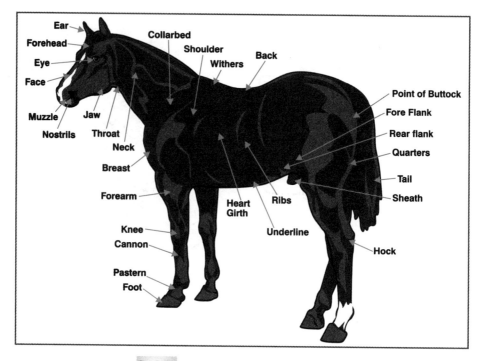

2-12. Major external parts of a horse.

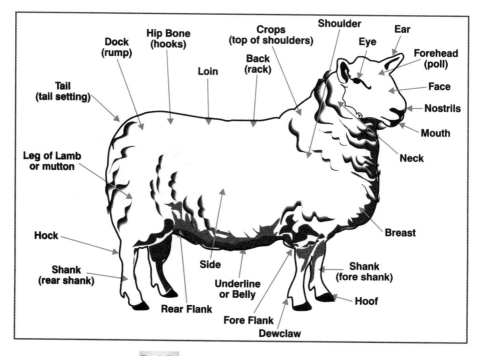

2-13. Major external parts of sheep.

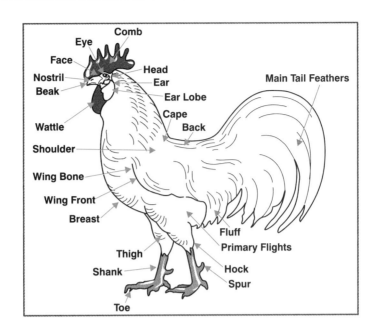

2-14. Major external parts of a chicken.

MAJOR ANIMAL SYSTEMS

As organisms, animals have several systems that make the living condition possible. These organ systems perform essential functions for an animal. The organ systems and the chief structures in the systems are shown in Table 2-3.

Table 2-3. Organ Systems and Structures	
System	**Chief Structures**
Circulatory	Heart
Digestive	Stomach and intestines
Integumentary	Skin
Endocrine	Ductless glands
Excretory	Kidneys and bladder
Muscular	Muscles
Nervous	Brain, spinal cord, nerves
Reproductive	Ovaries and testes
Respiratory	Lungs
Skeletal	Bones, joints

■ Muscular System—The ***muscular system*** acquires materials and energy. It creates bodily movements, maintains posture, supports, and produces heat. Figure 2-15 outlines parts of the hog's muscular system.

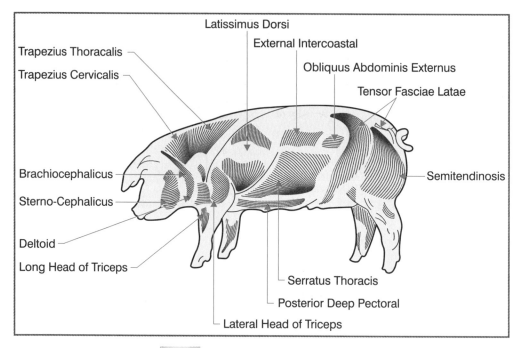

2-15. The muscular system of a hog.

■ Skeletal System—The ***skeletal system*** provides a framework for the body. The major components are bones and cartilage. The skeletal system supports and protects internal organs, stores minerals, and produces blood cells. The skeletal system of a chicken and a hog are shown in Figure 2-16.

■ Digestive System—The ***digestive system*** breaks food into smaller parts that are used by the body. Nutrients are made available from the food materials. Digestive systems vary. These differences allow animals to consume varying kinds of food materials. For example, cattle are ruminants. The polygastric system allows the animals to eat grass and other forages. But pigs are different. Pigs have monogastric or nonruminant digestive systems. They must be fed grain and other concentrated food materials. Figure 2-17 outlines the digestive systems of a hog (monogastric or nonruminant), a cow (polygastric or ruminant), and a chicken.

■ Respiratory System—The ***respiratory system*** governs gas exchange. Oxygen is taken in through the nostrils. It goes to the lungs where it passes into the blood. This system also maintains blood pH by expelling carbon dioxide. Fish

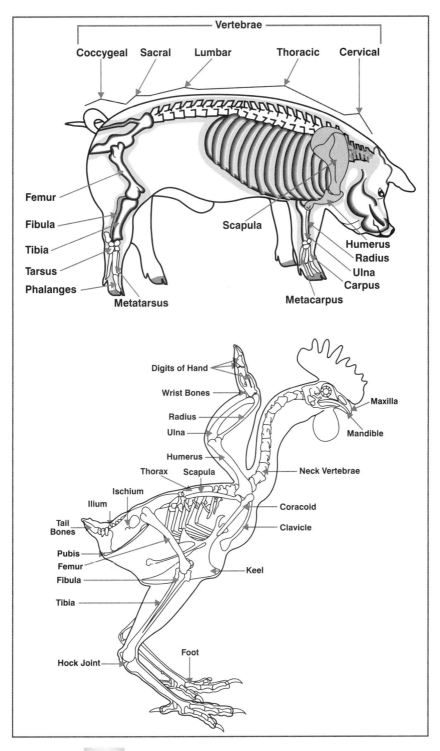

2-16. The skeletal systems of a hog and a chicken.

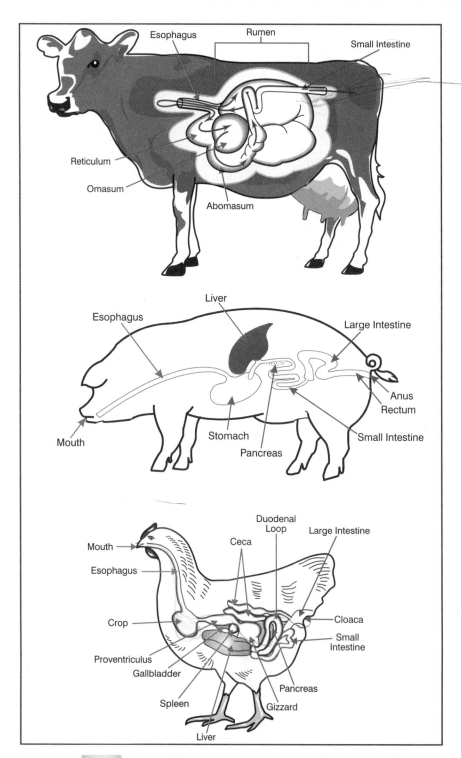

2-17. The digestive systems of a cow, hog, and chicken.

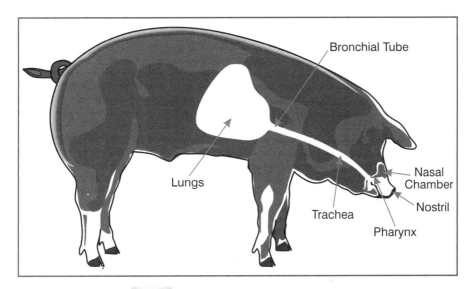

2-18. The respiratory system of a hog.

Connection

CAT PARTS

Knowledge of anatomy helps people care for animals. It helps in providing for their well-being. Most people use a preserved specimen in their study. Once opened, organ systems are observed. Skeletal, muscular, digestive, circulatory, and respiratory systems are the easiest to see.

These students at a New Hampshire high school are studying the anatomy of a cat as part of their small animal class. They are using a specimen that has been carefully pre-

served by a biological supply company. The students are following proper safety procedures with gloves, goggles, and aprons. The specimen has been properly opened and positioned on a dissection tray. They are in a well-ventilated laboratory.

The students will apply what they have learned in the school lab, with their jobs, and in caring for their own animals.

and some other animals are different. For example, fish have gills for removing oxygen from the water. The oxygen is exchanged in the gill filaments for carbon dioxide in the blood. Figure 2-18 shows the respiratory system of a hog.

■ Circulatory System—Nutrients, oxygen, and metabolic wastes are moved by the *circulatory system*. Oxygen and nutrients are transported to the cells, while wastes are removed. The circulatory system also moves hormones and protects against injury and microbes. The circulatory system of a horse is shown in Figure 2-19.

■ Excretory System—The *excretory system* rids the body of waste. It maintains the chemical composition and volume of blood and tissue fluid. Wastes are

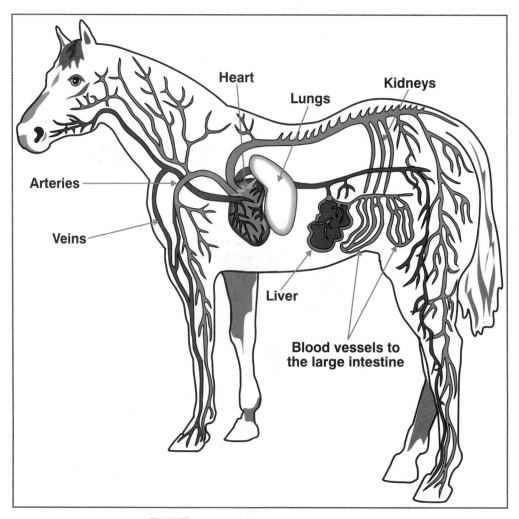

2-19. The circulatory system of a horse.

expelled through the lungs, skin, bladder, and anus as the body rids itself of undigested food. Figure 2-20 shows the general excretory system.

■ Lymphatic System—The *lymphatic system* circulates a clear fluid known as lymph. One role of lymph is to protect the body from disease. The lymphatic system transports excess tissue fluid to the blood, moves fat to the blood, and helps give immunity against diseases.

■ Nervous System—The *nervous system* coordinates body activities. In coordination with the endocrine system, nerve impulses react to stimuli. The system regulates other systems and controls learning and memory. The endocrine system secretes hormones that regulate the body's metabolism, growth, and reproductive systems. The nervous system of a horse is shown in Figure 2-21 and the endocrine system of a cow is shown in Figure 2-22.

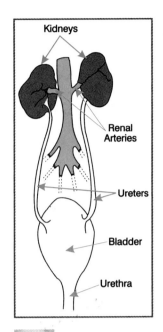

2-20. The general excretory system.

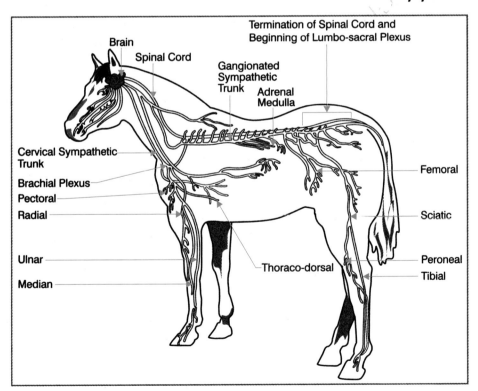

2-21. The nervous system of a horse.

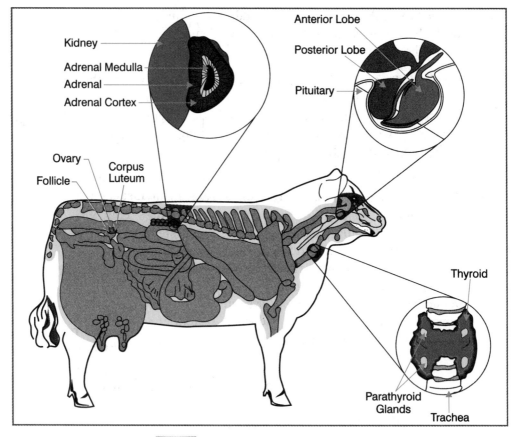

2-22. The endocrine system of a cow.

■ Integumentary System—The ***integumentary system*** is the skin. The skin protects internal body tissues from outside dangers. It keeps out foreign materials, such as bacteria and dust. It also helps regulate body temperature.

■ Reproductive System—The ***reproductive system*** produces offspring. New individuals of the same species are produced. The reproductive system is discussed in Chapter 5.

REVIEWING

MAIN IDEAS

The hierarchy of classification is a system for grouping animals. Kingdom is the broadest classification, with species the most specific classification.

The Kingdom Animalia includes all agricultural animals. Organisms in the Kingdom Animalia are made of cells, can move about on their own, and get their food from other sources. Many common animals belong to the phylum Chordata and the subphylum Verbrata. Birds, fish, and mammals are among the most common Chordata.

Animals carry out life processes. When these processes stop, the animal is no longer in the living condition. A structure is needed for an animal to survive. The structure begins with the cell. Cells specialize to carry out unique activities for an organism. Groups of specialized cells form tissues. Tissues form organs. Organs form organ systems. Animals have the following organ systems: circulatory, digestive, integumentary, endocrine, excretory, muscular, nervous, reproductive, respiratory, and skeletal.

A knowledge of anatomy and physiology helps animal producers understand their animals. They are better able to provide conditions that meet the needs of the animals. They can help animals grow efficiently and have healthy lives.

QUESTIONS

Answer the following questions using correct spelling and complete sentences.

1. What is taxonomy?

2. What are the three traits of organisms in the Kingdom Animalia?

3. What are the common characteristics of vertebrates?

4. What names are used with animals?

5. What is a scientific name?

6. Compare and contrast birds, bony fish, and mammals.

7. What are the life processes of animals?

8. What is a cell? Cell specialization?

9. What is an organ system?

10. Distinguish between anatomy and physiology.

11. What are the major systems of animals? What is the function of each?

12. Compare and contrast the digestive systems of a hog, a cow, and a chicken.

EVALUATING

Match the term with the correct definition. Write the letter by the term in the blank provided.

a. vertebrate   e. physiology   i. mitosis
b. anatomy   f. homeostasis   j. organ system
c. organism   g. taxonomy
d. invertebrate   h. mammal

_____ 1. Living things.

_____ 2. Animal with backbone.

_____ 3. Constancy of the internal environment of an organism.

_____ 4. Function of cells, tissues, organs, and systems.

_____ 5. Form, shape, and appearance of animals.

_____ 6. Females give birth to live young and secrete milk.

_____ 7. Animal that does not have a backbone.

_____ 8. Science of classifying organisms.

_____ 9. Cell division.

_____ 10. A group of organs that works together to carry out a specific function.

EXPLORING

1. Dissect a pig or other farm animal. Compare the various types of tissue and organ systems and their functions. Use proper procedures in your work. Follow all safety practices.

2. Visit a local farm or ranch that raises various agricultural animals or attend a livestock show. Compare and contrast the external parts of the animals. Use a camera to make photographs of the different animals. Prepare a poster or bulletin board that summarizes your observations.

3. Make a sketch of an animal that is raised on local farms or kept as a companion. Label the major external parts. The sketch can be made in the space below or on a separate sheet of paper.

3

ANIMAL NUTRITION AND FEEDING

Animals need the right feed to live and grow properly. Feeding animals often raises questions: What does an animal need? Why are these needs important? How are the needs met?

Selecting the right feed and method of feeding varies with nutrient needs and species of animal. Animals that do not get the right feed suffer from poor nutrition. Poor nutrition causes slow growth, poor reproduction, lowered production levels, and poor health. Sometimes, poor nutrition leads to the early death of an animal.

Producers want fast growth at the lowest cost when raising animals. They want animals that stay healthy and return a profit. Nutrition is a primary concern of all animal owners. Animals that are properly nourished are healthy and grow fast.

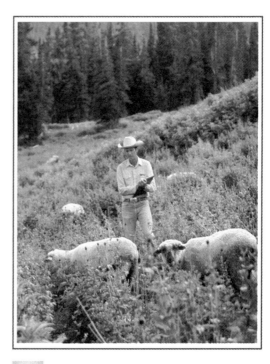

3-1. A range scientist is studying the kinds of plants sheep consume. (Courtesy, Agricultural Research Service, USDA)

OBJECTIVES

This chapter covers the nutrition and feeding of animals. It has the following objectives:

1 List the major nutrient needs of animals and describe the purpose of each

2 Contrast feed requirements with structure of digestive systems

3 Describe the ways animals use nutrients

4 Describe the types of feedstuffs

5 Explain how animals are fed

6 Describe how rations are formulated

TERMS

abomasum
amino acids
balanced ration
calorie
carbohydrate
carnivore
concentrate
diet
fat
feed
feed analysis
feedstuff
fiber
forage
free access
growth
herbivore
lactation ration
lipid

maintenance
mineral
nutrient
omasum
omnivore
palatability
Pearson Square method
permanent pasture
protein
reproduction ration
reticulum
roughage
rumen
scheduled feeding
stomach
supplement
temporary pasture
vitamin

NUTRITION AND NUTRIENTS

Nutrition is the process by which animals eat and use food. Food is needed by animals to live, grow, lactate, reproduce, and work. All animals must have the food they need for a balanced diet.

Proper animal nutrition increases feed efficiency and rate of gain. It decreases the number of days required to grow a meat animal to market size. Producers want animals to grow fast. They also want them to make good use of the feed. Keepers of companion and exotic animals want them to be healthy and content. Feeding too much or the wrong feed is a big waste!

Animals must be fed a diet that meets their needs. If their needs are not met, they will not grow, can get sick, and may die. What animals are fed depends upon many factors. Differences in digestive systems are important. For example, ruminants eat more grasses because their stomachs allow them to make use of these foods. The stages of life and the activities of the animal also affect feed needs.

3-2. Active animals need more energy than those that are not active. (Courtesy, American Quarter Horse Association)

NUTRIENT NEEDS OF ANIMALS

Feedstuffs contain nutrients. A *nutrient* is a substance that is necessary for an organism to live and grow. Nutrients make it possible for animals to carry out life processes. Nutritional requirements vary with age, stage of development, environmental conditions, activity, and genetic makeup.

A ration is the total amount of feed an animal has in a 24-hour period. The ration may be fed in amounts throughout the day or all at once. Feeding animals all of their ration at one time may not be best. A *balanced ration* is one that contains all the nutrients that the animal needs in the correct proportions. Too much of a nutrient is wasteful and, in some cases, can harm an animal. A nutrient deficiency results in stunted growth and low production.

The nutrients required by all animals include water, carbohydrates, fats, proteins, minerals, and vitamins.

Water

Water is necessary for an animal to live. Animals can live longer without food than without water. Water is found in all the cells of the body. It makes up about 75 percent of the weight of an animal's body. Water percent may be as high as 90 percent in newborn animals. Dairy cows may consume up to 50 gallons of water on a warm day. The amount needed is also related to activity, gestation, and lactation.

3-3. An automatic waterer is used in this "pet friendly" park.

Water has two basic functions:

■ Regulating body temperature—Water helps control body temperature because it accumulates, transfers, and loses heat through evaporation.

■ Promoting biochemical processes—All biochemical reactions require water. Water is the major component of cells, blood, and body tissues.

Animals get water in several ways. Most of it is by drinking. Water is also in feed and may be produced through biochemical reactions. Water is lost through urine, feces, sweat, and vapor from the lungs.

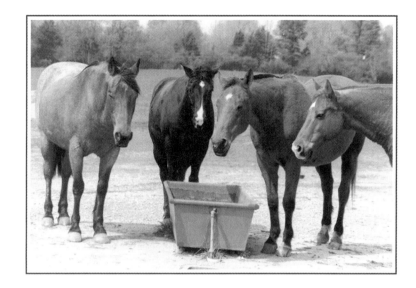

3-4. Horses have gathered around a trough with good water.

Water should be clean. It should be free of chemical pollutants, mud, and disease-causing organisms.

Special attention to water supplies is needed in weather extremes. In the winter, cold temperatures cause water to freeze. Animals cannot drink frozen water! Water consumption increases in warm weather. Plenty of water should be available all of the time. Animals should have free access to water. Companion animals, for example, should be able to drink at will from a bowl or other container.

Carbohydrates

Carbohydrates provide energy. They are a major component of plant tissues. Carbohydrates should make up about 75 percent of an animal's diet. Besides energy, carbohydrates aid in the use of proteins and fats. Carbohydrates provide energy for growth, maintenance, work, reproduction, and lactation.

Three types of carbohydrates are:

■ Sugars—Sugars are of two kinds: simple sugars (monosaccharides) and double sugars (disaccharides). Glucose and fructose are simple sugars. Sucrose is a double sugar and is used to make table sugar. Glucose is an excellent source of energy for most cells. Sugars are found in many foods, with fruit and milk being two examples. Sugars provide efficient energy.

3-5. Cereal grains that may be used to provide carbohydrates in livestock feed include oats, barley, and corn. These grains are concentrates in an animal's diet.

■ Starch—Starch is an important source of energy. It is found in grain, root crops, and other plant materials. Starch may be converted to glucose in the digestive process.

■ Fiber—*Fiber* is the material left after the food has been digested. It is made of plant cell walls and cellulose. Fiber helps the digestive system function smoothly. Fiber absorbs water and provides bulk. It also helps to increase the bacterial population in the rumen of cattle.

The major sources of carbohydrates are cereal grains, such as corn, wheat, barley, oats, hay, and rye. Abnormal carbohydrate metabolism can lead to ketosis and diabetes.

Carbohydrates are not stored in the body. They must be eaten every day. A small amount of glucose is always in the blood. Unused glucose and carbohydrates are converted to fats.

Lipids (Fats)

A *lipid* is a nutrient that can be dissolved with ether. Ether is a colorless, liquid solvent used in nutrition research. Using ether helps determine the nutrients in food materials.

Most lipids are fats or oils. A *fat* is a good source of energy. Fats contain the highest amounts of energy, with 2.25 times more energy than carbohydrates. Fats help supply energy for normal body maintenance. Fats also provide a healthy skin, keep the nervous system healthy, give food a good flavor, and carry fat soluble vitamins A, D, E, and K. Animals do not develop a fat deficiency for energy. However, they can develop a vitamin deficiency.

Sources of fats in animal diets (depending of the species) include meat, vegetable oil, cottonseed, and fish meal.

Obesity in animals is related to fat intake. Research has shown a genetic link between fat accumulation in an animal and fat intake. This indicates the importance of genetic selection in animal breeding programs. Keepers of companion and exotic animals should regulate fat intake and provide exercise for animals to reduce the likelihood of obesity.

Protein

Protein is a nutrient needed to grow new tissue and repair old tissue. Protein is especially important for weight gain, growth, and gestation. It is used only for energy when carbohydrates and fats are deficient.

Protein serves many functions. It is found in wool, feathers, horns, claws, beaks, DNA, RNA, skin, hair, hooves, blood plasma, enzymes, hormones, and immune antibodies. Protein is needed every day for basic functions. It is also needed to build and repair cells. Three to five percent of the body's proteins are rebuilt every day. The highest amounts of protein can be found in the muscles of animals.

3-6. Small companion animals, such as this cat, should receive balanced rations.

3-7. Broilers are fed specially formulated diets to assure rapid growth.

Protein contains amino acids. ***Amino acids*** are the building blocks of proteins. Organisms use amino acids to build proteins. Twenty-three amino acids have been found in protein. Ten of the amino acids are essential. An essential amino acid is one that must be fed. Essential amino acids are not synthesized by animals. The essential amino acids vary with the kind of digestive system, with ruminants having fewer essential amino acids. The essential amino acids are:

- arginine
- histidine
- isoleucine
- leucine
- lysine
- methionine
- phenylalanine
- threonine
- tryptophan
- valine

Animals with simple stomachs must eat a greater variety of food. Each food contributes only part of the essential amino acids.

Ruminants have microbes in their rumen. The microbes synthesize amino acids from feedstuffs. The microbes are then digested in the digestive

3-8. Baby pigs obtain needed nutrients from the milk of their mother. These pigs will soon begin eating high-protein feed and be weaned from nursing.

system. This provides the ruminant with needed amino acids. Because of this, much research has been done to increase the efficiency of ruminants.

Sources of protein include soybean meal, cottonseed meal, fish meal, tankage, skim milk, and alfalfa hay. Protein is the most common nutrient deficiency. Most feedstuffs are low in protein. Supplements may be needed. For example, corn is deficient in lysine and tryptophan (amino acids). The deficiency results in slow growth. By supplementing corn with soybean meal, this problem can be solved.

Symptoms of a protein deficiency include anorexia, slow growth rate, decreased feed efficiency, anemia, edema, low birth weight of young, and lower milk production. Protein consumption is like a chain. When one amino acid is deficient, the whole body suffers.

Young animals need diets with higher amounts of protein than older animals. This is because protein is essential for growth.

Minerals

A *mineral* is an inorganic element found in small amounts in the body. Minerals are essential in skeleton growth and necessary for body systems to function properly. Feedstuffs must provide needed minerals.

Soil deficient in a particular mineral may produce crops that are low in the mineral. This can lead to deficiencies in feeds made from the crops. Some of these include selenium, copper, manganese, cobalt, and iodine. Mineral supplements are often used.

Minerals are necessary for a strong skeletal system. Minerals also make up organic compounds needed by muscles, organs, blood cells, and other soft tissue. They regulate many body functions. Examples of sources of minerals are alfalfa hay, cereal grains, bone meal, molasses, and salt.

Macrominerals are minerals required in large amounts. Microminerals, or trace minerals, are minerals required in smaller amounts. They are just as important as macrominerals.

Here is a summary of several minerals:

- Calcium—Calcium is a macromineral required in the highest amounts. Lack of calcium causes bones and teeth to form improperly. When calcium is not consumed on a daily basis, the body removes it from the bones for use in other body processes.

- Phosphorus—Phosphorus is another macromineral. Phosphorus is a key ingredient in the body's use of protein.

3-9. Mineral blocks supplement forage for grazing animals. This block is in a holder that keeps it clean. A new block will soon be needed because this one has been nearly licked away.

- Iron—Iron deficiency leads to anemia and a tired feeling. Iron is needed to make hemoglobin for red blood cells.

- Iodine—Lack of iodine will lead to fatigue, increased appetite, and a rapid pulse.

- Magnesium—Magnesium deficiency causes muscle tremors and shaking.

- Sodium and potassium—Sodium and potassium are needed to maintain water balance. Sodium and potassium transfer nutrients and waste through the cell membrane.

Deaths due to mineral deficiencies are rare, but inadequate amounts can cause economic losses to the producer. Deficiencies may cause a poor rate of gain, feed inefficiency, decreased reproduction, and a decrease in production of meat, milk, eggs, and wool.

Trace mineralized salt should be made available to animals. This will give most animals all the minerals they need.

Appendix A presents the mineral needs of animals. The functions and deficiency symptoms are included.

Vitamins

A **vitamin** is an organic substance needed in small quantities to perform specific functions. Several vitamins are needed. They do not provide energy, but are necessary in using energy. Vitamins help to regulate body functions, keep the body healthy, and develop resistance to diseases. The deficiency of a vitamin can lead to disease and death.

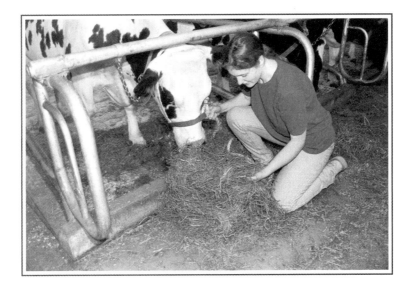

3-10. Alfalfa hay provides needed vitamins for this dairy cow.

Vitamins are either fat soluble or water soluble. Fat-soluble vitamins are stored in the fat and are released as they are needed by the body. These include vitamins A, D, E, and K. Water-soluble vitamins are dissolved by water and need to be consumed every day. They include vitamin C and the B vitamins.

Ruminants do not have any problems getting the necessary vitamins. Vitamin A may be the only one that is not readily available in most feeds. Vitamin D can easily be obtained when animals are exposed to the sun.

Appendix B lists the functions and deficiency symptoms of vitamins.

FOODS AND DIGESTIVE SYSTEMS

Animal digestive systems vary. Differences in the systems affect the foods they eat. Feedstuffs should be selected based on the digestive system. Using the wrong feedstuffs results in poor growth and can lead to malnutrition. Cattle may perform well on good quality hay but hogs, hamsters, and chickens would not!

FEEDING GROUPS

Animal species eat foods from different sources. The foods suit their digestive systems and natural feeding habits. The major differences are the

3-11. Sheep are herbivores. They eat grasses, legumes, and other succulent plants.

sources of the food: plant sources, animal sources, and combinations of plant and animal sources.

■ Herbivores—A *herbivore* is an animal that eats foods from plant sources. The food materials may be leaves, stems, twigs, flowers, roots, and seeds. Cattle are herbivores that eat leaves and stems. Other herbivores eat seeds, such as grain. Herbivores are the most efficient users of food materials.

■ Carnivores—A *carnivore* is an animal that eats foods from animal sources. The food is typically the flesh of other animals. Larger animals may eat smaller animals. In some cases, carnivores are fed meat foods that have been

3-12. A badger is a carnivore. In the wild, badgers prey on small animals, such as squirrels. In captivity, they are often fed horse meat.

3-13. Hogs are omnivores. Their diets often include grains as well as tankage (processed meat scraps) and supplements.

manufactured specifically for them. Carnivores need a meat-based diet. Many exotic animals in zoological parks are carnivores, such as badgers and lions. Dogs and cats are companion animals that are carnivores.

■ Omnivores—An *omnivore* is an animal that eats both plant and animal food materials. Omnivores eat grain, leaves, meat, and similar food materials. Fish are often omnivores. Hogs will eat meat as well as grain food materials.

SYSTEMS

Most animals have one of three types of digestive systems: monogastric, ruminant, and pseudoruminant.

Monogastric Systems

A monogastric digestive system has a simple stomach. The *stomach* is a muscular organ that stores ingested food and moves it into the small intestine. Some preparation for digestion occurs in the stomach.

Acid is secreted by the stomach. The acid results in a low pH of (1.5 to 2.5). The low pH destroys most bacteria and begins breaking down the food materials. Muscular movement of the stomach churns the contents softening food materials.

Dogs, cats, chickens, hogs, and rats have monogastric stomachs. These animals are fed feedstuffs that are high in nutrients and digestibility. They do not do well on feedstuffs that are high in roughage and fiber.

Ruminant Systems

A ruminant digestive system has a large stomach divided into compartments. The largest section of the stomach is the *rumen*. The rumen receives ingested food. It contains bacteria and other microbes that promote fermentation. Ruminant systems are designed for the food materials to be ingested, eructated (belched up), chewed, and swallowed again. The *reticulum* is the second segment of the stomach. It has many layer-like projections. The reticulum traps materials that are not food, such as nails or stones that have been ingested. The *omasum* follows the reticulum. The omasum is a small compartment that acts as a filter of materials for the next compartment–the *abomasum*. The abomasum secretes gastric juices that kill the microbes that have passed with the food materials from the rumen. The microbes are digested by the juices (acid) and other substances in the abomasum. The intestines serve in a similar manner to those of the monogastric animal in the absorption of nutrients.

Ruminant digestive systems use food materials high in roughage. This is why cattle are known as efficient animals on the grassland areas of the

Connection

BEAVERS GO TO DINNER

Exotic and wild animals have unique dietary needs. More study is needed on exact requirements. Watching animals in their native habitat provides useful information. Observing the kinds of food materials eaten helps in meeting dietary needs.

The beaver shown here are in an animal park near Seattle and are doing well in captivity. A natural setting has been made for them. A diet is provided that meets nutritional needs. The diet includes apples, carrots, and a manufactured rodent food pellet.

The rodent food provides protein, vitamins, and minerals. It is formed into hard pellets. Eating the hard pellet helps keep teeth clean and in good condition. The food is provided once each day.

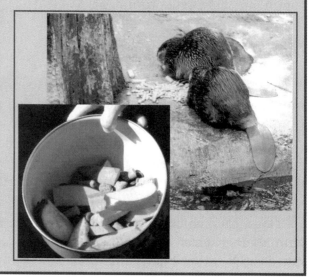

United States. The microbes that grow on the ingested forage in the rumen are sources of nutrients. Sheep, deer, cattle, and goats are examples of ruminants.

Pseudoruminant Systems

A pseudoruminant herbivore is an animal that eats large amounts of roughage but does not have a stomach with several compartments. Horses, rabbits, and guinea pigs are examples of pseudoruminant herbivores. The digestive systems

3-14. Rabbits are pseudoruminant herbivores.

of these animals do some of the same functions as those of ruminants. These animals often eat forage as well as grains and other concentrated feeds.

NUTRIENT USES AND REQUIREMENTS

Nutrient needs vary with the species, age, size, condition, and environment of an animal. Nutrients are used to maintain bodies, for growth, reproduction, lactation, and for work.

MAINTENANCE

Maintenance is keeping the body at a constant state. There is no loss or gain of weight. No matter what an animal is doing, nutrients are required. Animals require energy for internal workings of the body, heat to maintain body temperature, and small quantities of vitamins, minerals, and proteins.

A maintenance diet is usually high in carbohydrates and fats. It contains small amounts of protein, minerals, and vitamins. About one-half of the feed consumed is used for maintenance requirements. Animals on a maintenance diet are usually not growing, lactating, reproducing, or working.

GROWTH

In addition to maintenance, nutrients are needed for growth. *Growth* is an increase in the size of muscle, bones, and organs of the body. Meat production depends upon the growth of the animal. Growth requires mostly energy and smaller amounts of the other nutrients. A growth ration is high in protein, vitamins, and minerals. For example, chickens are fed a carefully prepared ration so they grow rapidly.

Young animals and breeding stock should receive a ration that provides plenty of protein. Young animals are expected to grow rapidly and efficiently.

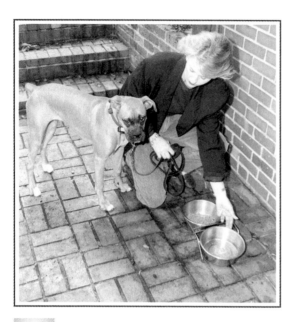

3-15. This maturing puppy is still on a growth diet.

REPRODUCTION

Proper nutrition is essential for efficient reproduction. Poor nutrition is a major reason for reproductive failures. During the first trimester of pregnancy, the fetus requires large amounts of protein, minerals, and vitamins. A *reproduction ration* is high in nutrients needed by breeding animals. Females that do not receive a proper ration will give birth to underweight babies or abort the fetus before birth.

Males also need additional nutrients for reproduction. Lack of nutrients can lower sperm production and fertility rates. Males with inadequate diets may be unable to breed the females.

LACTATION

Lactation is the production of milk. Females must have nutrients to produce large quantities of milk. A *lactation ration* is for females who are producing milk. Females that are pregnant and lactating require even higher levels of nutrients.

3-16. This mare with foal is on a lactation diet.

Since milk is high in protein, calcium, and phosphorus, milking animals need diets high in these nutrients. Dairy cows need a ration that has been specially prepared for them.

Work

Work and activity require energy. This includes all kinds of work and activity. Examples are draft animals that pull heavy loads, race horses, hunting dogs, and guard llamas. Increased carbohydrates and fats supply the needed nutrients. That is why draft horses receive additional grain in their diets.

TYPES OF FEEDS

A *feedstuff* is an ingredient used in making the feed for animals. Common feedstuffs include corn, soybean meal, and barley. Most feedstuffs have been analyzed to determine nutrient content. Some are high in certain nutrients and low in others.

Feed is what animals eat to get nutrients. Some feeds have few nutrients; others are high in nutrients. Most feeds provide more than one nutrient.

Feedstuffs can be fed to provide flavor, color, or texture to increase palatability. *Palatability* is how well an animal likes a feed. Some feedstuffs are made more palatable by adding molasses. Many animals like the sweet taste of molasses. A feed high in nutrients is of no benefit if animals refuse to eat it!

FEED CLASSIFICATION

Feeds are classified as forages, concentrates, or supplements.

Roughages (Forages)

A *roughage* is a feed that is mostly leaves and tender stems of plants. They are also known as forages. A *forage* is a bulky feedstuff that is high in fiber—roughage. They are lower in energy. Forages are fed because they are relatively inexpensive. Young, tender forage plants are more nutritious than older, tougher plants.

Grasses and legumes are grown in pastures as forage. Pastures may be permanent or temporary. A *permanent pasture* has plants that live from one year to the next. Crimson clover and white clover are examples of two le-

3-17. Corn being harvested as silage—an important roughage on dairy farms.

3-18. Crimson clover (left) and white clover (right) are used in some permanent pastures.

3-19. Hay being sun cured to maintain nutrient quality.

gumes used in permanent pastures. Most clovers are planted in mixtures with grass. Examples of common grasses are Bermudagrass, Kentucky blue grass, and bromegrass. Native grasses are predominant on rangelands. A *temporary pasture* is planted to forages for grazing during one season or year. Most temporary pastures are planted to high-yielding forages, such as millet or milo.

Some forages are harvested for feeding at a later day. Hay is forage that is cut, dried, and stored for feeding at a later date. Crop residue, green chop, and corn silage are other examples of forages. They contain high levels of fiber. Cattle often only eat forages. Some cattle in high-production uses are provided concentrates and supplements in addition to forage.

3-20. Round bales of hay are covered for protection from rain. Moisture causes hay to deteriorate and lose nutritional value.

Forages are harvested at the vegetative stage. As the plant matures, it loses energy because the energy is being put into the seed for reproduction. The younger the plant, the more energy and protein it will contain.

Concentrates

A *concentrate* is a feed that is high in energy. Concentrates have more energy per pound than forages and are usually lower in fiber. Examples of high-energy concentrates are corn, wheat, sorghum, barley, rye, and oats. Examples of high-protein concentrates are soybean oil meal, cottonseed oil meal, and sunflower meals. Higher producing animals need more nutrients from concentrates.

Ruminants are usually not fed concentrates in a maintenance ration. They are only fed if the forage is of very poor quality.

Many small animals need concentrates as a regular part of their diet. Cats, dogs, hogs, and similar animals are fed diets appropriate to their needs.

3-21. Three forms of manufactured feed are meal, small pellet, and range cube. Finely ground meal (left) is for baby animals, such as chicks or fish. The small pellets (center) are for calves, dogs, and growing fish. Large range cubes (right) are for cattle on pasture that need supplemental feed.

Supplements

High-producing animals are given supplements so they get the protein, minerals, and vitamins that they require. A *supplement* is a feed material high in a specific nutrient. Some contain several nutrients. Examples of protein supplements include soybean meal, cottonseed meal, corn gluten, sunflower meal, canola meal, and coconut meal.

Protein quality is usually less important to ruminants than nonruminants, except for high-producing dairy cows. Microbes present in the rumen produce protein. Nonprotein sources of nitrogen, such as urea or ammoniated molasses, are available for ruminants to eat.

Salt and mineral blocks can be placed in a feedlot or pasture for free-choice feeding.

GOOD FEEDSTUFFS

All animals require different amounts of nutrients. All feedstuffs provide differing amounts of nutrients.

Rations and Diets

A ration should provide the right amount and proportion of nutrients. A *diet* is the type and amount of feed an animal receives in its ration. Diet also includes water. Diets should be based on the needs of the animal being fed. Diets vary according to the nutrient content of the feed.

A good ration should be balanced, have variety, be succulent, be palatable, bulky, economical, and suitable. A balanced ration will increase feed gain, decrease expense, and increase profits. A variety of feeds will make ration-balancing easier and increase palatability. A succulent ration that is juicy and fresh will increase production. Bulky rations aid in digestibility because of the fiber. Economical rations should provide needed nutrients and maximize profits. A ration should be suitable to the type of animal.

3-22. The African hedgehog prefers a diet of living worms.

Some animals have diets that require live food. Snakes, lobster, hedge-hogs, and spiders are examples. A supply of the food must be available and provided to the animal, such as a mouse for a snake.

Feed Analysis

Feed analysis is the process of determining the nutrients in a feedstuff or prepared mixed feed. Chemical laboratory methods are used in analyzing feed. The information is important in selecting the diets of animals to assure a balanced ration. Labels on feed containers often report some of the information.

An analysis of a feed sample usually provides information in several areas. These are:

- Dry matter—Dry matter is the weight of feed materials after moisture has been driven out. Feed quality is based upon the proportion of water in feed. Silage has much higher water than No. 2 Yellow Corn.
- Crude protein—Crude protein is the nitrogen content of feed multiplied by 6.25 (a constant factor). Feedstuffs with higher crude protein are typically more nutritious.
- Fat—Fat content is determined by using an ether extract process. The fat is dissolved by the ether. The remaining feed material is weighed and a percentage of fat is calculated.
- Ash—Ash is the residue that remains after feed has been burned at a temperature of 600°C.
- Crude fiber—Crude fiber is determined by boiling the feed material in an acid and using laboratory procedures to dry the feed. The weight before and after drying is determined and used to calculate percent.
- Nitrogen-free extract (NFE)—The percentage of NFE is determined by subtracting the percentage of water, crude protein, ether extract, ash, and crude fiber from 100.
- Minerals—Feed materials can be analyzed for mineral content. Examples include calcium, phosphorus, and copper.

HOW ANIMALS ARE FED

How and when animals are fed is important in gaining production and growth. Animals need to consume the recommended amounts of each nutrient without overeating. They need feed for maintenance, growth, reproduction, lactation, and work. Feed can either be fed at scheduled times or be fed free access.

FREE ACCESS

Free access or free choice is allowing animals to eat feed when they want feed. The feed is provided in feeders and is always available. Hay is commonly fed free access to cattle. Hogs can be fed concentrates free access because they will not overeat. Cattle do not receive concentrates free access because they will overeat.

Livestock on pasture are commonly provided with salt or mineral blocks free access. Of course, the pasture is free access. Livestock can graze as they wish. Water is also usually provided free access.

3-23. A lamb with free access to hay and grain in a combination feeder.

3-24. A cow with free access to hay. The hay ring prevents the hay from being trampled by feeding animals.

SCHEDULED FEEDING

Many animals are fed at certain times of day. Feeding times are based on needs of the animal or management practices.

Scheduled feeding is providing feed at certain times of the day. It is commonly used with cattle because they will overeat. Dairy cows are usually fed at milking time. While they are in the barn, they will each be given an individual ration.

Baby fish, known as fry, are fed once every hour. Their digestive systems will not hold much food because they are small. They are unable to store much energy and thus need to be fed frequently.

Computer feeding systems are becoming more popular with cattle and hogs. Computerized chips can be placed in ear tags or under the skin in the ear. A computer will read the personal identification number and give an amount of feed into a feeding trough.

BALANCING RATIONS

Providing the feed animals need involves two major areas: knowing the nutrient needs of the animals and knowing the nutrient contents of feedstuffs. Scientists have studied the nutrient needs of animals. Major nutrient needs in most diets are protein and energy. Minerals and vitamins are also important.

USING NUTRITION INFORMATION

Protein and energy are the nutrients in greatest amounts in rations.

Protein is stated as crude protein and is given as percentage or grams. Protein needs are higher for young, lactating, and pregnant animals. The needs of an animal must be matched with its diet. Some feedstuffs are higher in protein than others. Most concentrates have higher energy than do roughages.

Energy is from carbohydrates, fats, and some proteins in feed. Energy is stated as total digestible nutrients (TDN). The calorie is the modern way of stating energy.

Energy is measured in calories. A *calorie* is the amount of heat needed to raise the temperature of 1 gram of water 1 degree C. Most animals require a large number of calories each day. Calories in feed or as requirements are

Table 3-1. Partial Daily Nutrient Needs of Selected Animals

Species/Condition	Weight (kg)	Selected Nutrient Needs		
		Digestible Energy	Calcium	Crude Protein
Mature working horse (moderate work such as roping cattle)	900	36.2 (Mcal)	44 (g)	10.4%
Colt (weanling, 6 months)	335	23.4 (Mcal)	44 (g)	14.5%
Swine (weaned pig)	20	3,230 (kcal)	—	18.0%
Swine (pregnant sow)	140	3,340 (kcal)	—	12.0%
Cow (dairy, lactating and pregnant)	450	19.9 (Mcal)	30 (g)	973 (g)
Rabbit (maintenance, adult)	2	4,200 (kcal)	—	12.0%
Chicken (4-week-old broiler)	0.5	—	0.9%	20.0%

Mcal = 1,000,000 calories
kcal = 1,000 calories

Source: Information in this table was adapted and summarized from multiple sources.

stated as kilocalorie (kcal) or megacalorie (Mcal). A kcal is 1,000 calories. An Mcal is 1,000,000 calories. Scientists have determined the energy in feedstuffs as well as the daily energy requirements of many animals.

The National Research Council (NRC) has prepared tables that provide the nutrient composition of feedstuff. All feedstuff has been given a number. That number is an international feed number (IFN). The number standardizes the identification of feedstuffs worldwide. For example, the IFN for #2 shelled corn is 4-02-931.

Nutritional information about feeds is used to formulate rations. The amount of each nutrient is figured into the ration. This is based on

```
7 16374 20070 9

CO-OP
FRY FISH
FEED

—Manufactured By—
ALABAMA FARMERS
COOPERATIVE, INC.
DECATUR, ALABAMA 35609
```

GUARANTEED ANALYSIS

Crude Protein, not less than 49.00%
Crude Fat, not less than 9.00%
Crude Fiber, not more than 2.00%

INGREDIENTS: Fish Meal, Meat Meal, Flash Dried Blood Meal, Dried Whey, Ground Corn, Fish Oil, Vitamin A Supplement, Vitamin D-3 Supplement, Vitamin E Supplement, Riboflavin Supplement, Calcium Pantothenate, Niacin Supplement, Vitamin B-12 Supplement, Choline Chloride, Menadione Sodium Bisulfite, Thiamine Mononitrate, Ascorbic Acid, Pyridoxine Hydrochloride, Folic Acid, Ethoxyquin (A preservative), Ground Limestone, and traces of Manganous Oxide, Calcium Iodate, Copper Oxide, Cobalt Carbonate, Zinc Oxide, Iron Carbonate and Sodium Selenite.

Tag Code: 021092

3-25. Sample feed label showing ingredients, analysis, and manufacturer.

Table 3-2. Examples of Nutrient Content of Selected Feedstuffs

Feed	Dry Matter	Crude Protein	Fat	Ash	NFE*	TDN**	DE***
	(%)	(%)	(%)	(%)	(%)	(%)	(Mcal/kg)
Roughages							
Alfalfa hay (mid-bloom, sun-cured)	91	17.1	3.3	7.8	37.4	52	2.46
Bermudagrass (fresh)	29	4.2	0.6	3.3	13.0	17	0.77
Clover (fresh Ladino	18	4.4	0.9	1.9	8.1	13	0.60
Millet (foxtail, fresh)	29	2.8	0.9	2.5	13.4	18	0.77
Sorghum fodder (with heads, sun-cured)	90	6.2	2.0	8.9	47.4	51	2.24
Wheat straw	90	3.2	1.8	6.9	40.4	40	1.90
Concentrates							
Barley grain (all analysis)	88	11.7	1.7	2.4	67.7	75	3.42
Corn (#2 grain)	87	8.9	4.0	1.2	71.3	80	3.47
Cotton Seed Meal (solvent)	93	41.2	4.7	6.1	28.9	70	3.27
Oats (grain, all analysis)	89	11.9	4.7	3.1	58.9	69	3.00
Soybean meal (solvent)	89	44.4	1.5	6.4	30.6	76	1.45
Other							
Molasses (black strap)	74	4.3	0.2	9.8	59.7	60	2.68

*NFE = nitrogen-free extract.
**Based on ruminant digestion.
***DE = digestible energy.

Source: Information in this table adapted from feed composition tables in M.E. Ensminger and R. C. Perry, *Beef Cattle Science*, Seventh Edition. Danville, Illinois: Interstate Publishers, Inc.

3-26. A mill that manufactures feed for a feedlot.

the nutrient requirements of the animal. The information tells how much roughage, concentrate, and supplement are needed. This is done with a calculator or with a computer program. Feed specialists formulate rations for specific needs.

In ration formulation, the feedstuffs are selected to give the least cost ration. Computer programs can also compute feed rations that are economical as well as nutrient balanced. For example, more corn may be in a ration if it costs less than barley.

3-27. Manufactured feed may be sold in bags.

3-28. Feed may be delivered in bulk using special trucks.

Manufactured feeds are made to contain specific nutrients. Nutrients are often stated as percentages. Labels attached to feed containers provide nutrient information. Labels give the amount of protein, the ingredients, and the name and address of the manufacturer. Labels may also have a bar code. Bar codes are electronically scanned and used in tracking a feed.

Using PEARSON SQUARE

The *Pearson Square method* is a simple way to calculate a ration for a specific animal. It can also be used to calculate ingredients for batches of feed. The method is satisfactory with some animal production. The calculated ration may be adequate in terms of protein but deficient in minerals and vitamins.

Using the Pearson Square method requires information on (1) the nutrient needs of the animal and (2) the nutrient content of feed. Information on the nutrient content of feed is provided in feed composition tables. The nutrient needs of animals are available in feeding standards tables.

The feedstuffs chosen must be appropriate and practical. It is possible to calculate a ration that is impractical, such as one based on an animal eating a greater quantity of low-quality roughage than it can possibly hold.

Use the example in Figure 3-29 as a guide for calculating other rations. Refer to nutrition books for additional information on feedstuffs and feeding standards.

STEPS	EXAMPLE
Step 1. Draw a 1- to 2-inch square. Place diagonal lines across the square. **Step 2.** Write the percentage of crude protein needed by the animal in the center of the square where the diagonal lines cross. **Step 3.** Write the feeds to be used at each left corner. Place the percent of crude protein in the feeds after the name of feed.	**Weaned pig weighing 20 kg (44 lbs.)** Corn (#2 grain) 8.9 Soybean meal 44.4 18.0
Step 4. Subtract the smaller of the numbers from the larger numbers. (This involves crude protein needed by the animal and that provided by the feed.) Write the difference at opposite corners.	8.9 26.4 parts corn 18.0 44.4 9.1 parts soybean meal
Step 5. The numbers at the two right corners are parts of the two feed ingredients that are needed. (Parts can be measured as weight or volume just so the proportion remains as was calculated.)	26.4 parts corn + 9.1 parts soybean meal **35.5 Total Parts**
Step 6. The percentage of each feed needed in the ration can be found by dividing the number of parts by the total parts.	$26.4 \div 35.5 = 74.4\%$ corn $9.1 \div 35.5 = 25.6\%$ soybean meal
Step 7. The amount of each feed ingredient for a large batch of feed is determined by multiplying the percentage of each by the total amount of feed desired.	If batch = 1 ton (2,000 lbs.): shelled corn = 1,488 lbs. soybean meal = 512 lbs. 2,000 lbs. Note: Vitamin and mineral supplements may be needed in the feed.

3-29. How to use the Pearson Square method.

REVIEWING

MAIN IDEAS

Nutrition is the process by which animals eat and use food. The food is used for maintenance, growth, reproduction, lactation, and activity. Nutrients are substances necessary for an organism to live, grow, and function. The nutrients that all animals require include water, carbohydrates, lipids (fats), proteins, minerals, and vitamins.

A maintenance diet is one in which the animal consumes enough to prevent loss or gain of the body. It is high in carbohydrates and fats. Growing animals need more protein, vitamins, and minerals. A reproduction ration includes extra protein, minerals, and vitamins for the fetus. Lactating animals require additional protein and minerals. Work utilizes high energy feed.

Feedstuffs must be appropriate to the feeding groups and digestive systems. The feeding groups are herbivores (eat plant foods), carnivores (eat animal flesh), and omnivores (eat both plant and flesh foods). Digestive systems are monogastric, ruminant, and nonruminant herbivores.

Forages, concentrates, and supplements are the three different types of feed. Forages are feeds made up of leaves and stalks of plants. Concentrates are feeds high in energy and low in fiber. Supplements are high in protein, minerals, and vitamins.

Animals can be fed free access at all times or they can be fed at scheduled times.

Feed analysis is used in balancing rations. Nutrition information from the National Research Council is needed. The nutrient content of feed and the needs of the animal are considered in balancing a ration. Feed manufacturing companies use computer programs to formulate rations. The Pearson Square method is a simple way to balance a ration.

QUESTIONS

Answer the following questions using correct spelling and complete sentences.

1. What is nutrition? Why is it important to understand the importance of nutrition?

2. What is a ration? Balanced ration?

3. What are nutrients? What are the six nutrients that all animals require?

4. Describe the functions of water.

5. What are carbohydrates? Describe the functions of carbohydrates.

6. What are lipids (fats)? What is the main function of fats?

7. What is the primary function of proteins?

8. What are the feeding groups? Briefly explain each.

9. What are the three types of digestive systems? Briefly explain each.

10. What are minerals? Why are minerals important?

11. Define vitamins. Why are vitamins required by animals?

12. Compare and contrast the five nutrient requirements of animals.

13. Compare and contrast the three types of feeds.

14. Describe the characteristics of good feedstuffs.

15. How are animals fed?

16. What is the Pearson Square method?

EVALUATING

Match the term with the correct definition. Write the letter by the term in the blank provided.

a. nutrition
b. ration
c. fiber
d. protein

e. lactation ration
f. feedstuff
g. concentrate
h. supplement

i. free access
j. scheduled feeding

_____ 1. Providing feed at a certain time.

_____ 2. A nutrient used to build new tissue and repair old tissue.

_____ 3. The process by which animals eat and use food.

_____ 4. The material left after food has been digested.

_____ 5. The amount of feed an animal has in a 24-hour period.

_____ 6. A ration for females producing milk.

_____ 7. Ingredient used to make feed.

_____ 8. Material fed animals for nutrients.

_____ 9. Feed material high in a specific nutrient.

_____ 10. When feed is available for animals when they want to eat.

EXPLORING

1. Visit a local feed mill. Determine the kinds of feeds sold and used by local animal farms. Look at how the feed is packaged. Observe where the ingredients are stored. Prepare a report on your observations.

2. Interview a local veterinarian concerning animal nutrition and the common problems that they deal with.

3. Interview a nutritionist at a feed company about animal feeding and rations and common problems with which they deal.

4. Tour a local farm. Compare the various rations and feedstuffs that are fed to different animals depending on their ages and uses.

5. Use the Pearson Square method and calculate the ration for an animal.

6. Get a label from an animal feed container, such as horse, cat, dog, or fish feed. Study the label and list the following information: kind of animal to be fed, manufacturer, major ingredients, percent protein, percent fat, calories, and percent fiber. What is your assessment of the quality of the food: good, poor, or unsure? Why?

4

ANIMAL HEALTH

People want animals to be healthy. Having animals with good health is easy to do by following practices that promote health.

Animals that do not have good health fail to perform as we want them to. They do not grow as well and are not as active as healthy animals. The health of animals is important in the production of meat, milk, eggs, and other products.

Some animal diseases are spread to humans. This makes disease control more important. Since people often have companion animals in their homes, they want to have animals that feel well and have a good appearance.

Producers need to be aware of the principles in having animals with good health. Each species and each animal within a species is unique. Sometimes, the assistance of a veterinarian may be needed. Regardless, it is far better to prevent health problems than to try to treat them.

4-1. Healthy animals are fun to raise and show.

OBJECTIVES

This chapter provides guidelines to help producers in areas of animal health. It has the following objectives:

1   Explain health and how it relates to normal and abnormal behavior

2   Explain the impact of environment on animal health

3   List and explain the economic losses caused by poor animal health

4   Describe some common diseases and parasites of animals

5   Explain how good health is maintained

TERMS

anaplasmosis	immunity
anthrax	implant
antibiotic	injection
antibodies	internal parasite
bacteria	isolation
balling gun	leptospirosis
behavior	lice
biological	mad cow disease
blackleg	mastitis
brucellosis	medication
coccidiosis	noncontagious disease
colostrum	oral medication
contagious disease	parasite
disease	pharmaceutical
disinfectant	preconditioning
dose syringe	protozoa
ectotherm	rabies
endotherm	sanitation
external parasite	shipping fever
fungi	sleeping sickness
grubs	topical medication
hog cholera	virus

HEALTH AND BEHAVIOR

Health is the condition of an animal in terms of how the functions of life are being performed. An animal is in good health when its life processes are normal. Behavior is related to health. *Behavior* is the reaction of an animal to certain stimuli. It is the manner in which an animal reacts to its environment.

Disease is a disturbance in the functions or structure of an animal. Sometimes, only one organ or body part may be affected. Other times, the entire animal may show disease. In short, disease is the lack of good health.

Healthy animals vary. Some animals that appear healthy may be suffering from pain or disease. Knowing each individual animal will help to easily detect and correct any ill health. An animal keeper should learn the signs of good and ill health.

4-2. A healthy ewe and lamb.

GOOD HEALTH

Good health is the absence of disease. Some signs of good health are: good appetite, alert and content behavior, bright eyes, shiny coat, normal feces and urine, normal vital signs, and normal reproduction.

Healthy animals will eagerly eat their food. A lingering or lazy animal may be in poor health. Healthy animals will stretch and watch the world around them. Eyes that are clear and alive are very important signs. A shiny

Table 4-1. Normal Signs of Selected Adult Animals

Species	Average Normal Temperature	Normal Pulse Rate	Normal Respiration Rate
	(rectal °F)	(rate/min.)	(rate/min.)
Cattle	101.5	60–70	10–30
Swine	102.6	60–80	8–15
Sheep	102.3	70–80	12–20
Goat	103.8	70–80	12–20
Horse	100.5	32–44	8–16
Chicken	106.0	200–400	15–30
Cat	103.0	140–240	20–40
Rabbit	102.5	140–150	50–60
Dog	101.3	60–120	10–30

coat is evidence of a proper diet. The presence of blood or mucus in the feces is a sure sign of a problem. Important vital signs to look for include temperature, pulse, and respiration. Table 4-1 lists the normal vital signs of various animals.

4-3. A swollen eyelid and discharge around the eye are signs of an eye disease.

ILL HEALTH

Animals with disease often have behavior that is not normal. If the signs of a healthy animal are not observed, then the animal is in ill health.

Signs of ill health may include one or more of the following: lack of appetite, sunken eyes or discharge from the eyes, discharge from the mouth or nostrils, inactivity, rapid breathing, rapid pulse rate, high temperature, full hair coat, lumps or protrusions on the body, open sores, seclusion, bloody urine or feces, and loss in production levels or weight.

IMPACT OF ENVIRONMENT ON ANIMAL HEALTH

Undue stress can be caused by an animal's environment. A change in the environment or an unsuitable environment will create stress. This causes the animal to become more susceptible to diseases and parasites.

Important factors affecting the health of an animal include temperature, light, moisture, moving, and pollution.

TEMPERATURE

Specific temperature ranges are best for different animal species. If the temperature is too high or too low, health problems may occur.

Connection

UNRAVELING HOLLOW TAIL

Hollow tail was once a problem that some cattle had in the late winter. Producers were concerned. The cattle were quite thin, sluggish, and had rough hair coats. Some got down and died. Examination revealed that the tail was hollow (hence the name hollow tail).

Producers began a treatment for hollow tail. They would cut into the tail with a knife and fill the hollow with salt. The animals would jump, run, and do most anything to try to escape the pain. No longer sluggish, the cattle were said to be recovering.

Hollow tail was unraveled by researchers. They found that the hollow in the tail was due to severe malnutrition. The animals were starving and had used fat and other tissue in the tail to support life processes. The animals were often wormy. What was being done to treat hollow tail was wrong. It did not reflect the real problem nor animal well-being. Producers now know that hollow tail results from not having enough food.

This photo shows a cow with the signs of hollow tail. With feed and care, she will overcome the condition.

4-4. As ectothermic animals, these discus fish require water that provides a good environment.

Animals are endothermic or ectothermic. An **endotherm** is an animal that strives to maintain a constant body temperature. Disease and extreme temperature conditions may make it impossible to do so. Dogs, cats, and horses are examples of endothermic animals.

An *ectotherm* is an animal that adjusts its body temperature to that of its environment. Fish, reptiles, and frogs are examples of ectotherms.

Endothermic animals adjust to temperature extremes as best possible. During hot weather they may sweat, pant, or wade in cool water. Confined animals need proper ventilation and fans to keep cool. Animals on pasture need plenty of fresh water and shade trees.

During cold weather, animals will burn more energy to produce body heat. They must be fed more carbohydrates and fats to produce this heat.

4-5. These new-born piglets have an environment designed just for them with a heat lamp and facility to protect them from being laid on by the sow.

The muscles of the animal will involuntarily contract or shiver to burn energy and keep the body warm.

Younger animals have a more difficult time maintaining body temperature than older animals. They require protection from weather extremes.

LIGHT

Light affects most animals. Chickens and lactating dairy cows are likely affected by light more than other animals. However, animals with light-colored skin, such as hogs, may become burned by too much sunshine. In laying hens, light causes the pituitary gland to release hormones into the bloodstream. This causes them to grow faster and lay more eggs. Electric lights are often on in broiler and laying houses to take advantage of this.

Color of the animal affects its reaction with light. This is because darker colors absorb light rays and lighter colors reflect light rays. Cattle with light colors around their eyes are more likely to have eye disease. Animals with light-colored skin may be more susceptible to disease. For example, white cats are more likely to get skin disease than those of darker colors.

4-6. These young chicks are in a lighted environment with nutritious feed, good water, clean litter, and heat.

MOISTURE

Moisture is due to humidity and precipitation. Air with high humidity (moisture in the air) will not cool animals as well as dry air. The perspiration on the skin of animals will not evaporate as fast in a humid location.

Diseases live and reproduce in moisture. Animals raised in confinement are more likely to have disease problems. Ventilation allows air to move about. It is needed to remove the moisture from a confined barn. Proper ventilation lowers the inside temperature and decreases the chance of a disease outbreak.

4-7. Companion animals should be lifted and handled to prevent injury. This shows a child enjoying a Yorkshire Terrier.

MOVING

Lifting and moving animals creates stress. Stress lowers resistance to disease. When animals are corralled, hauled, or chased onto trucks, their stress level increases. Loading and hauling animals exposes them to environmental conditions, such as rain, snow, wind, and hot or cold temperatures. Trailers and loading chutes should be free of protruding nails, broken boards, wire, and debris. Animals can be injured by sharp objects. Animals shipped on airplanes have stress. Water, temperature, and food needs must be met.

Animals in good physical condition are better able to cope with stress. Producers may precondition animals. This is done by gradually acclimating them to their new environment.

POLLUTION

Pollution of water sources, feed, and the environment may cause animals to develop a disease. Water and feed sources may be poisoned by pesticides or industrial waste. Storing or hauling feeds in areas that have contained chemicals may contaminate feedstuffs. Pastures sprayed with chemicals may pose a pollution problem.

LOSSES DUE TO POOR HEALTH

It is easy to see the economic loss to producers if animals die. The greatest losses due to poor animal health are less obvious. Diseases lower production and threaten humans.

DEATH

Death is the most obvious loss due to poor animal health. Dead animals are of no value. The producer has lost all that was invested. Costs go up because of the expense of disposing of the animals.

Dead animals should be buried, burned, or, with chickens, composted. Producers may need help with the proper methods of disposal. A veterinarian or health official is a good source of information. Improperly disposing of animals may create bad odors and attract coyotes or wild dogs; who may, in turn, spread disease to other animals and people.

LOWER PRODUCTION

Healthy animals mean increased profits for owners. Poor health may result in lower production levels. Some easy-to-see results of poor health are failure to reproduce or breed, slow growth rates, decreased milk production, decreased meat production and quality, and increased costs of production. All of these lead to decreased profits for producers.

HUMAN DISEASE

Diseased animals may spread diseases to humans. Examples are anthrax, brucellosis, leptospirosis, rabies, trichinosis, and tuberculosis.

Brucellosis or Bang's disease is found in cattle. Humans may get undulant fever from cattle infected with brucellosis. Drinking raw milk from an infected cow is one way humans may contract the disease. Another source of contamination is getting milk or other fluids from an infected animal on open wounds. Dairy cattle are regularly vaccinated and checked for this disease. Pasteurization makes milk safe for humans to drink.

Trichinosis is caused by a small worm in the flesh of infected hogs. When pork is not cooked thoroughly, the disease may be transferred to humans. Very few hogs have this disease. People have stopped feeding raw garbage with pork scraps in it to hogs. Properly cooked pork and clean cooking areas

4-8. A healthy Mexican Redhead Amazon Parrot is appealing.

have eliminated this disease. Always wash your hands after touching raw meat.

COMMON DISEASES AND PARASITES

Disease is any disorder that keeps animals from carrying out life processes in normal ways. Symptoms are not the disease itself, but a sign of a problem.

CLASSIFICATION OF DISEASES

Knowing how a disease is spread will help to better control and prevent it. Diseases are either contagious or noncontagious.

Contagious Diseases

A *contagious disease* is a disease spread by direct or indirect contact with other animals. Pathogens, such as viruses, bacteria, fungi, protozoa, and parasites, cause contagious diseases. Pathogens are organisms that can produce disease. They may produce a toxin or poison that upsets the normal metabolism of the body. They are small and reproduce very rapidly.

4-9. A blood sample is being taken from the tail of a cow for analysis for disease organisms or abnormalities.

Viral Diseases. A *virus* is a tiny disease-producing particle too small to be seen with an ordinary microscope. Viruses closely resemble the DNA of animal cells. This makes them hard to control. It is difficult to find a chemical that will kill the virus without killing the animal's DNA. Vaccinating animals is the most effective way of controlling viruses. Examples of viral diseases include rabies, hog cholera, smallpox, and distemper.

Bacteria. *Bacteria* are one-celled organisms. They are microbes that are sometimes called germs. Many pathogenic bacteria form spores that become resistant to control measures. Examples of diseases caused by bacteria include tuberculosis, brucellosis, leptospirosis, mastitis, and tetanus.

Fungal Diseases. *Fungi* are unicellular organisms that generally cause dis-

4-10. A microscope is used to check blood samples for disease.

eases on the outside of the body. Conditions caused by fungal diseases may result in secondary infections. Bacteria may be responsible for secondary infections. Examples of fungal diseases are ringworm and fescue foot.

Protozoa. *Protozoa* are unicellular organisms. They are the simplest forms of animal life. Examples of diseases caused by protozoa include malaria and anaplasmosis.

Parasites. A *parasite* is a multicellular organism that lives in or on another animal. Parasites receive their nutrients from their host animal.

An ***internal parasite*** lives inside the host. These parasites are found in the digestive system, muscles, or other tissues. Internal parasites enter the host when it is eating contaminated feed or when eating feed on a dirty floor. Most internal parasites are obtained by feces being in food. Pastures with short grass increase the likelihood of ingesting parasites. Examples of internal parasites are tapeworms, trichina, roundworms, cattle grubs, and hookworms.

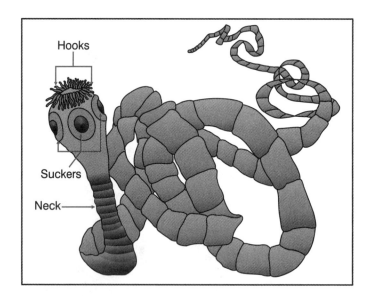

4-11. The tapeworm is an internal parasite. It is a flat, segmented worm with suckers and hooks for attaching to the intestines. Tapeworms can grow to several feet long in large animals.

An ***external parasite*** lives on the external parts of the animal. They obtain their food by biting or sucking the blood and tissue of the animal. External parasites can transfer contagious diseases or other parasites to other animals. Examples of external parasites include ticks, fleas, lice, mites, and leeches.

4-12. A cow is being checked for external parasites. (Courtesy, Agricultural Research Service, USDA)

Noncontagious Diseases

A ***noncontagious disease*** is a disease that is not spread by casual contact. It is caused by a nutritional, physiological, or morphological problem.

Nutritional. Most nutritional diseases result from an unbalanced diet. Animals require specific amounts of fats, carbohydrates, proteins, minerals, and vitamins. Too high or too low amounts of any of these may result in a nutritional disease and decrease profitability. Mineral and protein deficiencies are common. (Chapter 3 discusses the nutrition of animals.)

Physiological. This type of disease results from a defect of a tissue, organ, or organ system. When an organ causes the body's metabolism to be off balance, other problems occur. Examples of physiological diseases include milk fever, heart failure, birth defects, and acetonemia.

Morphological. Morphological diseases are birth defects and physical injuries to animals. Poor management can result in accident or injury to animals. Possible injuries include cuts, scrapes, scratches, bruises, and broken bones. These decrease the efficiency of the animal. Good management practices can decrease the likelihood of these problems. Pick up loose wire, old lumber, and equipment. Remove protruding nails and broken boards. Cattle may be given magnets to collect nails and wire that they eat. Feed mills often run feed over magnets to remove any metal particles that may be in the ingredients.

4-13. A knee problem on a dairy cow likely resulting from an injury.

SELECTED DISEASES OF ANIMALS

Some diseases affect only certain species while others affect several species. Roundworms are found in hogs, cattle, sheep, horses, poultry, and fish. Blackleg primarily affects cattle.

Several common animal diseases are briefly described here. Proper diagnosis and treatment of a disease may require a veterinarian.

- Anaplasmosis—*Anaplasmosis* is a parasitic disease caused by a protozoan that attacks red blood corpuscles. Cattle are primarily affected. In chronic anaplasmosis, animals become anemic and may overcome the disease on their own. In acute forms, animals have rapid heartbeat, muscular tremors, and loss of appetite. Sick animals may become aggressive and want to fight. Death may occur in a few days. The disease is prevented by immunizing cattle. Some treatments are available, if the disease is caught early.

- Anthrax—*Anthrax* is an acute infectious disease that affects most endothermic animals. It most frequently affects cattle during the summer when they are on pasture. Affected animals have a fever, rapid respiration, and swelling on the neck. Animals will die suddenly. Prevention of anthrax includes vaccinating animals, controlling flies, and sanitizing their environment. Penicillin may be effective if given in large doses.

- Bang's (Brucellosis)—Bang's or *brucellosis* affects cattle, sheep, goats, and hogs. The reproductive tract of the female is infected. Developing fetuses may be aborted. Cows may have to be bred several times before they become pregnant. When they do become pregnant, abortion may end pregnancy in only a

4-14. The paw of a captive wolf is being examined. (Courtesy, U.S. Fish and Wildlife Service)

few weeks. Bang's is prevented by vaccinating heifers. Sanitation, testing cattle, and bringing only Bang's-free cattle into a herd help to control this disease. Humans can get this disease from infected animals or their milk. In humans, this is known as undulant fever. If humans survive undulant fever, they may suffer permanent crippling and disability.

- Blackleg—*Blackleg* is an acute, highly infectious disease that usually results in death. It primarily affects cattle, although sheep and goats can also get it. Symptoms include high fever; swelling in the neck and shoulder; and muscles in the neck, shoulder, and thighs crackle when mashed. Eventually animals lose their appetite and die. Few animals recover from blackleg. Blackleg is prevented by vaccinating calves with a bacterin, which is a vaccine made from killed or inactive bacteria.

4-15. The teeth of an animal are being examined. (Courtesy, U.S. Fish and Wildlife Service)

■ Hog Cholera—**Hog cholera** is a viral disease that is highly contagious. Hogs get high fevers, lose their appetite, get weak, and drink lots of water. Considerable effort has been spent eradicating hog cholera in the United States through federal government programs. There is no known treatment. Any hogs with hog cholera are destroyed.

■ Coccidiosis—**Coccidiosis** is a parasitic disease affecting chickens, turkeys, ducks, geese, and game birds. It is caused by a protozoan. Symptoms include bloody feces, ruffled feathers, unthrifty or sick appearance, and pale coloring. Medications known as anticoccidials are put in feed and water to prevent and treat this disease. Since it is transmitted by infected birds, wild birds should be kept away from poultry flocks and infected areas should be sanitized.

■ Equine Sleeping Sickness—Horses are affected by **sleeping sickness**. It is caused by a virus transmitted by insects that bite horses, mules, and wild rodents. Affected animals walk aimlessly crashing into things; later they appear sleepy, cannot swallow, grind their teeth, and possibly go blind. Some horses will recover, but they will usually die within two to four days. No effective treatment has been developed, but horses can be vaccinated to develop immunity.

■ Mad Cow Disease (Bovine Spongiform Encephalopathy or BSE)—**Mad cow disease** became prominent in the mid-1980s because of outbreaks in Great Britain and Europe. The disease has not been diagnosed in the United States, but it has affected the importation of animals. Mad cow is a serious neurological disease with forms likely transmissable to humans, sheep, and goats. In humans it is known as Creutzfeldt-Jacob Disease (CJD). Outbreaks of mad cow have been traced to feeding cattle meat meals made from sheep carcasses where the sheep had a disease known as scrapie. There are no effective treatments.

■ Grubs (Warbles)—Cattle **grubs** are internal parasites caused by heel flies. The fly lays its eggs in the summer around the heels of cattle. When they hatch as larva, they go into the skin and move through the body until they reach the back in a few months. In the back, the grubs cause bumps to swell in late winter and spring. The larva (about $3/4$ inch or 2 cm long) will come out of the bumps and grow into an adult fly. A good fly control program will prevent

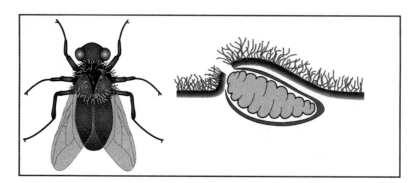

4-16. Drawing of a cattle grub fly and larva under the skin. Larva damage the hide and meat.

grubs. Several treatments are available and should be followed according to their directions. Grubs damage the hide by making holes in it as well as the flesh around it. Cattle grubs cause more injury to cattle than any other insect.

■ Leptospirosis—*Leptospirosis* is a bacterial disease that affects cattle, dogs, sheep, and most other farm and companion animals. Symptoms include high fever, poor appetite, and bloody urine. Females will abort their fetuses. Antibiotics are sometimes used to treat leptospirosis. Animals may be vaccinated against this disease.

■ Lice—*Lice* are external parasites that attack cattle, hogs, and other species. Lice are small insects that suck their host's blood. Animals may become anemic. Lice also cause animals to itch; they will often be seen scratching or rubbing against trees or posts. Lice usually appear in the winter. Back rubbers and other means can be used to administer insecticides to control lice.

■ Mastitis—*Mastitis* is the leading cause of profit loss in dairy cattle. Mastitis is a bacterial disease that affects female cattle, sheep, goats, and swine. It is an inflammation in the udder that interferes with milk production. Chronic mastitis causes the milk to be thick or lumpy. Acute mastitis is indicated by a fever and a hard, warm udder. Laboratory tests are used to detect most mastitis. Chronic mastitis that is not treated can cause the death of the animal.

■ Poisonous Plants—Poisonous plants are sometimes problems. Examples of poisonous plants include low larkspur, oak, tall larkspur, timber milk vetch, and water hemlock. Most poisonous plants cannot be totally eliminated. Possible loss is reduced with good pasture management. Routine weed control will kill most poisonous plants. Knowing the common poisonous plants in the area prevents losses. Study the symptoms of poisoned animals. Provide good pasture and feed for range animals. Promptly treat animals that may have been poisoned.

4-17. A poisonous plant garden is maintained by the College of Veterinary Medicine at Mississippi State University. Veterinary medical students are trained in plant identification.

■ Rabies—*Rabies* is a disease that occurs in nearly all endothermic (warm-blooded) animals. It can be transferred to humans. Caused by a virus, rabies results in death. Signs include irritable and aggressive behavior, avoiding light, fever, vomiting, and diarrhea. Some animals become frenzied, vicious, and attack everything that moves. Lips may be drawn to expose the teeth. Once the disease develops, it is fatal. Incubation is three to eight weeks. Autopsy of the brain is a sure method of diagnosis. Vaccination can prevent rabies. Dogs, cats, bats, foxes, skunks, and raccoons are examples of animals where rabies may be a problem. In recent years, skunks have had 40 percent of the diagnosed cases. Never touch or pet an animal that you do not know. Humans bitten or scratched by an infected animal should receive immediate medical attention.

■ Roundworms and Tapeworms—Roundworms, tapeworms, and other worms are parasites that can infect all animals. They are found in the intestines and the stomach. Tapeworms can grow several feet long. When they become this large, they use much of the feed that the animal has eaten. Animals with low infestations of small-sized worms may not show symptoms. As the number and size of worms in the digestive tract increases, the animal loses weight, becomes anemic, and may have diarrhea. Sanitation is a good control measure.

■ Shipping Fever—Cattle and sheep of all ages can get *shipping fever*, though it is more of a problem for younger animals. It is an environmental disease caused by conditions animals encounter when hauled or sold. Animals will develop a high temperature, discharge from the eyes and nostrils, and may cough. Most will have difficulty breathing. It is more likely to affect thin, underfed animals that are hauled long distances. Cattle should be vaccinated three to four weeks before being moved.

■ Poisonous feed—Animals sometimes eat poisonous food. Feed may contain substances that poison animals. Molded feedstuffs should be avoided. Poisonous plants may grow in pastures and create problems with animals that graze. Feed contamination is sometimes a source of poisoning. Some animals are allergic to specific foods. Companion animals typically do not do well on human foods. For example, dogs should not be fed chocolate.

MAINTAINING GOOD HEALTH

A key to success with animals is keeping them healthy. Using good management will increase production and decrease losses due to poor health. An understanding of the animal's bodily defenses will lead to healthy management practices.

4-18. Many animal owners have routine health exams of their animals. (Courtesy, College of Veterinary Medicine, Texas A&M University)

Body Defenses

The first line of defense of an animal is the epithelial tissue or skin and mucous membranes. Once a foreign substance has gotten past the epithelial tissue the body uses other mechanisms.

The digestive tract is very high in acidity. This depresses bacterial growth. Tissue fluids (lymph) contain leukocytes that neutralize or engulf pathogens. The liver and other lymph organs will trap pathogens until they can be destroyed.

When an inflammatory reaction occurs, redness will be present. Redness is due to blood flow to the infected area. The area will feel hot and swelling or edema will occur. Increased sensitivity or pain is felt by the animal. All of these are the body's attempt at destroying pathogens. In addition, metabolic activity and immune reaction (development of antibodies) will increase.

Health Management Practices

Animal keepers should use practices to maintain good animal health. Several practices are described here.

Environmental Conditions

The environmental conditions that the animal lives in must be checked regularly. Space requirements should be met. Allow enough room for animals; do not overcrowd. Ventilation is needed to have a continuous supply of fresh and clean air. Providing a clean and dry place to lay helps to decrease the disease potential.

Sanitation

Keeping areas where animals are raised and fed clean is *sanitation*. Filth carries disease. Sanitation decreases the chances of animals contracting disease.

Dead animals should be removed and disposed of properly. Barns and facilities should be cleaned regularly. Waste and manure should be disposed of properly and regularly. Many farms have fenced lagoons or pits to temporarily store animal waste.

A *disinfectant* is a substance that destroys the causes of disease. Facilities, especially where young are raised, should be cleaned with a disinfectant regularly. Disinfectants include alcohol, iodine, lime, chlorine bleach, and soap. People should always follow safety precautions in using disinfectants.

Clean facilities increase the natural beauty of the surroundings and lessen susceptibility of disease.

Proper Nutrition

Animals that have proper nutrition are more resistant to disease and stress. Providing the proper ration is important. Poorly fed, weak animals do not have the resistance needed to fend off diseases.

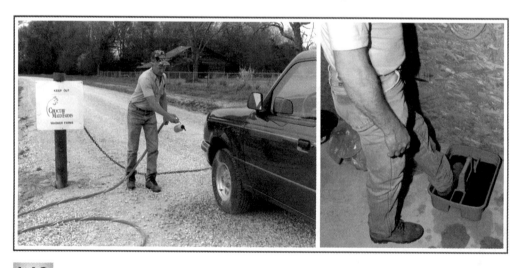

4-19. Two practices in sanitation are using a sanitizing solution on the tires of vehicles that come onto a farm and using a dip for shoe soles.

Isolation

Isolation is separating diseased and non-diseased animals. Diseases are often spread by animal contact. Isolating diseased animals decreases the risk of spreading a disease.

New animals brought to a farm should be isolated for three to four weeks. If the animals have a disease, the signs should be visible in that time. If no diseases are seen, the animals may be turned in with the herd or flock.

Restrict Truck and Equipment Traffic

Diseases can be brought to a farm by trucks, farm equipment, and other vehicles. Trucks and equipment should be disinfected between trips to other farms or feedlots. Mud, manure, sawdust, and straw can harbor disease.

Restrict Human Access

People may bring diseases with them as they travel from farm to farm. Diseases can be transferred on shoes or boots. Some farms require everyone to walk through a boot tub of disinfectant solution. Other places restrict access to facilities to only a few individuals. Breeding farms may require visitors to wear plastic slip-on boots or to view the facility with video cameras.

Preconditioning

Preparing animals for stress is **preconditioning**. Stocker calves and other animals that are to be hauled are often preconditioned. Preconditioning is important with any activity that might cause stress. Castration and dehorning should be done by the time cattle are two months old, so as not to coincide with transfer to another lot or weaning. Calves should be weaned and started on feed thirty days before being sold. The amount of handling and moving time should be kept to a minimum.

Immunization

Immunity means that an animal resists disease. An animal often develops immunity when it has a disease. Artificial immunization is used to protect animals from disease.

Antibodies are immune substances produced in the body. Immunity can be developed to many different diseases, thus allowing animals to withstand exposure to disease even when coming in contact with it.

4-20. Teats of a dairy cow are cleaned before the milkers are attached to assure wholesome milk. (Courtesy, Ann Clark, Kentucky)

Vaccines, serums, bacterins, and toxoids are used to help animals develop artificial immunity to diseases. Animals should be vaccinated to prevent common diseases.

Mothers often pass immunity to newborn young. Young mammals get immunity by antibody transfer from the mother through the placenta and colostrum. The placenta is the organ that unites a fetus with the uterus in which it is developing. *Colostrum* is a substance found in milk shortly after giving birth. The presence of colostrum quickly drops a few hours after birth.

Animals may be fed vaccines in their feed or water or given shots. Most vaccines contain dead or modified living organisms that cause the disease.

TREATING DISEASE

Animals that get a disease need medication to help them in trying to overcome the disease. A *medication* is a substance that is used to prevent and control disease. Animal producers are well aware that prevention is preferred to treatment. Treatments for some diseases may not be effective.

The assistance of a qualified person may be needed to identify and treat a disease. A disease must be properly identified before a treatment is selected.

KINDS OF MEDICATIONS

Several types of medications are available. Some can only be used by a licensed veterinarian. Proper use is essential. The animal being treated can be injured or killed by improper use.

Two general kinds of medications are used: biologicals and pharmaceuticals. A **biological** is a medicine primarily used to prevent disease. The purpose is often to develop immunity in an animal. A **pharmaceutical** is a medicine used to treat a diseased animal. Pharmaceuticals reduce or prevent the continued growth of microbes that cause disease.

- Antibiotics—An **antibiotic** is a substance produced by one organism that will inhibit or kill another organism. Antibiotics are used in diluted form. Antibiotics are used to treat diseases caused by microbes. Penicillin, Terramycin, and streptomycin are the best known antibiotics. These should be used only when needed. Overusing antibiotics may result in the disease pathogens developing resistance to them.

4-21. An antibiotic that is given with a syringe and hypodermic needle.

- Pesticides—Pesticides are used on parasites. These should be carefully used. Both internal and external parasites are controlled with pesticides. Wormers, insecticides, and miticides arc cxamplcs. Wormcrs arc also known as anthelmintics.

- Dietary supplements—A dietary supplement is a substance that provides vitamins, minerals, or other nutrients that help animals overcome disease.

- Sulfonamides—The sulfonamides are commonly known as sulfa drugs. Several sulfa drugs are used. These agents stop the growth of infecting germs without poisoning the animal. The sulfas have been used for many years.

4-22. A topical medicine used to treat fungal diseases.

■ Others—Many other medications are available. These often use destroyed or altered germs. Vaccines, serums, bacterins, and toxoids are most common. In addition to treatment, most of these are primarily to prevent disease by developing immunity.

ADMINISTERING MEDICATIONS

Medications must reach sources of infection or need. This requires proper administration. Several methods are used. Select the method based on the type of medication and kind of disease. Be sure the medication is approved for the intended use.

Additives

Additives are materials placed in food and water. A variety of medications are suited to this method. The materials must be added at the proper rate. As the animal eats, medication is ingested. The feed and water must be used only for the animals as approved. The use of feed and water with additives must usually be stopped for a period before animals or their products are consumed.

Injections

An *injection* is a medication given directly into the bloodstream, tissues, muscle, or body cavity. Injections are known as shots. A hypodermic needle

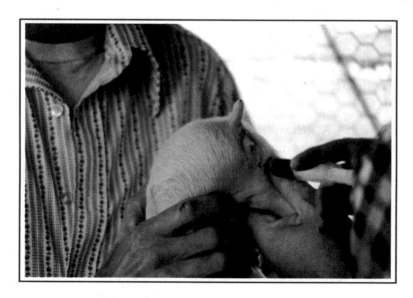

4-23. A piglet is being injected with a vaccine to develop immunity to disease.

and syringe are typically used. Sanitation around injection sites is important. This may involve washing the area with a cotton ball containing isopropyl alcohol. Careful attention to dose size is needed. Doses are measured in cubic centimeters (cc). Read the instructions for dose size.

Several kinds or sites of injections are used. Medications vary in where they are given. This information is given on the label that accompanies injectable materials. The kinds are:

■ Intradermal—An intradermal injection is made into the skin. Since the skin is often thin, care is needed. Few drugs are given this way. Tests for disease (such as TB) may involve skin injections.

■ Subcutaneous—A subcutaneous injection is made just beneath the skin. These injections are given in easy-to-get-to loose skin. With cattle, the site of the injection is often on the neck or behind the shoulder.

■ Intramuscular—An intramuscular injection is made into a muscle. The most frequent sites are in the neck and shoulders. Penicillin, for example, is given with an intramuscular injection.

■ Intravenous—An intravenous (often known as IV) injection is made into the veins of an animal. These injections are used when a quick response is needed, such as to save the life of an animal.

4-24. Wrong and correct way of placing an IV needle in a vein.

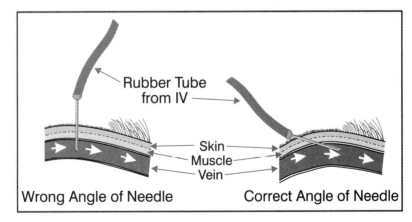

Rubber Tube from IV

Skin
Muscle
Vein

Wrong Angle of Needle   Correct Angle of Needle

■ Intranasal—An intranasal injection is given through the nose. It involves spraying the medication into the nostrils. Intranasal injections are often used with innoculants and do not involve a hypodermic needle and syringe.

■ Intraperitoneal—An intraperitoneal injection is made into the body cavity. It is most commonly used with cattle. The site is in the hollow of the flank. These are used when a quick response is needed.

■ Intramammary—An intramammary injection is made into the udder through the opening in the teat. It is used to treat mammary infections, such as mastitis in dairy cattle.

Implants

An *implant* is a small pellet that is placed underneath the skin. Most implants are made in fairly loose skin behind the ears of animals. The medication is slowly released. Implants are more widely used to promote growth or control reproduction.

Topicals

A *topical medication* is one that is placed on the skin or surface area of an animal. These medications may be sprayed, poured, or dusted on animals. In some cases, the animals are dipped into tubs or vats of solution. Backrubbers or other devices that animals rub against can be used as topical applicators. Topical medications are typically for abrasions (cuts or wounds) or to control external parasites.

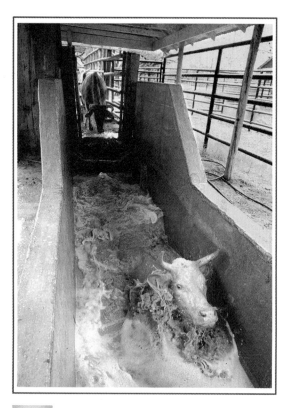

4-25. Cattle moving through a dipping vat to treat for external parasites. (Courtesy, Agricultural Research Service, USDA)

Oral

An *oral medication* is given through the mouth. The forms of medication may be liquid, pill, bolus, or as an additive to the feed or water. These medications are often for digestive problems or to control internal parasites. The sulfa drugs are also given orally. Liquids are given with a *dose syringe*. A pill or bolus may be given with a *balling gun*.

Some animals are particularly choosy about what they eat. For example, a small pill can be concealed in a piece of hotdog wiener for a dog.

4-26. A balling gun with bolus. (The bolus is placed in the end of the gun and released into the animal's throat.)

4-27. Using a dose syringe to administer liquid medicine.

With a cat, one person restrains the animal and another opens its mouth and quickly places the pill as far back as possible. The throat is stroked to promote swallowing. Be careful! Cats bite and scratch.

As with all medications, proper dosage is essential. Animals used for food or milk production must be off the medication for a while before its products are used for food.

Other

Medications may occasionally be given in other ways. Some are in capsule or liquid forms and inserted into the rectum. Medications for some re-

4-28. A beef animal is being given a vaccination by a veterinarian. (Courtesy, College of Veterinary Medicine, Texas A&M University)

productive problems may be placed in the reproductive tract. Ear disease may involve placing medications in the ear canal. Eye problems may involve placing medications directly into the eye. Always read and follow directions on medicine. Get the assistance of a veterinarian or other qualified person.

REVIEWING

MAIN IDEAS

Health is the condition of the body and a measure of how the functions of life are being performed. Health affects the behavior of the animal. Poor health causes personal and economic losses to the owner or producer. Death, low production, and human contraction of disease may result.

Signs of good health include good appetite, alert and content behavior, bright eyes, shiny coat, normal feces and urine, normal vital signs, and normal reproduction.

Signs of ill health include lack of appetite, sunken eyes or discharge from the eyes, discharge from the mouth or nostrils, inactivity, rapid breathing, rapid pulse rate, high temperature, full hair coat, lumps or protrusions on the body, open sores, seclusion, bloody urine or feces, and loss in production levels or weight.

Diseases may be contagious or noncontagious. Contagious diseases may come from viruses, bacteria, fungi, protozoa, and parasites. They are spread by direct or indirect contact with other animals. Noncontagious diseases are due to poor management practices.

Maintaining good animal health is easy if the well-being of the animals is considered. Animals must be placed in a good, clean environment that is free of debris. They should not be overcrowded and must have a clean, fresh, and nutritious supply of feed and water. Dead animals must be disposed of properly. Diseased and sick animals should be isolated. A disinfectant should be used to clean stalls, feeders, and waterers. Animals may be preconditioned for stress and vaccinated to develop artificial immunity. Use appropriate methods of administering medications.

QUESTIONS

Answer the following questions using complete sentences and correct spelling.

1. Compare and contrast health and disease.

2. How does health affect behavior?

3. Give four signs of good health.

4. Give four signs of ill health.

5. How does the environment affect health?

6. How can the environment be manipulated for the benefit of animals?

7. What are some results of poor health? Explain each.

8. Explain the difference between contagious and noncontagious diseases.

9. List three diseases of animals and describe each.

10. How does the body fight off disease?

11. List and explain several methods for maintaining good health.

12. What kinds of medications are used?

EVALUATING

Match the term with the correct definition. Write the letter by the term in the blank provided.

a. disease
b. health
c. immunity
d. disinfectant

e. contagious disease
f. isolation
g. shipping fever
h. preconditioning

i. balling gun
j. injection

_____ 1. Disease spread by direct or indirect contact among animals.

_____ 2. Substance that destroys the causes of disease.

_____ 3. Separating diseased animals from healthy animals.

_____ 4. The ability to resist disease.

_____ 5. Preparing animals for stress.

_____ 6. Caused by conditions animals encounter when hauled.

_____ 7. Disturbance in the functions of the body.

_____ 8. Condition of the body and how well life functions are being carried out.

_____ 9. Using hypodermic needles and syringes to get a substance into a body system.

_____ 10. Equipment for administering pills to large animals.

EXPLORING

1. Visit a local veterinary medical clinic and tour the facilities. Ask about the most common health problems that they deal with, including symptoms, causes, preventions, and treatment. Observe the treatment of an animal. Write a report on what you saw.

2. Conduct a job-shadowing project and follow a veterinarian, rancher, small animal producer, or animal health technician for at least a day. Write a report on your experiences.

3. Check the vital signs of an animal. Record its pulse, respiration rate, and temperature. Compare the data to normal animals. Check the vital signs over a period and observe the fluctuations. Be sure to follow the proper procedures so the animal is not injured.

4. Observe an animal that is in good health. List the signs that this animal exhibits. Observe an animal in ill health and list the signs observed that indicate ill health. Compare the two lists.

5

ANIMAL REPRODUCTION

Baby animals create excitement! People enjoy baby animals. Pet owners like a new litter of puppies or hatch of fish. Animal producers like baby animals, too, but from a different point of view. Producing new individuals is a big part of raising animals.

Animals are used for many purposes. People want a wide range of animal products. The goal of producers is to provide plenty of animals to meet the demand. If there are not enough animals, people will not have the products they want. Shortages develop and cause prices to go up.

Animal producers must raise the kinds of animals that give the products people want. The number of animals available is tied to reproduction. Producers need to understand the processes involved in reproduction.

5-1. A cow with a nursing calf represents the reproduction process of cattle.

OBJECTIVES

This chapter covers the fundamentals of animal reproduction. It has the following objectives:

1 Explain the role of animal reproduction

2 Name and describe the functions of the major reproductive organs

3 Describe the phases of the estrous cycle

4 Explain the phases of reproductive development in the life of an animal

5 Describe the role of animal reproduction technology

TERMS

accelerated lambing
anestrous
artificial insemination
breed
breeding
cloning
conception
copulation
crossbreeding
egg
embryo
embryo transfer
estrous cycle
estrous synchronization
estrus
fertilization
fetus
gamete
genetic engineering

gestation
incubation
insemination
lactation
natural insemination
ovary
ovulation
parturition
pregnant
puberty
purebred
reproductive efficiency
scrotum
sexed semen
sexual reproduction
spawning
sperm
testicle
zygote

ROLE OF ANIMAL REPRODUCTION

Reproduction is the process by which offspring are produced. The offspring are of the same species and have traits of their parents. For example, rabbits reproduce other rabbits though the offspring are not completely identical to their parents. We know them as rabbits when we see them!

Reproduction is not essential for an organism to live. It is essential if a species is to stay in existence. Without reproduction, there would be no new animals.

SEXUAL REPRODUCTION

Sexual reproduction is the union of a sperm and an egg. A new animal begins at the time of this union. Two parents are required: male and female. The *sperm* (spermatozoa) is the sex cell produced by male animals. The *egg* or ovum is the sex cell produced by female animals.

The union of sperm and egg is fertilization. Fertilization takes place inside the reproductive tract of female mammals, such as dogs, cattle, and hogs. With some animals fertilization is outside the reproductive tract, such as fish. In either case, time is needed for the union to grow into a new individual.

5-2. Sexual reproduction involves the union of a sperm and an egg.

With mammals, the fertilization of an egg to create new life is *conception*. The new developing organism is formed in the reproductive tract of the female. A female that has a developing unborn baby in her reproductive tract is *pregnant*. The period of pregnancy varies by species.

Insemination

Insemination is placing sperm in the reproductive tract of the female. Animal producers sometimes regulate insemination. Male animals may be kept separately from the females. New technologies are being used in animal reproduction to assure desired offspring.

Natural insemination is the process of the male depositing semen (fluid containing sperm) in the reproductive tract of the female. It occurs during copulation (mating). *Copulation* is the sexual union of a male and female animal. During copulation, the male releases semen in the reproductive tract of the female near the eggs.

Artificial insemination is collecting semen from a male and placing it in the reproductive tract of a female of the same species. Special equipment and methods are used in the process. Artificial insemination gives the producer control over the reproductive process.

Breeding

Breeding is promoting animal reproduction so the desired offspring result. Producers get the kinds of animals they want to meet market demand. Without some control over breeding, animal quality would not meet the demands.

Producers want animals that meet their needs. Animal breeds have been developed. A *breed* is a group of animals of the same species that share common traits. For example, Angus is a breed of beef cattle. Angus are black and polled (some Angus are red and known as Red Angus). Breed traits are inherited. They are passed from one generation to the next. Common traits offspring inherit are color, milk capacity, size, type, and presence of horns (in some species).

A *purebred animal* is one that is eligible for registry in a breed association. Its parents must meet the standards for the breed. The offspring of two purebred animals of the same breed will qualify for registration as a purebred. Some animal producers raise purebred animals.

To produce animals of a certain type, purebred animals of the same species, but of different breeds, are mated. This is *crossbreeding*. For example,

5-3. A purebred Hampshire boar meets all of the standards of the Hampshire Swine Registry. (Courtesy, Hampshire Swine Registry)

an Angus bull could be mated with a Hereford cow. Crossbreeding is used to improve the quality of the products produced by the offspring. In some cases, crossbred animals are mated to make offspring even more desirable. Other systems of breeding are also used.

REPRODUCTIVE EFFICIENCY

Reproductive efficiency is the timely and prolific replacement of a species. It may result in some species flourishing and others declining. Reproductive efficiency is the difference between success and failure in animal production.

The union of a sperm and an egg may not always produce a new, healthy individual. The developing animal may die before birth or hatching. With mammals, the death of a fetus is natural abortion. Some are born or hatched at low birth weights or defective in some way. Losses to abortion and other problems lower reproductive efficiency. For example, a cow that does not have a calf is not productive. She should be culled from the herd.

Research can influence a species or breed so the cost of products is reduced. Artificial insemination is an important example. More new technology is being developed.

A good example of new technology is accelerated lambing. *Accelerated lambing* is a method that uses out-of-season lambing techniques to get three lamb crops in two years from a ewe. Most ewes have one lamb crop a year. Accelerated lambing allows sheep farmers to increase their productivity by 30 to 50 percent.

5-4. This ewe with the newborn quadruplet lambs represents reproductive efficiency. Most ewes produce two young at each lambing.

REPRODUCTIVE ORGANS AND SYSTEMS

Reproductive processes must work properly. When they do not, new animals are not produced. A new calf being born or a chick hatching depends on the proper function of reproductive organs. Anyone planning to breed animals needs to know about the reproductive organs of animals and their functions.

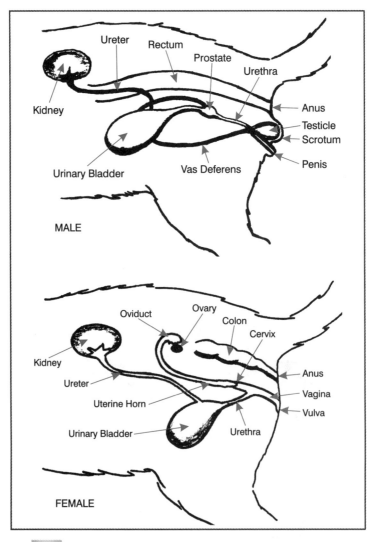

5-5. The male and female reproductive systems of the cat.

FEMALE REPRODUCTIVE ORGANS AND SYSTEMS

The primary reproductive organ of the female is the **ovary**. The ovary produces female gametes. A **gamete** is a sex cell that can unite with other sex cells. It is a mature sex cell capable of a union that produces an individual organism.

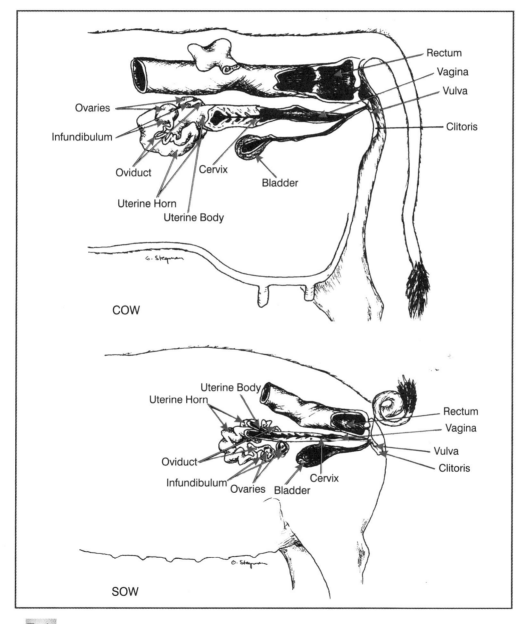

5-6. Cow and sow reproductive organs. (Courtesy, Gary Stegman, Crookston, Minnesota)

Female mammals have two ovaries. Some variation exists in other animals. Only the left ovary fully develops in chickens. Ovaries contain follicles, which are tiny structures that produce ova. A heifer has approximately 75,000 primary follicles in her ovaries at birth. A chick has about 4,000 miniature ova when hatched.

The maximum ova are present at birth or hatching. Ova are slowly used throughout a lifetime. An older cow may have only 2,500 follicles containing ova when culled from the herd.

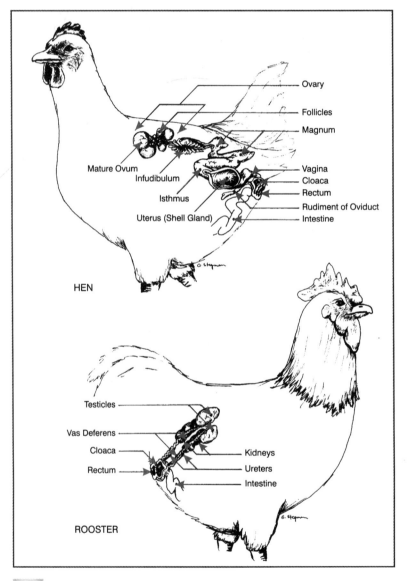

5-7. Male and female chicken reproductive organs. (Courtesy, Gary Stegman, Crookston, Minnesota)

As an ovum matures, it moves to the edge of the ovary for release. Releasing an egg is ***ovulation***. The egg is normally caught by the infundibulum (first segment of oviduct) and sent through the oviduct (fallopian tube).

In mammals, the oviduct moves the ova and spermatozoa (if present) for fertilization and early cell division. The embryo (developing young) remains in the oviduct an additional three to five days.

Fertilization in a hen occurs in the infundibulum (sometimes called a funnel). The oviduct of the hen has four segments. A developed ovum (egg yolk) goes through these segments. The magnum secretes the albumen (thick white) of the egg. As the egg travels through the isthmus, it receives shell membranes. The uterus adds the thin white, the outer shell, and the shell pigment. The completed egg then passes through the vagina and is laid through the cloaca. The cloaca is the opening for both the digestive and re-

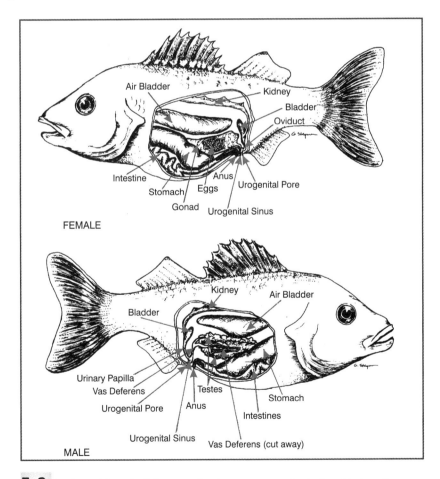

5-8. Male and female fish reproductive organs and relationship to other organs. (Courtesy, Gary Stegman, Crookston, Minnesota)

productive tracts. The entire time from ovulation to laying is slightly more than 24 hours.

In mammals, the embryo goes to the uterus. The uterus has two parts: uterine horn and uterine body. A placenta forms and attaches itself to the uterine wall. Cattle and swine differ from horses. A mare carries a foal in the uterine body, while a pig or calf is carried in the uterine horn.

The cervix is a thick-walled, inelastic organ (technically part of the uterus) that connects the uterus and the vagina. The cervix contains annular rings and cervical mucus that seal the uterus to keep out contaminants. During pregnancy, the mucus thickens to form a gel-like plug. Just before giving birth, hormones (chemical substances produced by the body) relax the cervix to allow the young to pass through.

The vagina is the passageway for reproduction and urine excretion. The vulva is the external opening of the reproductive and urinary systems. The clitoris is located just inside the vulva. It is a highly sensitive organ.

The reproductive organs of fish differ from mammals and poultry. The ovary of most fish species is a single large sac of eggs that appears as two ovaries that are joined. The eggs travel via the oviduct to the urogenital opening where they are excreted. Most fish eggs are fertilized outside the female's body.

MALE REPRODUCTIVE SYSTEM

The primary reproductive organ of the male is the ***testicle***. Testicles produce male gametes or sperm. Two testicles are held externally in the scrotum of mammals and internally in the body cavity of male poultry and fish.

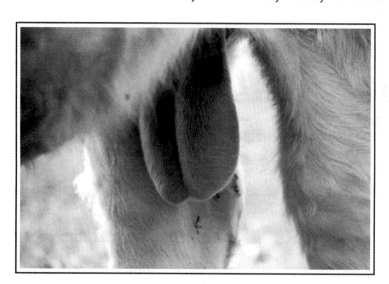

5-9. Scrotum of young bull with properly developing testicles.

5-10. Normal sperm (greatly enlarged). (Courtesy, Elite Genetics, Waukon, Iowa)

Males begin producing sperm at puberty and do so throughout life. The *scrotum* is a two-lobed sac in mammals that protects the testicles and helps control the temperature. Sperm production is best if testicles are cooler than the body.

Sperm are formed in the seminiferous tubules of the testicles. Sperm pass through the epididymis into the vas deferens. The vas deferens carries the sperm and fluid (semen) for release. In mammals, it goes to the urethra. With poultry, sperm go to the cloaca (papillae). Sperm go to the urogenital sinus in fish. Male poultry do not have a true penis and transfer sperm to an undeveloped copulatory organ into the oviduct of the female. From the urogenital sinus of fish, sperm are excreted via the urogenital pore. Near the urethra of some mammals (bulls, stallion, and ram), sperm pool in an enlarged end of the vas deferens called an ampulla.

At ejaculation in mammals, semen passes through the penis and is released in the reproductive tract of the female. The penis is normally relaxed but is erect during sexual arousal. The free end of the penis (known as glans penis) serves as the sensory organ similar to the clitoris in females. It also retracts beneath the protective sheath (fold of skin) in most animals when not aroused.

THE ESTROUS CYCLE

The *estrous cycle* is the time between periods of estrus. The estrous cycle of female mammals has four periods: estrus, metestrus, diestrus and proestrus. These periods are cyclical (except during pregnancy) for many ani-

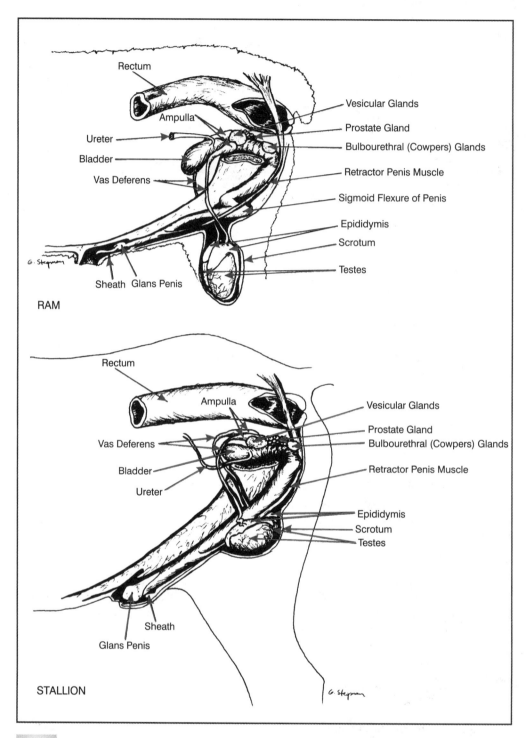

5-11. Ram and stallion reproductive organs. (Courtesy, Gary Stegman, Crookston, Minnesota)

mals, such as cows or sows. Some animals are seasonal breeders. They may go through periods of cycling and periods of *anestrus*. Anestrus is the absence of cycling. Anestrus is often related to the number of hours of light in a day. Examples of seasonal breeders include ewes, queens, does, and mares.

The estrous cycle length of common animals is shown in Table 5-1.

- Estrus—*Estrus* (heat) is the period when the female is receptive to the male and will stand for mating. The length of estrus varies between species. Estrus ranges from 12 hours to six days. It is longer in some small animals.

  Periods of heat or estrus are triggered by the hormone estrogen. Many changes take place, such as restlessness, mucus discharge, a swollen vulva, and standing to be ridden by other animals. Ovulation takes place during estrus for the ewe, sow, mare, and some companion animals. The cat only ovulates after mating. The number of eggs ovulated varies. Cows and mares typically ovulate one egg. Dogs, hogs, and cats may ovulate several eggs—15 or more.

- Metestrus—The period following heat or estrus is metestrus. Ovulation occurs during metestrus in the cow and doe. It is the time when LH (luteinizing hormone) triggers the corpus lutea (CL) to develop from follicular tissue that remains after release of the ova. The corpus luteum (yellow body) is important in maintaining pregnancy.

  After ovulation, some capillaries break and release small amounts of blood. The blood is occasionally seen on the tail of a cow a day or so after estrus. This blood is not a sign that a cow is or is not pregnant.

- Diestrus—Diestrus is the period in each estrous cycle in which the system assumes pregnancy. It is characterized by a fully functional corpus luteum (yel-

Table 5-1. Length of Estrus, Estrous, and Time of Ovulation

Species	Estrous Cycle (days)	Length of Estrus (heat)	Ovulation
Cow	21	12–18 hourse	10–14 hours after estrus
Mare	22	6–8 days	1–2 days before estrus ends
Doe (goat)	21	30–40 hours	at the end of or just after estrus
Doe (rabbit)	Constant	Constant	8–10 hours after mating
Sow	20–21	40–72 hours	mid estrus
Ewe	17	24–36 hours	late estrus
Bitch (dog)	—	9 days	1–2 days after estrus begins
Queen (cat)	14–21*	5 days	24 hours after mating

*Estrous cycle influenced by length of daylight, with mating season typically when there are more than 12 hours of daylight in one day.

5-12. These three-week-old Siberian Husky puppies are evidence of successful sexual reproduction.

low body) that releases high levels of progesterone (the hormone that maintains pregnancy). Diestrus is often 9 to 12 days in length. It is during diestrus that the uterus is prepared for pregnancy.

■ Proestrus—Proestrus begins with the regression of the corpus luteum and a drop in the hormone progesterone. FSH (follicle stimulating hormone) causes rapid follicle growth in preparation for estrus and ovulation. Late in proestrus, changes in behavior may occur as estrus approaches.

PHASES OF REPRODUCTIVE DEVELOPMENT

Reproduction involves a series of events that must be properly timed. Its success is measured in various ways, such as the number of pigs per litter or calves per 100 cows. Hatchability and liveability are important to poultry and fish operations.

PUBERTY

Puberty is the time at which animals reach a level of sexual development that makes them capable of reproduction. Puberty in female animals is the age of the first estrus with ovulation. Puberty in males is the first ejaculate with fertile sperm.

Connection

TWO HEADS ARE NOT BETTER THAN ONE

This day-old calf was born with two heads. Each head was joined by a neck to the body. Both heads appear to be well developed. The body was near normal except where the necks joined. The calf did not live.

Why was the calf born this way? No one knows for sure. Several reasons have been suggested. The two heads could have been the result of a mutation. Another reason is that the birth defect was due to pollution in the pasture where the calf's mother lived, such as a range with radioactive wastes present. Still another reason could be that the animal was from a diseased parent.

Two-headed calves do not usually live. They rarely occur. When they do, they represent reproductive failure. Producers prefer normal calves. An abnormal calf has cost the owner feed for its mother and delayed income from a marketable calf for about two years.

Neither males nor females are sexually mature at puberty. The female is often too small to bear young. The male is not highly fertile nor capable of breeding regularly.

Both environmental and genetic factors affect the age at which puberty occurs. Weight at puberty is affected by genetic factors, such as size of the parents. The primary environmental factor is anything that affects growth. Differences between breeds are the primary genetic factors within species. It is important to not breed animals until they reach the recommended weight.

FERTILIZATION

Fertilization is the union of a sperm and an egg. The sperm penetrates the egg and pairs of genetic material are formed. The fertilized ovum is a *zygote*.

5-13. Normal presentation of a calf in the uterus of a cow is needed to assure giving birth without complications.

Gestation

Gestation is the period of pregnancy. It begins with conception and lasts until parturition. The average length of gestation varies from 114 days for a sow to 337 days for a mare. Fertilization is the beginning of gestation. A series of cell divisions without growth (cleavage) occurs while the embryo slowly migrates toward the uterus over a three-to-four day period.

The new animal is known as an *embryo* while the organs in the developing animal are forming, which is differentiation. Growth is rapid during this period. Differentiation is completed by day 28 in a sow and day 45 in a cow. After differentiation and until birth, the young animal is a *fetus*. The fetus gains weight and matures in preparation for birth.

Parturition

Parturition is the process of giving birth. Several hormone levels change and initiate this process. These initiate uterine contractions and milk production.

Signs of approaching parturition can be observed. Females begin to exhibit a "nesting" behavior. They become restless and attempt to separate from the herd or flock.

The first stage of parturition includes dilation of the cervix and entry of the fetus into the cervix. This usually is the longest stage; ranging from one to 12 hours, depending on the species. The second stage completes the birth of an animal through strong uterine contractions.

The birthing process and the first hour after birth are critical. Animal producers may observe if the fetal position is correct. They observe the delivery for problems. They check to see that respiration has started and help the newborn nurse to get colostrum—the first milk that is high in antibodies.

The last stage of parturition is the expulsion of the afterbirth (placenta) from the uterus. This normally occurs shortly after giving birth. Sometimes the placenta is retained. This can interfere with future reproduction. The next gestation depends on a return to normal estrous cycling. The uterus must return to normal size and condition.

LACTATION

Lactation is the production of milk. Young mammals must have milk for food. The second purpose of lactation is to provide early disease resistance for the newborn.

Lactation is a part of the reproductive cycle. Hormones that trigger the onset of lactation also play a role in parturition. Lactation requires nutrients that extend periods of anestrus following parturition. Weaning pigs or temporarily removing calves from cows can trigger estrus.

MATING BEHAVIOR

Conditions for domestic animals must be such that mating occurs. In addition to a properly cycling female, the male must be in proper condition and have the proper status (social rank). Social rank refers to status of the male within a herd or group. Males need both the desire to mate (libido) and the ability to mate. Both can be enhanced with proper nutrition that keeps the males from becoming overweight and with an exercise area/breeding pen that allows animals to move about. Males should be kept disease and injury free so mating is possible.

Management of rams in a sheep flock is an example of stimulating proper estrous cycling and mating behavior. Rams should be kept away from ewes (out of sight and smelling range) until near breeding time. Reintroducing rams near breeding time will stimulate cycling in ewes. Males of any species that are very dominant can keep other males from mating. Pasture and breeding pen design help males that are lower in the social order breed their portion of females. This insures a high pregnancy rate in the total herd/flock.

INCUBATION

Incubation is the development of a new animal in the fertile egg of poultry, fish, and other egg-laying species. Incubation occurs outside the body of the female. Proper incubation is needed to hatch eggs. Eggs are tended by species in different ways, such as birds "sit" on the nest. Eggs can and are artificially incubated in hatcheries.

Four important factors in incubating eggs are temperature, humidity, oxygen, and egg rotation. Temperature should be maintained at 99 to 103°F (37 to 39°C). The humidity should be about 60 percent during the first 18 days and 70 percent during the last 3 days (assuming a 21-day incubation period). Eggs should be slightly higher in the large end and be rotated two to five

5-14. Eggs being removed from a trout for fertilizing and hatching.

5-15. Catfish eggs being incubated in a hatching trough. The eggs turn pink to red as they near hatching. The incubation period is about seven days under proper conditions.

times daily for the first 18 days. Sufficient air exchange to prevent carbon dioxide buildup, while maintaining a 21 percent oxygen level, will enhance hatching results. Incubation times range from 21 days for chickens to 42 days for ostriches.

Early embryo development takes place in the body of the hen (107°F or 42°C) before the egg is laid. Several cell divisions take place. After laying, the egg cools and growth stops. Fertile eggs should be incubated shortly after laying. Some must be held for a while before incubation. Eggs stored below 80°F (27°C) are more likely to hatch when incubated.

Table 5-2. Incubation Times for Selected Poultry Species	
Common Species Name	Incubation Period (days)
Chicken	21
Pheasant	24
Duck	28
Turkey	28
Goose	28–32
Ostrich	42

Incubation continues embryo development. Organ formation (cell differentiation) is completed in the first six to eight days. Two key changes take place near the end of incubation. The beak turns toward the air cell and the yolk sac enters the body cavity. This yolk sac provides nutrition and water for the first several hours of life after hatching. The yolk is gradually used during the first 10 days of life.

SPAWNING

Spawning is the releasing of eggs by a female fish and the subsequent fertilization by the male. A female may lay thousands of eggs at one spawning, depending on the species. The male fish then fertilizes the eggs by releasing sperm on the mass of eggs. The eggs of some species are protected only by a gravel covering. The males or females of some species provide the eggs with protection or incubation. Spawning is covered in more detail in Chapter 12.

ANIMAL REPRODUCTION TECHNOLOGY

The use of artificial insemination (AI) has increased. Today, AI is more common than natural service in some species, such as dairy cattle and turkeys. Synchronization and embryo transfer are also used to a lesser extent.

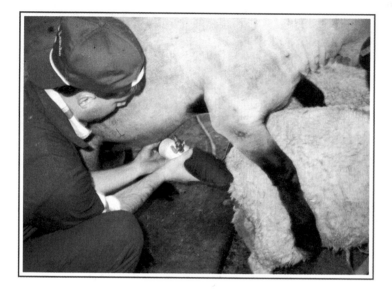

5-16. Ram semen is being collected with an artificial vagina and teaser ewe. (Courtesy, Elite Genetics, Waukon, Iowa)

ARTIFICIAL INSEMINATION

Artificial insemination is placing semen in the female reproductive tract by artificial techniques. AI has the advantage of using a quality male to breed many females. Superior traits can be extended to a large population. This has improved many species. Desired qualities can be obtained in companion animal offspring. Meat animals can be produced that have high yields. Cows can be produced that give more milk. Hens can be produced that lay more eggs.

5-17. Sperm motility (movement) is being examined before processing. (Courtesy, American Breeders Service)

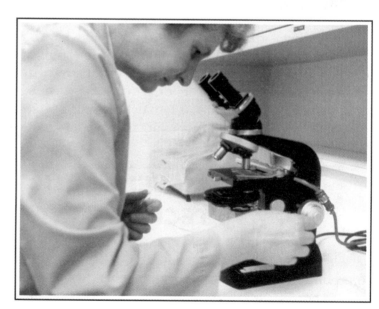

SEMEN COLLECTING

Artificial insemination requires that semen be collected from the male. A tom turkey or rooster requires manual stimulation and semen removal (sometimes called "milking of semen"). Sperm from a male fish can be obtained by applying gentle pressure on their abdomen. Bulls, rams, and stallions are usually collected with an artificial vagina (AV).

Semen collection is improved if time is given for the male to become stimulated. The collected semen is evaluated, cooled slowly, and processed in preparation for freezing at temperatures of –320°F (–196°C). Semen stored in liquid nitrogen at –320°F (–196°C) can be thawed and used 30 to 40 years later.

INSEMINATION

High conception rates require much more than quality semen. The female must be cycling and in estrus (heat). The semen should be thawed and inseminated after heat detection.

Detecting estrus usually requires keen observation. Failure to detect heat is the most common cause of AI failure.

Estrus signs vary between species. The best indication of estrus for most species is standing heat. Standing heat is the stage of estrus when a female stands when mounted by another animal. Other behavioral patterns change when an animal is in estrus. Most become more active. Cows try to mount other animals and are generally restless. Mares urinate frequently and expose their clitoris in a process called "winking." Many animals display extra mucus and redness in the vulva. Sows may have redness in the vulva. The ewe requires that a male be present before she will show signs of estrus. A ram that has had a vasectomy (made infertile) is often used to detect ewes in heat.

The timing and placement of semen vary depending on the species. All require that frozen semen be

5-18. Artificial insemination kit. (Courtesy, Auburn University)

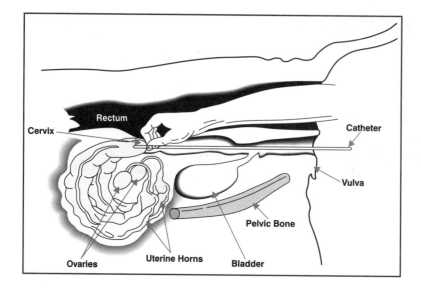

5-19. Using the recto-vaginal method to artificially inseminate a cow.

thawed properly (95 to 98°F or 33 to 34°C). Fresh semen should be used within an acceptable time, depending on the species.

Most turkeys must be artificially inseminated due to their large size. Insemination does not need to be on a daily basis. A week is usually suggested between the first two inseminations and a slightly longer interval on subsequent inseminations. This is possible in poultry because the hen has storage glands for semen.

Cows should be inseminated in the last two-thirds of estrus (heat) or the first few hours after the end of heat. Heat detection is most often done early in the morning or late at night, since, most riding activity occurs during or

5-20. Artificially inseminating a cow.

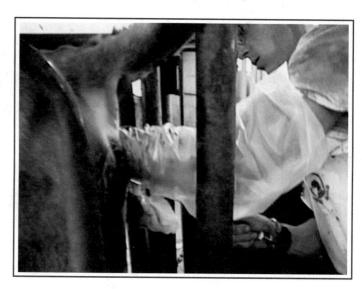

near the night hours. An approved practice is to breed cows detected in heat in the morning during the afternoon of the same day. Cows in heat in the evening should be bred early the next day. Insemination of cattle usually involves the recto-vaginal method where a gloved arm is placed in the rectum to palpate (feel) the reproductive organs.

Mares may need to be inseminated several times. This is because it is difficult to detect ovulation. Inseminations are made on the third, fifth, and seventh days of estrus. The AI procedure for mares differs from that for cows. Mares have a delicate rectum and other anatomical differences. The arm and syringe are placed in the vagina with a finger through the cervix. This requires special emphasis on cleanliness. The semen is deposited in the body of the uterus.

Sows should be bred 24 hours after the onset of estrus. Some people prefer to breed a second time, 40 to 48 hours after the onset of estrus. An inseminating tube is inserted into the cervix. The semen is ejected with a plastic squeeze bottle syringe.

Artificial insemination of ewes is less common. Two inseminations will increase conception and multiple lambs. The size of a ewe requires smaller equipment or the use of lapriscopic artificial insemination. In lapriscopic artificial insemination, semen is placed in the uterus through the ewe's abdominal wall.

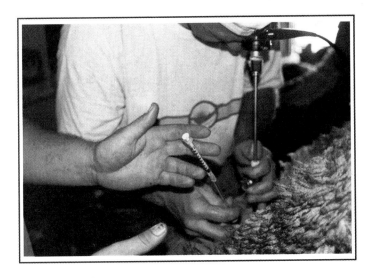

5-21. Lapriscopic artificial insemination is being used on this ewe. The semen is being injected from a syringe through a pipette into upper one-third of the uterine horn. (Courtesy, Elite Genetics, Waukon, Iowa)

SEXED SEMEN

The sex of offspring can be controlled with sexed semen. *Sexed semen* is semen that has been prepared to produce all male or all female offspring. It is collected in the manner as other semen used in artificial insemination. Sexed semen will predict sex with about 90 percent accuracy. The cost of sexed semen will likely be about four times the cost of unsexed semen.

ESTROUS SYNCHRONIZATION

Estrous synchronization is bringing a group of animals into heat simultaneously. This helps schedule animal breeding and birthing.

Synchronization usually involves the use of prostaglandin, progestin, or a combination of the two. Prostaglandin causes the corpus luteum to stop production of progesterone, allowing the animal to come into estrus. Progestin has the effect of keeping progesterone levels high, holding animals in an "extended" diestrus. When the progestin source is removed, the animal quickly comes into estrus.

CIDR (controlled internal drug release) dispensers are used with sheep to promote synchronization within a flock. CIDRs are placed in the vagina of ewes. They are removed and ewes cycle shortly afterward. This is used to breed ewes during times of the year when they do not cycle. This helps sheep producers get three lamb crops every two years.

5-22. A controlled internal drug release (CIDR) dispenser for use in synchronizing ewes. (Courtesy, Northwest Experiment Station, University of Minnesota, Crookston)

EMBRYO TRANSFER

Embryo transfer is moving embryos from one female (known as a donor) to the reproductive tract of another female (known as a recipient). Donor females usually carry extraordinary genetics. Recipient animals have far less worth and are used as surrogate mothers.

Most embryo transfer work is done with beef and dairy cattle. Embryo transfer is used following superovulation. This assures a larger than normal number of eggs. Multiple breedings and careful administration of extra follicle stimulating hormone (FSH) will cause a donor to produce an average of five or six transferable embryos. Transfer of embryos can be done surgically or nonsurgically. The success rate is higher when transferring fresh embryos. Embryos can be frozen in liquid nitrogen and transferred later.

CLONING

Cloning is the production of one or more exact genetic copies of an animal. Cloning has been expanded and may be widely available soon. There are several methods of cloning animals.

One method of cloning in animals involves letting embryos grow to the 32-cell stage before splitting into 32 identical embryos. Scientists feel that this could be repeated producing 1,024 identical copies, 13,088 identical copies, etc.

A second method of cloning in animals involves taking a cell from an adult animal. This method resulted in the creation of Dolly from mammary cells of a six-year-old sheep (ewe).

Still another variation of the cloning technique involves taking cells from primordial germ cells during fetal development. These cells are more stable and can be cultured and frozen for indefinite, but longer periods of time. Bovine (cattle) calves have been born using this technique.

5-23. Cloning technology.

GENETIC ENGINEERING

Genetic engineering is removing, modifying, or adding genes to DNA. Genetic engineering using recombinant DNA (gene-splicing) along with other reproductive technology has the potential to greatly change animal science.

5-24. Turkeys have been bred to give large amounts of meat. Many are so fleshy that they cannot mate naturally. Some people view breeding turkeys so they cannot mate as an ethical issue. What do you think?

Genetic marker technology is being used to detect the presence of certain genes. This technology, along with cloning, has the potential to further increase genetic progress. (Chapter 6 presents additional information on genetic engineering.)

REPRODUCTIVE TECHNOLOGY ETHICS

Reproductive technology is rapidly increasing. This raises ethical questions. Technology can be good or bad depending on how it is used. People must be honest. Fear of genetic engineering abuse is not a good reason to stop the potential benefits.

People need to be educated about the use of reproductive technology. If not, poor decisions may be made.

REVIEWING

MAIN IDEAS

Reproduction is the process by which offspring are produced. Animals naturally reproduce sexually. Sperm and egg unite to begin a new individual. Producers also use artificial insemination (AI) to help assure productive animals.

Reproductive efficiency is important to animal producers. Animals must breed and regularly produce offspring. If they do not, the animals should be culled from the herd or flock.

Male and female animals have distinctive reproductive organs and systems. The testicle produces sperm and is the primary reproductive organ in the male. The ovary produces eggs and is the primary reproductive organ in the female.

Reproductive development goes through several phases. Puberty is the phase of development when animals can reproduce. Fertilization is complete when there is a union between the sperm and the egg. Gestation is the time between conception and parturition in the female. Parturition is the birthing process.

New technology is used in reproduction. Artificial insemination is widely used. Research is developing cloning, genetic engineering, and other areas. People must consider the ethics of the newer technology.

QUESTIONS

Answer the following questions using complete sentences and correct spelling.

1. What is reproductive efficiency and why is it important?
2. Describe the structure of the female reproductive system.
3. Describe the structure of the male reproductive system.
4. Where does mammal fertilization take place? Hen fertilization? Fish fertilization?
5. What are the four phases of the estrous cycle? What happens in each phase?
6. How is puberty defined in males? In females?
7. What four factors are essential to successful incubation?
8. What reproductive technology has had the largest impact on improving animal efficiency? Why is it useful?
9. Discuss two ways that reproductive technology of the future may influence animal and poultry production.

EVALUATING

Match the term with the correct definition. Write the letter by the term in the blank provided.

a. sperm   e. reproduction   i. estrus
b. egg   f. gestation   j. estrous cycle
c. testicle   g. parturition
d. ovary   h. conception

_____ 1. The process of producing offspring.

_____ 2. A male sex cell.

_____ 3. The time between periods of estrus.

_____ 4. Primary female reproductive organ.

_____ 5. The union of sperm and egg.

_____ 6. Birthing process.

_____ 7. Heat or the period when females of some species are receptive to mating.

_____ 8. Time between conception and parturition.

_____ 9. Female sex cell.

_____ 10. Primary male reproductive organ.

EXPLORING

1. Dissect the reproductive tract of a cow. Find all of the important reproductive organs. Animal slaughter houses may provide the reproductive tracts to schools for educational purposes. Be sure to follow proper safety procedures and properly dispose of the reproductive tract after the dissection. Caution: Be sure to follow all safety precautions with this activity.

2. Tour an artificial insemination company and observe the correct procedure used to collect semen from a bull. Observe how the semen is handled in preparation for storage and shipment.

3. Hold a debate in class on reproductive technology. Consider the ethics of the technology and other areas that are important.

4. Observe heat signs in your herd or a friend's herd or flock. Determine the kind of animal and previous breeding history. What change in behavior did you see?

5. Producers feed non-productive animals even when they do not reproduce. How much does a rancher lose when cows do not calve? Assume that ZYX Ranch has 1,000 cows. How much income is lost if 5 percent of the cows do not have calves in a year? In your calculations, use a weaning weight of 500 pounds for each calf and a per pound selling price of 70 cents. (Clues: first determine the total number of calves not born and multiply by 500 pounds. This is the total weight of all lost calves. Multiply the total weight by 70 cents to get the number of dollars lost.) How would you overcome this loss if you owned ZYX Ranch?

6

ANIMAL BIOTECHNOLOGY

Animal producers want better ways of raising animals. They also want animals that grow fast and meet the needs of people. For example, people want turkeys with more white meat. Research has developed new kinds of turkeys to meet the demand for more white meat.

Some scientists feel that most of our food will soon come from new kinds of plants and animals. Discoveries in biotechnology will change animal agriculture. Animal production will have many changes in the years ahead.

People have different opinions about biotechnology. Some people feel that it should be used. They feel that the benefits outweigh the possible problems. Other people feel that biotechnology should not be used. What do you think?

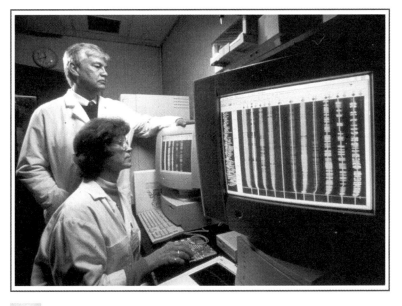

6-1. Scientists examining DNA sequences of a virus disease in poultry. (Courtesy, Agricultural Research Service, USDA)

OBJECTIVES

This chapter introduces biotechnology in animal production. It has the following objectives:

1   Explain biotechnology and how it is used

2   Explain the role of genetics in animal production and biotechnology

3   Distinguish between organismic biotechnology and molecular biotechnology in animals

4   Describe molecular biotechnology, including genetic engineering and recombinant DNA processes in animals

5   Explain examples of organismic biotechnology

6   Identify issues associated with animal biotechnology

TERMS

allele
animal biotechnology
atom
biotechnology
chromosome
deoxyribonucleic acid (DNA)
dominant trait
DNA sequencing
gene
gene transfer
genetic code
genetics
genome
genotype
heredity
heterozygous

homozygous
microinjection
molecular biotechnology
molecule
mutation
organismic biotechnology
particle injection
phenotype
probability
Punnett square
recessive trait
recombinant DNA
superovulation
synthetic biology
transgenic animal

BIOTECHNOLOGY:
AGRICULTURE FOR THE FUTURE

Biotechnology is the management of biological systems for the benefit of people and their environment. This broad definition includes many areas of animal production. Biotechnology is often viewed as the application of science in food and fiber production. All plants and animals are included.

Biotechnology has been used for hundreds of years. Using yeast to make bread and bacteria to make cheese is biotechnology. However, today's biotechnology refers to new technologies developed since 1977. It is based on a deep and thorough understanding of life and the mechanics of living things.

Animal biotechnology is the application of biotechnology methods to improve animals. Livestock, companion animals, exotic animals, and other animals are included. The products are important to animal producers, companion animal keepers, and consumers.

MULTIPLE TECHNOLOGIES

Biotechnology involves many fields. Examples of important areas are biology, microbiology, genetics, biochemistry, and computer technology. These fields have technology that enables biotechnology to be carried out.

In animal biotechnology, specific fields related to animals are involved. Animal science, veterinary medicine, agricultural engineering, and other areas are important.

People working in biotechnology must be well educated in a specific area and must have broad preparation across related fields. They must be able to work with other people. Often, a team approach is used. This allows several people with in-depth education to bring their areas of specialization together to solve a problem.

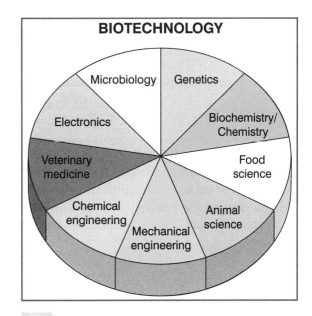

6-2. Many fields of study are involved in animal biotechnology.

6-3. Using microscopy with attached video and computer systems to study synthetic biology.

SYNTHETIC BIOLOGY

Synthetic biology goes beyond biotechnology. It is the use of chemicals to create systems with some characteristics of living organisms. Now confined to laboratories, synthetic biology uses vesicles to create lifelike conditions. Vesicles are tiny cell-like structures that have external membranes.

All synthetic biology work is carried out using powerful microscopes with attached computer networks. Emory University in Atlanta, Georgia, has the only laboratory in the world doing this research.

Physical organic chemists are studying the use of synthetic biology in animal health and production. The interaction of vesicles and living organisms is a top priority. Could the creation of living organisms from nonliving substances be next?

GENETICS

Genetics is the study of the laws and processes of biological inheritance. It is important to animal producers as well as in biotechnology. Parents have traits that are passed to their offspring. The passing of traits from parents to offspring is *heredity*. The heredity of animals is important to producers.

CHROMOSOMES

All organisms are made of cells. A living cell consists of a cell membrane and the cytoplasm, which contains the nucleus. The nucleus has chromosomes. A *chromosome* is a tiny threadlike part in a cell that contains the ge-

6-4. The study of genetics can explain why this black cow has a red calf.

netic material. This genetic material is known as the *genome* of the organism.

Chromosomes are the link between parents and offspring. When organisms reproduce sexually, the genome is the combination of the traits from the mother and father. The offspring receive chromosomes from both parents. This gives them pairs of "like" chromosomes. All of the cells within the organism are genetically identical. Each cell contains identical numbers of chromosomes. Table 6-1 lists the number of chromosomes for a few species.

Chromosomes are made of genes that consist of deoxyribonucleic acid (DNA). DNA forms genes that make up chromosomes. Chromosomes make

Table 6-1. Number of Chromosomes for Selected Animal Species			
Species	**Number of Chromosomes**	**Species**	**Number of Chromosomes**
Cat	38	Horse	64
Cattle	60	Human	46
Chicken	78	Mule	63
Dog	78	Sheep	54
Donkey	62	Swine	38

up genetic information for cells. Cells make tissues that form organs. Organs form organ systems. These organ systems make up the organism.

Genes

A **gene** is a segment of a chromosome that contains the hereditary traits of an organism. Since chromosomes come in pairs, genes also come in pairs.

An **allele** is a different form of a gene. Forms of alleles may be similar or they may be different. An organism having similar alleles is said to be **homozygous** for a trait. A **heterozygous** organism is one having different alleles for a particular trait.

Genotype and Phenotype

The genetic makeup of an organism is its **genotype**. The **phenotype** is the organism's physical or outward appearance. In basic genetics, a homozygous allele is called "AA" or "aa". A heterozygous allele is commonly noted as "Aa."

The transmission of traits by parents to offspring is important in breeding animals. Breeding practices use the concept of alleles to produce various phenotypes or outcomes. Outcrossing is used to develop heterozygous offspring and inbreeding produces homozygosity. The genotype and the environment determine phenotype.

6-5. Although they are from the same breed, these two whippets illustrate different phenotypes.

Dominant and Recessive

Some traits are dominant while others are recessive. A ***dominant trait*** is one that covers up or masks the alleles for recessive traits. A ***recessive trait*** is a trait or gene that is masked by dominant traits. If an animal has a dominant allele, it will phenotypically show the dominant trait in complete dominance. If the animal receives two recessive alleles, then it will be homozygous recessive.

An example of dominant and recessive traits is the polled trait in cattle. The polled (PP) trait is dominant over the recessive (pp) trait of having horns. If a calf is homozygous dominant (PP) or heterozygous (Pp), they will be polled. If they receive an allele for the horned trait from both parents, they are homozygous recessive (pp) and will grow horns.

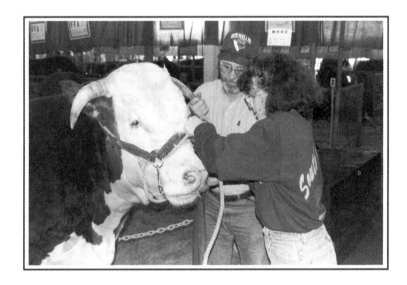

6-6. This Hereford bull received the genetic trait for horns from both parents.

Red Angus cattle are another example showing dominant and recessive alleles. Angus cattle have been black for hundreds of years. Occasionally, a red calf is born. Black is the dominant allele and red is the recessive allele. When two red calves are crossed, they will always produce a red calf. These red animals are homozygous recessive.

Other examples of dominant and recessive traits are:

- Color is dominant to the absence of color in most animals. (The absence of color is known as albinism.)
- A white face is dominant to a colored face in cattle.
- Black is dominant to brown in horses.
- Rose and pea comb are dominant to single combs in chickens.

Connection

KITTENS WHEN YOU WANT KITTENS

Kittens are cute and cuddly. They must be fed, watered, trained, and otherwise cared for. Kittens grow up to be cats! Their upkeep requires effort. Unwanted kittens may be abused and not receive proper care. Responsible owners prevent unwanted kittens. Cat birth control is used.

Neutering is the most effective method. However, a queen that has been spayed or a tom that has been castrated can never be used for breeding. Tubal ligations with queens and vasectomies with toms are options that may be reversible.

Cat birth control drugs may become available in the United States. Two examples are megestrol acetate and delmadinone acetate. These products are for females. They alter the estrous cycle and delay heat. Once removed from the drugs, a queen can cycle into heat and be bred. Biotechnology will help make good animal birth control possible.

Cats can be kept in isolation as a means of birth control so toms and queens are separate. Breeding is not possible. This may not always be practical.

Incomplete Dominance

Most traits are not products of complete dominance. They are usually influenced by several alleles and by the environment. An example is incomplete dominance. Shorthorn cattle are either red (RR), roan (Rr), or white (rr). The dominant allele does not completely mask the white color. The Shorthorn animal appears to have red and white hairs.

Predicting Genotype

Prediction is based on probability. **Probability** is the likelihood or chance that a trait will occur. It is used by geneticists to predict the outcome of mating. Mating animals of particular traits does not guarantee that the trait will be expressed in offspring.

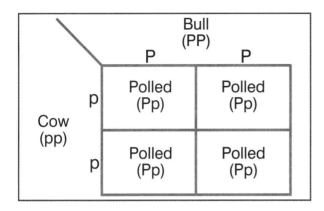

6-7. An example of the Punnett square prediction of the polled trait. A homozygous (dominant) bull is mated with a heterozygous (recessive) cow. This is noted as a **PP** x **pp** cross. All the F_1 have the genotype **Pp** (heterozygous) and will be polled.

The **Punnett square** is a technique for predicting genotype. It considers the dominant and recessive genes of the male and female parents for one trait. In a Punnett square, a parent is represented by the letter P and the offspring are represented by the letter F. Subscript numerals are used to designate the generation involved, such as F_1 is the first generation and F_2 is the second generation. Homozygous and heterozygous genes are considered. Figure 6-7 shows an example using the Punnett square.

ROLE OF DNA

Deoxyribonucleic acid (DNA) is a protein-like nucleic acid on genes that controls inheritance. Each DNA molecule consists of two strands shaped as a double helix or spiral structure. These strands are nucleotides bonded together by pairs of nitrogen bases.

This double helix is similar to a tiny twisted ladder. The supports or backbone are made up of sugar molecules held together by phosphates. The rungs consist of four nitrogen materials: cytosine (C), guanine (G), adenine (A), and thymine (T). The rung structure allows DNA segments to be cut out and new ones inserted.

Genetic Code

The **genetic code** is the sequence of nitrogen bases in the DNA molecule. This sequence codes for amino acids and proteins. The unique ability of DNA to replicate itself

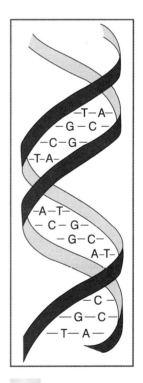

6-8. DNA structure consists of two nucleotide strands, which are bonded by the nitrogen bases cytosine, guanine, adenine, and thymine.

6-9. This DNA has been extracted from the thymus gland of a bovine using a simple process.

allows for the molecule to pass genetic information from one cell generation to the next.

When a chromosome divides during mitosis, the two strands unravel and separate. Each strand contains the complementary base and will immediately become a new double strand identical to the original. A gene is structurally a triplet sequence of nitrogen bases (C, G, A, and T).

Genetic code determines the nature of an organism. Hogs have a different genetic code than do cattle or humans. Research is identifying the genetic code of animals.

DNA Sequencing

A tool being used in biotechnology is DNA sequencing. ***DNA sequencing*** is determining the order of nucleotides on a DNA fragment. It is being used to unravel the mystery of heredity and allow genetic engineering to be used. A good understanding of the DNA double helix is essential.

6-10. Sequencing gel apparatus used in electrophoresis.

A C G T

Lane

6-11. Sample DNA sequencing ladder obtained with gel electrophoresis.

A process known as chain termination sequencing is used. Three steps are followed:

Step 1. DNA fragments are made using DNA synthesis.

Step 2. DNA fragments are separated according to length using gel electrophoresis. (Gel electrophoresis is a technique for separating substances, typically nucleic acids. An electrical field is used to move the substances through the gel. Separation is by the size of the molecule.)

Step 3. DNA fragments are made visible by using a stain. (Once visible, these fragments can be studied and used in determining if an organism is likely to have a genetic disease, such as dwarfism.)

6-12. The DNA sequence of a beef animal is being studied. (Courtesy, Agricultural Research Service, USDA)

DNA sequencing is being used to understand the genetic makeup of animals and other organisms. The genetic information in one organism is huge. Large projects are being used to research the information. These are known as genome projects. Among the best known is the Human Genome Project.

With DNA sequencing information, heredity can be used in animal selection. The information will provide a genetic map of the characteristics of an animal. Methods of genetic engineering can be used to modify the genetic makeup of an organism. The use of DNA sequencing information is now largely limited to scientists. This is because of its complexity and the ethical issues surrounding some uses of biotechnology.

MOLECULAR AND ORGANISMIC BIOTECHNOLOGY

The two major areas of biotechnology in animal science are molecular and organismic.

MOLECULAR BIOTECHNOLOGY

Molecular biotechnology is changing the structure and parts of cells to change the organism. It begins with the *atom*, which is the smallest unit of an element. Elements are such things as hydrogen, oxygen, nitrogen, phos-

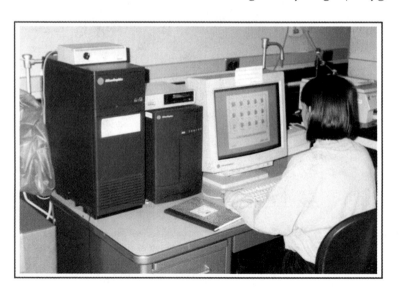

6-13. A silicon graphics computer is being used for molecular modeling.

phorus, and sulfur. Two or more atoms make up a *molecule*, which is the smallest unit of a substance. Molecules form cells in plants and animals.

Each cell of an organism has several molecules of water within it. Water is a molecule represented by the chemical symbol of H_2O. The elements in water are hydrogen (H) and oxygen (O). Water is made up of three atoms, two hydrogen and one oxygen.

Molecular biotechnology often changes the physical appearance of an organism. In some cases, undesirable traits may develop along with those that are desired. Thorough research and testing are always needed. Today's livestock operations are much more productive than earlier domestic animals. This is because of genetics and heredity information.

Genetics and molecular biology are important in both molecular and organismic biotechnology. Both areas are covered in more detail in this chapter.

ORGANISMIC BIOTECHNOLOGY

Organismic biotechnology deals with intact or complete organisms. The genetic makeup of the organism is not artificially changed. Organismic biotechnology is used to improve animals for the benefit of humans. This is the most widely used type of biotechnology.

Much genetic variation exists within a species. Hogs will always have piglets and cows will always have calves, but each piglet or calf is unique. Agricultural researchers have used this variation to improve animals using organismic biotechnology.

MUTATIONS

A *mutation* is a change that naturally occurs in the genetic material of an organism. The changes are not predictable. Mutations include changes in the chromosome number or chromosome structure. As animals grow and develop, thousands of cells are constantly dividing and millions of genes are being reproduced. This increases the risk of change resulting in a mutation.

Most mutations result in death of the cells, but occasionally mutations will result in a viable offspring that is genetically different. Mutations may be thought of as birth defects. It is likely that the first Polled Hereford beef animal resulted from a mutation.

6-14. The Polled Hereford breed probably originated as a mutation in the genetic material of horned Herefords.

MOLECULAR METHODS AND APPLICATIONS

Molecular methods have increased in the last few years. These have provided new applications to improve animal production.

METHODS

Much of the initial work in molecular biotechnology has involved previously unknown processes. The scientists were pioneers in their fields. Three areas are included here.

Genetic Engineering

Genetic engineering is a molecular form of biotechnology. The genetic information is changed to make a new product. Through this process, sections of the DNA strand are cut out and new sections are inserted.

Chromosomes contain thousands of genes. This makes it difficult to find the right gene to remove or add. A genetic map is used to locate the desired genes. This is needed for successful genetic engineering. *Gene transfer* is the moving of a gene from one organism to another. Recombinant DNA methods are used.

Genetic engineering is a means of complementing traditional breeding programs. It is not a replacement. Genetic engineering of plants is less controversial than of animals. The first release of a genetically engineered product was a tomato. This tomato has an altered gene to increase the time it is ripe without spoiling. This makes a higher-quality final product at a lower price. Improvements in animals will be made in the future using genetic engineering.

Recombinant DNA

Recombinant DNA is gene splicing. Genes are cut out of a DNA strand with a restriction enzyme that works like a scalpel. They are then inserted into circular DNA molecules in bacteria plasmids. Plasmids are circular pieces of DNA found outside the nucleus in bacteria. The plasmid is inserted into the

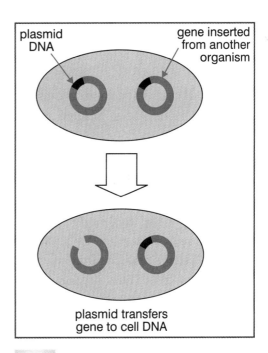

6-15. Recombinant DNA is moving a tiny part of one chromosome to another using plasmid DNA.

6-16. A microprojectile unit is used for particle injection. It transfers genetic material from one cell to another by shooting it the DNA on a small particle of gold.

cell that is to be altered. This allows the DNA of two different organisms to be combined. Gene splicing is a challenging process because DNA is so small!

All cells have membranes to protect them. Adding DNA to the nucleus of a cell is not easy. Several techniques may be used in an attempt to insert DNA into the nucleus. DNA may be mixed in a solution and injected into the cell with microinjection. Most cells die in the process. Many surviving cells do not have the "new" DNA present in the nucleus.

Transgenic Animals

A **transgenic animal** is an animal that has incorporated a "foreign" gene into its cells. This organism can pass to its offspring this transgene (altered gene). All of the cells within the transgenic animal contain this transgene.

Transgenic methods involve microinjection and particle injection. **Microinjection** is a common method of producing transgenic animals. It is injecting DNA into a cell using a fine diameter glass needle and a microscope. Microinjecting a bovine growth hormone has increased growth rates in fish, chickens, sheep, cows, hogs, and rabbits. **Particle injection** is using a microprojectile unit to shoot tiny particles coated with DNA into cells. Microprojectile units are sometimes known as particle guns. These methods have relatively low success rates. Many attempts may be needed for one successful genetic transfer to occur.

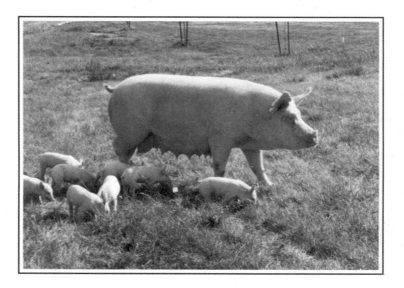

6-17. This sow is a transgenic animal named Genie. She was developed through genetic engineering. Her milk contains a human blood protein needed in human medicine. She has produced a litter of normal piglets with the same trait. (Courtesy, Virginia Tech University)

APPLICATIONS

Molecular methods have been used to genetically alter animals. Results with animals can often be applied to humans. Medicines, nutrients, and animal quality are improved with molecular biotechnology.

Animals are used to study diseases in humans, such as Lou Gehrig's disease, sickle cell anemia, and cancer. The growth and development of animals, diseases, mutations, and the influence of male and female chromosomes can be studied in similar ways.

Insulin was once extracted from the pancreas of slaughtered cattle and hogs. This process is expensive and would sometimes cause allergic reactions. Today, high-quality insulin is artificially made in laboratories.

Milk composition is suited to using molecular biotechnology. Increases of beta casein decrease the time required for rennet coagulation and whey expulsion in making cheese. This reduces the amount of time needed to make cheese. Fat content in milk can be reduced to make fat-free cheeses and ice cream. If a substance known as kappa casein is increased 5 percent, milk is more stable and easier to ship.

Livestock quality is improving. Growth rate, efficiency, and disease resistance are increasing. Reproductive performance is being enhanced. The high birth rate among Chinese hogs is being studied to increase the proliferation of American breeds.

Transgenic and other engineered animals are expensive to produce. Research is often with mice or other small animals because of relatively lower cost.

ORGANISMIC BIOTECHNOLOGY

Organismic biotechnology deals with improving organisms without artificially altering their genetic makeup. Most animal management practices are in this area of biotechnology. Several examples of organismic biotechnology are presented here.

GREATER FERTILITY

Increasing the reproductive capacity of top animals is important to producers. Sexually reproducing more offspring has been a major limitation.

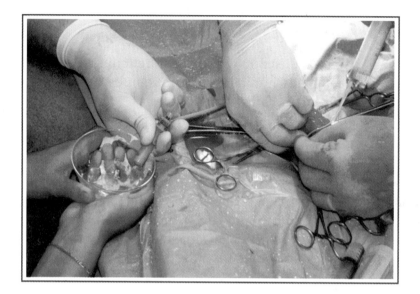

6-18. Harvesting embryos from a donor ewe. The embryos are collected in a small round dish at the end of the tube. (Courtesy, Elite Genetics, Waukon, Iowa)

Superovulation and embryo transfer are methods of increasing the number of young from a genetically superior individual.

■ Superovulation—*Superovulation* is getting a female to release more than the usual number of eggs during a single estrous cycle. Hormones are injected to assure more eggs. Superovulation is used with sheep, cattle, and a few other animals.

  Here is how superovulation is used with cows: Cows normally produce only one egg every 21 days. By injecting the cow with a hormone, such as go-

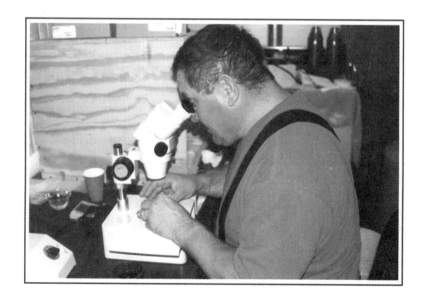

6-19. A microscope is used to search for harvested embryos in a dish. (Courtesy, Elite Genetics, Waukon, Iowa)

6-20. This shows the magnified appearance of harvested embryos. Unfertilized ova are identified and removed. Unfertilized ova show no cell division, with the cytoplasm filling the entire egg. Unfertilized ova have a lighter color. Fertilized ova show cell division, are darker, and have cell mass. Embryos are evaluated by stage of development and spherical shape. (Courtesy, Elite Genetics, Waukon, Iowa)

nadotropin, the cow can release up to 20 eggs. When in heat, the cow is inseminated and the eggs are fertilized. The embryos are flushed out of the cow seven days after breeding. A saline solution (saltwater) is used to wash the embryos out of the uterus. The embryos are no bigger than a match tip. They are either frozen or immediately placed in a recipient cow. Embryos can be frozen in straws similar to semen. After transfer, the embryos are carried to full term by the recipient. They do not receive any genetic material from their recipient mothers, only nutrients.

■ Embryo Transfer—Embryo transfer is taking an embryo from its mother (donor) and implanting it in another female (recipient). The embryo completes

6-21. Potential recipient ewes are waiting to receive embryos. (Courtesy, Elite Genetics, Waukon, Iowa)

6-22. Recipient ewes are sedated, scrubbed, and ready for embryos. Their well-being is considered throughout the process. (Courtesy, Elite Genetics, Waukon, Iowa)

6-23. Embryos are being transferred into the uterine horns of recipient ewes. (Courtesy, Elite Genetics, Waukon, Iowa)

development in the recipient. Embryo transfer usually follows superovulation, but superovulation is not necessary. With cattle, the valuable donor cow can continue estrus cycling and produce more eggs. Beef producers may implant embryos in dairy cows because they are larger and give more milk. These traits help the newborn calf grow fast.

Increased Production

Production can be increased by using milk and meat hormones and growth implants.

■ Milk Hormones—Bovine somatotropin (bST) is an approved hormone product used to promote milk production. Since bST is a naturally occurring hormone, it is acceptable to the vast majority of producers and consumers. Injections cause the cow's mammary system to become more productive. The dairy industry has adopted bST on a wide basis since its introduction in 1994.

■ Meat Hormones—Larger pork chops and hams are produced by injecting hogs with porcine somatotropin (pST). PST is produced in the pituitary gland of hogs. It causes them to produce more muscle cells resulting in larger and leaner cuts of valuable pork. Research is being done on mixing PST with feed.

■ Growth Implants—Growth implants have been used for years to promote growth. Implants are small pellets placed under the skin—usually behind the ear. They promote growth by making the animal more efficient in using feed. Implants slowly release the materials over several weeks. The materials come from hormones or molds grown on corn products.

6-24. A growth enhancer is being implanted into a calf. (Courtesy, Mississippi State University)

Enhancing Animal Nutrition

Greater results have been in animal nutrition than any other area. Enhanced nutrition has increased feed efficiency, decreased time to market weight, increased milk production, and decreased overall production problems.

■ Digestibility Testing—Digestibility of feeds varies. Tests are made by getting samples from the digestive system. Cattle, which are ruminants, have been used for most of this research. A small opening is made in the side of the cow

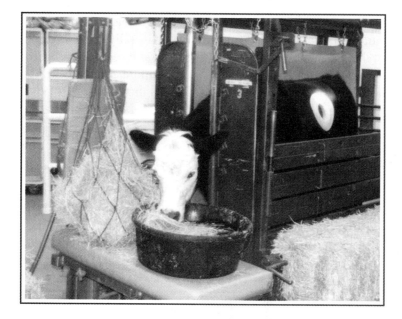

6-25. A steer with a rubber fistula in its stomach is in a feed digestibility study.

and into the rumen (large compartment of the stomach). A rubber fistula is put in the opening. This allows taking partially digested feed samples from the stomach. Cannulas or small tubes can be placed into the duodenum (part of the small intestine) of cattle. The animal has no pain during these processes. Fistulas make it easy to monitor feed intake and digestion. This information is used to improve feed use and animal growth.

■ Controlled Feeding—In a feedlot or pasture, many animals will over or under eat to meet their nutritional demands. One method of regulating this is with

6-26. Sensors on chains around the necks of these steers are used in a study that monitors feed consumption.

computer chips. The chips can be put in a monitor around the animal's neck, in an ear tag, or just under the skin in their ear. A computerized feeding station will read this chip and regulate the amount of feed that the animal gets. A daily feed intake sheet is used to detect health problems.

BIOTECHNOLOGY ISSUES

Some people are concerned about the use of biotechnology. They feel that it should not be used. Other people feel that biotechnology is tested and used in safe ways. This gives rise to controversy.

The major areas of biotechnology where issues have developed are listed here.

- Threats to Animals—Molecular biotechnology alters the natural genetics of an animal. Some people fear that this unnatural intrusion is wrong. Other people feel that animals should be improved as long as it does not threaten the well-being of the animals.

- Threats to People—The risks with new, unknown products need careful study. Biotechnology has many benefits. It is not always dangerous. New products always have some risk to workers and users. Proper safety standards are set to protect people. The safety of food products and the effects on humans are questions that consumers may have. Some people are reluctant to accept biotechnology because it represents change. Change brings uncertainty.

- Environmental Concerns—Some people fear the release of altered life forms into the environment. That is not likely to happen. Researchers are careful with new life forms. Other people fear that approved or-

6-27. Safety procedures are followed in biotechnology research. This shows a micropipette with a disposable tip being used to assure safety and sanitation.

ganisms will have an unknown bad trait. The possibility exists for the organism to upset the ecosystem.

■ Economic Gains—New feedstuffs and altered organisms present questions about patented products. Gaining a patent on an item allows a company to make money. On the down side, patents can slow the process of scientific knowledge.

■ Benefits to Developing Countries—Developing countries may not have the resources to benefit from new technologies. They may struggle to sustain a way of life. On the other hand, new technologies may make it possible for developing countries to increase production.

■ Possible Threats to Personal Beliefs—Some people are concerned about biotechnology. They often need more information to help understand issues. This could change views about creation and evolution theories. New technology could change our view of the "perfect" child. Should new plants or animals be patented? Is it right to genetically engineer an organism?

REVIEWING

MAIN IDEAS

Biotechnology is using biological systems for the benefit of humans. This uses molecular and organismic techniques.

Genetics is important in many areas of biotechnology. It is useful because it deals with heredity and how genes can be altered to change heredity.

Molecular biotechnology involves changing the structure and parts of a cell to change the organism. Genetics is a key to successful molecular biotechnology. A cell's nucleus contains chromosomes. Chromosomes are made of genes that are in turn made up of DNA. Alleles are the different forms of genes that code for a particular trait. DNA is the ultimate controller of inheritance. Recombinant DNA involves "cutting and splicing" new genetic material into a cell.

Methods of molecular biotechnology include genetic engineering, recombinant DNA, and transgenic animals. These are expensive techniques, but have great potential for helping society.

Organismic biotechnology deals with improving organisms as they are. Examples of organismic biotechnology include superovulation, embryo transfer, milk hormones, meat hormones, growth implants, digestibility testing, and controlled feeding.

Biotechnology also raises many questions. What damages could occur to individuals? How could biotechnology affect our environment? What economic gains will occur? Who will benefit from biotechnology? Will third world countries suffer because of biotechnology? Are people's personal beliefs threatened?

QUESTIONS

Answer the following questions using correct spelling and complete sentences.

1. Distinguish between the two main types of biotechnology.

2. Compare and contrast elements, atoms, and molecules.

3. What is genetics? Why is genetics important in understanding biotechnology?

4. Describe the relationship between DNA, genes, and chromosomes.

5. Distinguish between genotype and phenotype.

6. Distinguish between dominant and recessive.

7. What are mutations? How are mutations helpful?

8. What is genetic engineering?

9. What is recombinant DNA? How can recombinant DNA technology be used?

10. How is the genetic code related to recombinant DNA?

EVALUATING

Match the term with the correct definition. Write the letter by the term in the blank provided.

a. biotechnology   e. chromosome   i. phenotype
b. DNA   f. transgenic animal   j. Punnett square
c. synthetic biology   g. mutation
d. gene   h. molecular biotechnology

_____   1. Changing the structure and parts of cells to change an organism.

_____   2. The part of a chromosome that contains heredity material.

_____ 3. A protein that controls inheritance.

_____ 4. Changes that occur naturally in the genetic material of an organism.

_____ 5. Using biological systems for the benefit of humans.

_____ 6. An animal that has a foreign gene in its cells.

_____ 7. Threadlike parts of a cell that contain genetic material.

_____ 8. Using chemicals to create systems with the characteristics of living things.

_____ 9. A technique for predicting genotype.

_____ 10. The physical features or appearance of an animal.

EXPLORING

1. Invite an animal scientist or veterinarian to serve as a resource person on biotechnology in animal production. Ask the person to review superovulation and embryo transfer. Write a report that summarizes what you have learned.

2. Conduct a debate in class addressing the question "Is animal biotechnology good for society?"

3. Take a field trip to a research or biotechnology facility to study the work that is underway. Write a report on your observations.

4. Survey students in your class to determine their opinions and perceptions about biotechnology. Carefully write questions and collect data. (This could be a science fair project.)

7

BEEF CATTLE PRODUCTION

Beef! Many people enjoy eating beef. Steak and other beef foods are favorites in North America. Social gatherings often feature beef on the menu. A backyard barbecue with ribs, hamburger, and loin is a popular summertime event.

People like the excellent taste and ease of cooking beef. When properly cooked, beef products are among the safest to eat. Beef helps meet important nutritional needs because it is a major source of protein. In recent years, the fat in beef has been reduced. This has made it more appealing to health-conscious consumers.

Raising beef animals is big business. Feed and other supplies are needed. Marketing and preparing the products form a large meat industry. Many people work in the industry.

Beef animals also provide much fun. Raising a calf is a favorite project for many youth. Feeding, grooming, and caring for animals is rewarding. Youth learn about animal science and other areas of life from activities with beef animals.

7-1. Beef cattle must be regularly observed on the range. Horses are frequently used. (Courtesy, Beefmasters Breeders Universal and B. E. Fichte, Rosebud Communications, Inc.)

OBJECTIVES

This chapter introduces beef cattle production. It has the following objectives:

1   Explain the importance of beef production

2   Explain how to select beef animals

3   Name and identify the common breeds of beef animals

4   List the advantages and disadvantages of beef cattle production

5   Describe the types of beef production systems

6   Describe the promotion of herd health

7   Assess methods of animal identification

8   Describe two common herd management practices

9   Describe facility and equipment needs with beef cattle

10   Identify grooming and training practices with show animals

TERMS

anthelmintic	dehorning
backgrounding	dual-purpose breed
Bos indicus	elastrator
Bos taurus	finishing system
brand	forager
commercial cattle	halter broken
conformation	marbling
cow-calf system	polled
creep feeding	steer
cutability	

IMPORTANCE OF
BEEF CATTLE PRODUCTION

Beef cattle are raised for their meat, known as beef. The animals are selected to yield a high amount of good meat. This is contrasted with dairy cattle, which are used for milk production.

Beef production is a large part of agriculture in North America. Providing quality beef requires good animals and efficient marketing. Products must reach the consumer in wholesome condition.

ORIGIN OF BEEF CATTLE

The origin of cattle is not fully known. It is assumed that they were first domesticated in Europe during the Stone Age. Today's cattle are believed to have originated from two species: *Bos taurus* and *Bos indicus.*

Bos taurus are cattle common to the temperate regions, such as the midwestern United States. They are often known as the European breeds. Examples include Angus and Hereford.

Bos indicus cattle are those cattle that are descendants of the Zebu cattle from India with humps on their necks. They also have large, droopy ears and loose skin. Bos indicus cattle are common to warmer tropical areas, such as the southern part of the United States.

7-2. Young people enjoy caring for beef animals. This Charolais heifer is being fitted for showing.

Christopher Columbus introduced beef cattle in the United States. Beef animals were on his second voyage to America. Early settlers from the British Isles and other European countries brought cattle with them as they settled new land in America. As the settlers moved west, so did the cattle. Abundant pasture was found in the great plains areas.

THE BEEF CATTLE INDUSTRY

Today, the cattle business is an important part of American agriculture. Sales of cattle amount to a large portion of all farm markets. There are 101 million head of beef cattle on the farms and ranches in the United States.

7-3. Steaks are a favorite food for many people. (Courtesy, Missouri Cattlemen Association)

There is also a global market for beef. The United States exports beef to countries such as Japan, Canada, Republic of Korea, Mexico, and Hong Kong. Japan is by far the largest foreign customer of the United States' beef industry, including imports of variety meats and hides.

Consumption

The consumption of beef is part of a balanced diet designed for good nutrition and health. Beef continues to

7-4. Locations of the beef cuts on an animal. (An asterisk * indicates the most desirable cuts that sell for a higher price.)

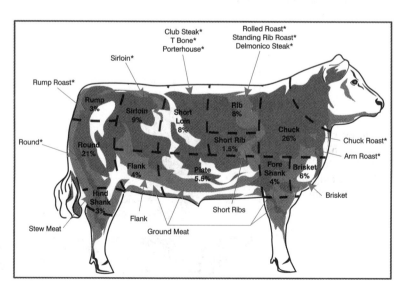

be one of America's favorite foods. The per capita consumption of beef is 66.6 pounds per year in the United States.

Beef helps consumers meet daily dietary requirements in protein, vitamin B-12, iron, zinc, and other essential nutrients. In recent years, beef has received criticism because of the amount of fat and cholesterol it can add to the diet. New research, improved genetics, and better management techniques are reducing these concerns. Moreover, the key to a healthy diet is in the balance, variety, and moderation of consumption.

Consumer Demands

Consumers want good beef. They will pay top prices for certain cuts, such as steak and loin. Consumers pay less for organs, bony pieces, and the parts made into hamburger and similar products.

Producers need to be aware of where the more valuable cuts are located on an animal. They also need to produce the type of animal that has larger amounts of meat in these cuts.

BEEF CATTLE SELECTION

Selecting beef cattle is an important decision. Some people want a specific breed. Other people are not concerned about specific breeding. Much money and other resources are involved. The strengths and weaknesses of the breeds should be considered. This will allow producers to meet their goals.

PUREBRED AND COMMERCIAL CATTLE

A purebred is an animal of a particular breed. It is entitled to be registered as a purebred animal. Breed associations set rules purebred cattle must meet. Producers must follow the rules and certify breeding.

Some cattle producers decide to raise cattle that are not purebreds (entitled to registration). They may have crosses between the breeds to get the desired traits. For example, bulls of one breed may be bred to cows of another. The animals that result are crosses. Other producers raise commercial animals. *Commercial cattle* are not registered. Some have a purebred background, but most are mixed breeds. They are selected for production potential.

7-5. Commercial beef cows on a North Carolina school farm. What breeds are involved in these crosses?

Most meat production in the United States is from commercial beef animals. Many purebred producers raise animals to improve the quality of the breed and to breed to commercial animals.

SELECTION GUIDELINES

Eight guidelines to follow in selecting a breed are:

1. No one breed is the best for all traits. Producers need to have priorities on the traits that are most important to them.
2. All breeds of beef cattle will have variation within their genetics. A decision has to be made by the producer on the genetic traits that are most important.
3. Selection of superior animals and sound breeding practices are as important as the breed. The cattle will only be as good as their selection and management.
4. Select a breed that has the desired conformation as a beef animal. *Conformation* is the type, shape, and form of an animal. It is related to muscle (meat) in the valuable cuts and the ability of the animal to be productive. Over the years, the desired conformation has changed to moderate-framed animals that can provide much meat.
5. The breed needs to be selected on its production capabilities in the specific environment where it will be raised. The producer needs to consider climate, feed, and forage resources. Some breeding animals are per-

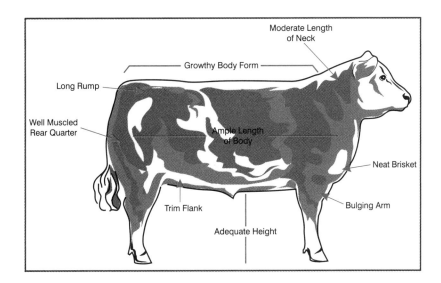

7-6. Conformation of a desirable beef animal.

(Figure labels: Moderate Length of Neck, Growthy Body Form, Long Rump, Well Muscled Rear Quarter, Ample Length of Body, Neat Brisket, Bulging Arm, Trim Flank, Adequate Height)

formance tested. This involves careful growth and feeding evaluations of young animals, particularly bulls.

6. Markets for the breed should be evaluated. Can the offspring be readily sold in the area in which they are produced? Breeding stock needs to be available at a price that makes production economically feasible. Buy the best animals at the lowest price.

7. The producer should consider personal preference and select a breed with which they are comfortable and familiar. This is probably the most commonly considered criterion in the selection process.

8. Buy animals from established producers with a good reputation.

BEEF BREEDS

The breeds of beef cattle common in the United States are briefly described here. These were developed in Europe, Great Britain, or America. Some genetics of Indian Zebu cattle may be included. Animals in the breeds conform to standards set by the breed associations. Many are purebred and registered. Others may be purebred but have not been registered.

ANGUS

The first Angus cattle were imported from Scotland by George Grant around 1873. Angus are the most popular purebred cattle. Nearly 20,000 are

7-7. A modern Angus bull. (Courtesy, American Angus Association)

registered in the United States with the American Angus Association each year.

The breed is black in color and has a smooth hair coat and black skin. A few Angus cattle carry a recessive gene for the color red, which are described below as Red Angus. Nearly all purebred Angus carry the dominant gene for being polled. *Polled* means that the animal is naturally without horns.

Angus are cattle with moderate frames. They produce a high-quality carcass that has good marbling. *Marbling* is the presence of intramuscular fat. Marbling increases the quality of the carcass and makes the meat cuts tastier when we eat them.

Red Angus

The Red Angus breed developed from Black Angus in the United States about 1945. Black Angus carrying the recessive red gene for color were mated to produce red offspring.

Red Angus are similar to Black Angus in many respects, except for color. However, red does not absorb as much heat as black, so Red Angus are somewhat more heat tolerant.

Ankole Watusi

The Ankole Watusi breed originated in the Ankole district of Uganda, Africa. It is not wide spread in the United States but has striking features.

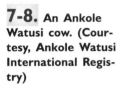

7-8. An Ankole Watusi cow. (Courtesy, Ankole Watusi International Registry)

Colors include dark red, black, white, gray, and yellow. This breed has the largest horns of any beef cattle. The horns may reach 8 feet in total length, with a diameter of 9 inches at the base. The horns are said to serve the animal by casting-off excess heat in hot climates. The animals have a slight hump. Many recessive genes result in a wide variety of animals, including some with dwarfism and some that are polled.

CHAROLAIS

The Charolais is an old breed developed in central France. The first Charolais were introduced in the United States in 1936 by the King Ranch in Texas. The Charolais has continued to increase in popularity since its introduction.

7-9. A Charolais cow and calf. (Courtesy, Gary and Pam Naylor, Missouri)

Charolais cattle are white or off-white in color with pink skin. They are a large-framed, heavy-muscled breed. Charolais may be either polled or horned. This breed is commonly used in crossbreeding programs to increase frame size and muscle.

*C*HIANINA

The Chianina breed originated in the Chiana Valley of Italy. It is, perhaps, one of the oldest breeds in the world. Chianina were introduced to the United States in 1971, when semen was first imported.

Chianina are large framed and are commonly used in crossbreeding programs. They have good maternal traits and generally improve the growth rate of the offspring. The color of Chianina cattle varies because of the influx of crossbreeding. Original Chianina cattle were white with a black tail switch and black skin pigment.

7-10. A Chianina bull. (Courtesy, American Chianina Association)

*G*ELBVIEH

The Gelbvieh breed was developed by crossing four yellow breeds of cattle. These four breeds were Glan-Donnersburg, Yello Franconian, Limburg, and Lahn. Together they form the Gelbvieh breed, which originated in 1920.

Gelbvieh cattle vary in color from cream to reddish yellow. In recent years, black Gelbvieh lines have been developed. These cattle are medium framed, produce an acceptable carcass, and have a high milking ability, when compared to other beef breeds.

7-11. A Gelbvieh cow and calf. (Courtesy, American Gelbvieh Association)

HEREFORD

Hereford cattle originated in Hereford County, England. The first Hereford cattle were imported by Henry Clay to Kentucky. These cattle were mated to other native cattle and the first purebred herd was believed to have been established in 1840 in New York.

Hereford cattle have white faces and red bodies. Additionally, they may have a white feather or stripe on the top of their necks, which is commonly called a "feather neck." They are predominately white on the belly, legs, and

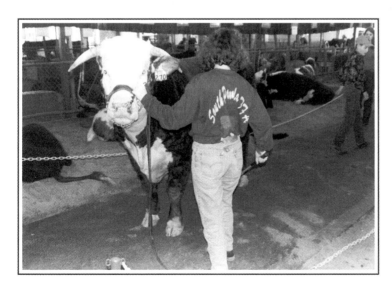

7-12. A Hereford bull being trained for showing.

switch. Herefords are horned, usually very docile and easy to handle, and are moderate in frame size.

Hereford cattle are extremely good foragers. A *forager* is an animal that makes especially good use of grasses and plants for food. They will search out food plants on rangeland. Furthermore, this breed is very hardy. The foraging ability and hardiness make this breed popular to the grazing regions of the western United States.

LONGHORN

Longhorn cattle were brought to the United States from Mexico by Spanish explorers in the 1800s. The fertility of Longhorn cattle is usually better than many other breeds. They are commonly used in crossbreeding because of their light birth weights.

Longhorn cattle do not grow as fast as some other breeds, but are very durable and adapt well to rough range conditions.

7-13. A Longhorn.

POLLED HEREFORD

The Polled Hereford breed originated in Iowa in 1901. An Iowa breeder named Warren Gammon was instrumental in developing the breed. Gammon located four Hereford bulls and ten Hereford cows that were carrying the polled gene and mated them to develop the foundation for the Polled Her-

7-14. A Polled Hereford cow.

eford breed. Polled Herefords have the same basic traits as Herefords, except they do not have horns.

LIMOUSIN

Limousin cattle originated in France. Limousin cattle were introduced in the United States in 1968, when semen was imported from Canada.

7-15. A modern Limousin bull. (Courtesy, North American Limousin Foundation)

The Limousin will be either a light yellow or an orange color. The area around the muzzle will generally be a lighter version of the body color. Recently, black lines of Limousin cattle have been developed. Both horned and polled cattle are prevalent in this breed.

Limousins have a moderate, heavy-muscled frame. The breed is noted for its carcass leanness and large loin areas. Therefore, this breed's carcass is usually very high in cutability. *Cutability* is the amount and quality of saleable retail cuts that may be obtained from a carcass.

Maine-Anjou

The Maine-Anjou breed originated in France in the early 1800s. It is thought to have developed through crossbreeding English Shorthorns and French Mancelle breeds. Maine-Anjous were originally used as draft animals. Beef traits were developed in the breed through selective breeding. Maine-Anjou cattle were introduced to the United States in 1970, when semen was imported from Canada.

Maine-Anjou cattle are dark cherry red and white in color and have light pigmentation in the skin. They are commonly used in crossbreeding. From this crossbreeding, black Maine-Anjous have been developed. The breed is usually horned and moderate framed. They are generally docile and easy to handle, grow quickly, and have good marbling in their carcasses.

7-16. A young Maine-Anjou bull. (Courtesy, Hinds Community College)

Connection

BREED ADAPTATION

Cattle breeds are adapted to certain climates. Most breeds are best adapted in climates similar to where they originated. The breeds with *Bos indicus* breeding are best suited to hot climates with high moisture. The *Bos taurus* breeds are best suited to cooler climates.

One example of a breed adapted in a cool climate is the Belted Galloway. This breed was developed in Scotland. Scotland has a cool climate that can be harshly cold in the winter. Northern areas of the United States may raise the Belted Galloway. This photograph shows Belted Galloway cattle grazing in Northern Maine on an early autumn day. They have the ability to make good use of the kind of forage that grows in that area and withstand cold winters.

One factor in selecting a breed is the environmental conditions under which it is to be raised.

SALERS

The Salers breed was developed in central France and recently brought to the United States. Initial herds were established in 1973, when semen was imported.

7-17. Salers bull. (Courtesy, American Salers Association)

Salers are a deep cherry red, and sometimes, will have white on their bellies. Recently, black Salers cattle have been developed. Salers cattle are noted for their calving ease, primarily because of their smaller heads and long slender necks and bodies. Most Salers females milk quite well.

SHORTHORN

Shorthorns originated in northern England around 1600. Shorthorns were originally developed as a ***dual-purpose breed***, meaning they were used for both milk and meat production. The breed was introduced to the United States in 1783, in Virginia.

Shorthorn cattle are red, white, or roan. Shorthorns may be either polled or horned. Frame size of this breed is moderate. The females are good mothers and milk exceptionally well. Shorthorns are generally gentle and have good dispositions.

7-18. A Shorthorn bull. (Courtesy, *Shorthorn Country*)

SIMMENTAL

The Simmental breed was developed in western Switzerland and is still quite popular in Europe today. Originally, they had many purposes, including milk and meat production, and use as draft animals. Simmentals were brought to the United States in 1969, from Canada.

Simmental cattle usually have a cream-colored face with a light yellow, red, or spotted body. Recently, black strains of the breed have been developed.

7-19. A Simmental cow with her calf. (Courtesy, American Simmental Association)

Simmental cattle may have some color pigment around their eyes. Horned and polled cattle will be found in this breed as well. The Simmental is large-framed, heavy-muscled, and produces a lean carcass.

SOUTH DEVON

South Devon cattle were developed in the western region of England. This breed of cattle is believed to have originated from a large, red breed in

7-20. A South Devon cow and calf. (Courtesy, North American South Devon Association)

France, which was brought to England during the Norman Invasion. The South Devon breed is related to the Devon breed. This breed was developed for dual purposes of meat and milk production.

South Devons were first imported into the United States in the late 1930s and early 1940s. However, the popularity of the breed increased in 1970, when a group from Stillwater, Minnesota, started using South Devons for crossbreeding purposes. The South Devon breed will be a medium-red color and usually be horned.

Brahman

Brahman cattle were developed in the southwestern part of the United States. The breed is thought to have developed between the mid 1800s and early 1900s and originated from the cattle from India. *Bos indicus* cattle have a hump on the shoulders and usually have long floppy ears and loose hide. These cattle may also be called Zebu.

Breeders select the Brahman breed because of their hardiness, resistance to insects and diseases, and their heat tolerance, which makes them very popular in the warmer climates of the south and southwestern United States. Furthermore, the females are good mothers and Brahman cattle do quite well on poor-quality forages common to range conditions.

7-21. A modern Brahman cow. (Courtesy, American Brahman Breeders Association)

Brahman cattle are somewhat unpredictable in their temperament. The Brahman breed is earlier in skeletal maturity and often is used in crossing with other breeds of cattle. Brahman cattle will vary in color from gray to red to black.

BRANGUS

The Brangus breed is a cross between the Brahman and Angus. This breed was developed at a U.S. Department of Agriculture Experiment Station in Louisiana in 1912.

Brangus cattle will have many of the same characteristics as the Angus and Brahman. They are solid black in color and are polled. (There is also a Red Brangus, which is a different breed from the Brangus.) They also have long floppy ears (not as long as the Brahman) and loose hide. Brangus cattle are very adaptable to various climates and have good mothering abilities.

7-22. A modern Brangus bull. (Courtesy, International Brangus Breeders Association, Inc.)

SANTA GERTRUDIS

Santa Gertrudis cattle were developed around 1918, on the King Ranch in Texas. This breed was developed by crossing Brahman and Shorthorn cattle.

Santa Gertrudis are a deep cherry red and may be horned or polled. Because the Santa Gertrudis are part ancestors of the Brahman breed, they have

7-23. A Santa Gertrudis bull, cow, and calf. (Courtesy, King Ranch)

many of the same characteristics, including loose skin and long ears. They are also very adaptable to the warmer climates and are more resistant to disease and insects.

BEEFMASTER

Beefmaster cattle were developed in 1931 in Texas. The breed is the result of crossing Herefords, Shorthorns, and Brahmans. The breed varies in color and reds and yellowish-whites are common. Beefmaster cattle are hardy and milk well.

7-24. A young Beefmaster bull. (Courtesy, Hinds Community College)

TARENTAISE

The Tarentaise originated in the French Alps in 1859. It was brought to North America in 1973 with importations into Canada. They are a hardy breed. Tarentaise are relatively free of eye cancer and udder burn because of dark pigment around the eyes and on the udder.

Tarentaise are wheat-colored, ranging from cherry to dark blonde. Black pigment is found on the muzzle, eyes, and udder. Females calve easier than most breeds. Bulls are well suited for use with first-calf heifers. Calves are somewhat smaller and vigorous at birth. Bulls may weigh up to 1,800 pounds and cows up to

7-25. A Tarentaise bull. (Courtesy, American Tarentaise Association)

1,150 pounds. Adults are more active than other breeds. The breed is growing in popularity. Research by Montana State University animal scientists has found that Tarentaise show particularly good performance at higher mountain elevations.

ADVANTAGES AND DISADVANTAGES OF BEEF CATTLE PRODUCTION

Producing beef cattle has advantages and disadvantages. The producer has to analyze these and decide if beef production is the right thing to do.

ADVANTAGES

Producing beef cattle may have the following advantages:

- Beef cattle can consume roughages, including most grasses and several weeds, which other livestock may not eat or be able to use. Therefore, little forage is wasted with beef cattle.

■ Labor requirements are usually not as great with beef cattle as with other livestock production. Because beef cattle are good foragers, they do not necessarily have to be fed daily.

■ Death losses are relatively low, if good management practices are followed.

■ Beef cattle adapt well to various sizes of operations. Productive beef herds can be from one or two head to several thousand head.

■ There is a good demand for meat. Beef is a mainstay in the American diet. Additionally, meat is an important export to other countries.

■ The beef industry creates many jobs. It relies on services, such as herdsmen, trucking firms, processors, researchers, and feed suppliers.

■ Beef cattle may provide avocational interests, as well as youth projects through FFA and 4-H, which allow new experiences for both young and old.

DISADVANTAGES

Producing beef cattle has the following disadvantages:

■ If cattle are fed to slaughter weight, the producer incurs risk. Grain prices tend to fluctuate and available forage is dependent on the precipitation for a specific area. An example of this would be whenever a region suffers a drought. Grain costs more because yields are lower and forages do not grow readily because of the lack of precipitation.

■ Beef cattle do not convert feed and forages into meat as efficiently as do other animals.

■ Because of the 283-day gestation period of beef cattle, it takes longer to increase herd size. The producer will usually only get one offspring a year from each cow.

■ Capital investment is high to start a modern beef operation. Not only does the producer have to invest in the cattle, but also equipment, facilities, and herd management.

TYPES OF
BEEF PRODUCTION SYSTEMS

Beef cattle production varies among producers. The kind of system depends on the producer's goals. It is also influenced by the resources available and the market in the local area.

COW-CALF SYSTEMS

A *cow-calf system* is keeping adult cattle to raise calves for sale to other growers. Most cow-calf systems involve high-quality commercial cattle. The cows are bred to calve in the late winter or early spring. The calves are sold in the fall. This takes advantage of the warm summer growing season. Most of the calves go to feedlots for finishing into meat animals.

Natural breeding is typically used with commercial cow-calf systems. The bull may be put in the pasture with the cows at breeding time. A producer schedules breeding to assure calving at the desired time. One three-year-old bull will breed 25 to 40 cows a year.

Some beef producers use artificial insemination (AI). AI is more important with expensive, purebred females. This allows the producer to have access to a wider variety of bulls.

Heifers need to be the proper size and age before breeding. Most heifers will have their first calf at two years of age. Since the gestation period of a cow is approximately 283 days, a cow should have one calf a year. Occasionally, cows will have twins or multiple births, but this is not prevalent in beef.

7-26. This cow-calf producer has a calf crop underway with commercial Charolais.

The cow-calf system uses forages for feed. Several kinds of plants are grown in pastures for forage. Both grasses and legumes are used. Pasture grasses include fescue, orchard grass, or Bermuda grass. Legumes include several kinds of clovers, lespedeza, vetch, and alfalfa. Pastures should be kept free of weeds, especially those that are poisonous. If pasture is not available, hay or silage can be fed as a forage, but this is not typical with beef cattle production.

An additional mineral supplement should be fed free-choice to meet the need for minerals, which are vital to beef cattle. Forage alone may not provide sufficient nutrients. Calves that are a few months old may be supplementally fed with grain. A method known as creep feeding is used. *Creep feeding* is using equipment that only allows small animals to enter. Creep feeding will allow the calf to gain more weight while it is still nursing the cow. Creep feeding is usually done when the calves are not receiving enough milk from the cow, the supply of forages is limited, or when grain prices are low and it is economically feasible.

Producers of purebred cattle follow a somewhat different cycle. They raise herd replacements and animals for other cattle breeders. Purebred cattle are not raised to go to a feedlot.

Backgrounding Systems

A *backgrounding system* is an approach that takes a calf from the time it is weaned from the cow to the feedlot phase. Backgrounding is essential to add more weight to the calf before it enters the final feeding phase.

During the backgrounding, forages, such as annual grasses, are commonly used to add more weight to the calf. Supplemental grain may also be fed to the calves to increase the rate of gain and prepare the calves to enter the finishing phase earlier. Calves that are backgrounded will also need a free-choice mineral supplement.

Finishing Systems

A *finishing system* is an approach that takes a calf through the complete grow-out phase of beef production. The calf is fed to slaughter weight. Slaughter weight is usually 1,100 to 1,300 pounds, depending on the breed.

7-27. The goal of finishing is to produce a beef carcass with desired quality.

Some beef producers feed calves to slaughter weight, but most cattle will be fin-

7-28. A California feedlot.

ished by commercial feedlots. The feedlots are located in the western and southwestern United States. During the finishing period, calves are fed feed concentrates. These are feeds high in energy from grains, such as corn, milo, and oats. Free-choice minerals may be provided if the minerals are not part of the finishing ration.

HERD HEALTH

A good herd health program is essential in all phases of beef production. Prevention, including good sanitation and vaccination programs, is key to maintaining good herd health. Diseases and parasites greatly reduce the profitability for the beef producer. A good health program will also aid in marketing beef animals. (Chapter 4 covers animal health in more detail.)

PARASITES

Beef cattle are affected by both external and internal parasites.

Common external parasites are flies, mosquitoes, lice, mites, and ticks. Good sanitation practices with facilities and equipment will reduce problems with these external parasites. In addition, external parasites may be controlled by using insecticides.

Only approved products should be used in controlling external parasites. Some producers consult veterinarians for parasite control. In general, insecticides may be applied by spraying or dipping the cattle. Backrubbers may also be used when spraying or dipping is not an option. A backrubber should be placed in a high-traffic area, such as near feeders and waterers. The insecticide is poured on the backrubber. As the cattle walk under the backrubber, the insecticide is applied.

Internal parasites, such as roundworms and flatworms, must be controlled. Common symptoms of internal parasites are weight loss, poor weight gain, rough hair, and diarrhea.

Anthelmintics are chemical compounds used for deworming animals. Anthelmintics may be administered to cattle orally, by injection, by pour-on, or by adding to the feed or mineral. A regular worming schedule should be followed. Best results are achieved if cattle are dewormed once or twice a year. A veterinarian may be consulted to decide which anthelmintics to use.

PATHOGENIC DISEASES

Controlling disease is important in a profitable beef operation. Many diseases affect beef cattle. Some producers use a veterinarian to help manage herd health. A few examples of disease problems with beef cattle are included here.

Bovine virus diarrhea (BVD) symptoms may include fever, coughing, discharge from the nose, poor gains, rapid breathing, diarrhea, dehydration, rough hair, and lameness. In addition, BVD may cause pregnant animals to abort.

7-29. Baby calf receiving its first injection. (Courtesy, Thunderbolt Angus)

Brucellosis (Bangs) is a reproductive disease. The symptoms may include an aborted fetus, retained afterbirth, and infertility. Government programs have helped control the disease, but it is still a risk to cattle. There is no known cure for brucellosis, and infected animals should be destroyed. All female calves should be vaccinated for brucellosis while they are young. Replacement heifers should be from brucellosis-free herds. Brucellosis may also be spread to humans, causing undulant fever.

Blackleg is a bacterial disease that can live without oxygen. The bacteria commonly live in the soil and may be present for many years. Symptoms of blackleg are sudden death, high fever, swollen stiff muscles, and lameness. Calves should be vaccinated when young and again at weaning.

Infectious bovine rhinotrachetis (IBR, Red Nose, IPV) has several forms. Symptoms may include fever, discharge from the nose, rough breathing, coughing, weight loss, inflammation of the genitals, watery eyes, convulsions, and incoordination. Since the symptoms are quite varied, a veterinarian should be consulted if IBR is suspected.

Leptospirosis is caused by several strains of bacteria. Symptoms may include a rise in temperature, rapid breathing, stiffness, loss of appetite, bloody urine, and abortion. All animals should be vaccinated twice a year to control the disease.

Campylobacteriosis or vibriosis may occur in intestinal or venereal form. The intestinal form usually does not have harmful effects on beef cattle. The venereal form is more serious. The symptoms may include infertility, abortion, erratic heat periods, and poor conception rates due to early embryonic death. Animals should be vaccinated before breeding and the vaccination should be repeated annually.

ANIMAL IDENTIFICATION

Most beef animals need to be identified in some way. Proper identification is important in managing the beef herd, as well as specifying ownership.

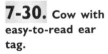

7-30. Cow with easy-to-read ear tag.

7-31. An electronic ear tag has information stored inside. (Courtesy, Allflex USA)

It allows the producer to identify all animals and keep records on production. Identification helps follow an animal from the producer through marketing. Newer methods are being developed for this purpose.

■ Tags—Tags are usually placed in the ears of animals. Some tags are on chains around the necks of animals. Tags are typically made of plastic materials. Some tags have identifying numbers or letters. Each animal has a different number. Records are kept by numbers. Newer tags allow electronic tracking of animals. Electronic codes in tags are sensed by a scanning device. This can be used in association with a computer that immediately identifies the animal. Detailed records can be kept via computer. These include the date of birth, weight, and vaccination and worming schedule. Tags are sometimes torn out of ears and lost.

7-32. Using an applicator to insert an electronic ear tag. (Courtesy, Allflex USA)

■ Branding—A *brand* is a permanent method of identification. Branding may be with a hot iron or by freeze branding. The producer needs to select a brand design that will represent ownership of the cattle. Brands are usually regis-

7-33. A hand-held reader emits a radio signal that detects the identification number programmed into an electronic ear tag. (Courtesy, Allflex USA)

tered with a state agency. A hot iron brand will burn the surface of the skin, leaving a permanent mark. When freeze branding, liquid nitrogen is used with the iron, which freezes the skin pigment. This depigments the hair follicles and allows the hair to grow back white in color. In white-haired cattle, the liquid nitrogen is left on a longer time. This produces a bald spot the shape of the branding iron.

7-34. Hot brand. (Courtesy, Tommy Perkins, Southwest Missouri State University)

■ Earmarks—Earmarks may be as simple as using a knife or an earnotcher to cut a notch in the ear. Tatoos may also be placed in the ears of some animals. With earnotches, a system is developed based on location in the ear using the left and right ears for the notches.

- Neckbands—Neckbands or chains may also be placed on animals for identification. Again, animals may tear the bands or chains or someone other than the owner may remove them. In rare cases, an animal may catch the chain or band on an object, which could possibly choke the animal to death.

- Tatoos—Using a tattoo is a common method of identification used among purebred breeders. An instrument applies letters and numbers by piercing the skin. Ink is then rubbed into the area. A common place for tattoos on beef cattle is the inside of the ear. Tattoos are frequently used. A drawback is the animal must be confined for the tattoo to be read. Some tattoos are hard to read on dark-skinned cattle.

- Other Methods—Several other methods of identification are available, including nose prints, DNA testing, and microchips. Nose prints and DNA testing are used to establish the precise identity of an animal. These procedures are often used with show animals. Microchips may be placed beneath the skin of an animal. A scanner will decode the signal from the microchip. The information identifies the animal. Microchips are not appropriate in animals to be slaughtered. Removing the microchip from the carcass may pose a problem.

COMMON HERD MANAGEMENT PRACTICES

Producing quality beef animals is more than turning them out in a pasture. The animals must receive the proper care. Care is known as management practices. Several areas of care include providing feed and water, breeding, disease control, worming, and animal identification. Animals should be observed each day. This is to check for injury and disease as well as calving by pregnant females. Two additional practices are included here.

DEHORNING

Dehorning is used with horned breeds of cattle to remove horns. Large horns can be dangerous and injure other animals, as well as people.

Several dehorning methods are used. It is preferable to dehorn young calves. Dehorning larger animals is more difficult as well as traumatic to the animal.

With calves, caustic chemicals may be placed on the location where horns grow. This prevents horn growth by destroying tissue. It is widely used on younger animals.

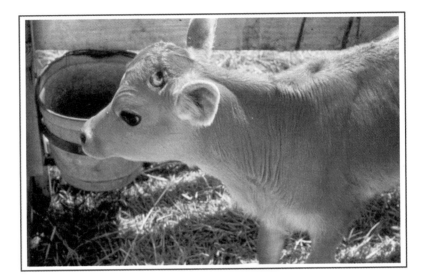

7-35. A calf dehorned with a chemical that prevents the growth of horn tissues and reduces the stress of cutting the horns off.

Horns may also be sawed or clipped off. This is a surgical procedure that removes the horns. When using this method, all bleeding needs to be stopped afterward. A repellent should be used to keep flies off the wound.

CASTRATION

The beef producer will usually not want to keep all male offspring as bulls. These calves are castrated.

Castration is removing the testicles or destroying their development. Castration makes male calves easier to handle and manage. A calf castrated at a young age is a *steer*. Except for breeding animals, steers often bring higher prices than bulls. Castration also prevents bulls from breeding any females that are present once puberty has been reached.

A common method of castrating cattle is with an *elastrator*. This stretches a specially made rubber band over the scrotum, which will cut off circulation. The scrotum will essentially dry up and fall off. This method should be done on younger, smaller calves.

A knife or a scalpel may be used to cut the testicles from the scrotum. This is a popular method. Bleeding must be stopped and flies controlled. Additionally, castrated calves need to be watched closely for infection.

A Burdizzo clamp is also used. No bleeding occurs with this method. The clamp crushes and severs the cords and blood vessels going to the testicles. This causes them to dry up. When using the Burdizzo, it is important that the cord does not slip out of the clamp and miss being crushed.

FACILITIES AND EQUIPMENT

The facilities and equipment needed for beef cattle depend on the system of production and climate. Factors that assist in planning include the number of cattle, amount of space, location, kind of facility, herd and property security, environmental conditions, amount of money to invest, and availability of other facilities and equipment.

FACILITIES

Facilities may range from simple to elaborate. Producers should allow for flexibility so they can expand or change beef production systems.

Because beef cattle are primarily range animals, it is not necessary to have a great amount of shelter. However, barns and loafing sheds should be available to provide shade in the summer, if trees or other shade sources are not available. Additionally, barns and loafing sheds may be used to protect cattle from harsh weather.

All beef cattle producers need fences to keep the animals confined. These fences must be kept in good condition. Some fences are of barbed wire while others use a meshed wire. Electrically charged fences are used on some farms and ranches. Board and vinyl material are also used, especially in public areas.

7-36. A properly designed corral. (Courtesy, University of Missouri Extension Service)

EQUIPMENT

Cattle production requires various equipment. The kind needed depends on the type of animal produced. Water tanks and water fountains need to be available for all cattle.

Because most producers supply salt and mineral free-choice to beef cattle, feeders need to be available. These feeders should be large enough to allow several head to consume salt and mineral mix at the same time.

Feed storage is needed to assure high-quality feedstuff and reduce the amount of feed wasted by spoilage and pest damage. Feed may be stored in upright silos, trench silos, metal storage bins, or feed rooms inside barns.

Feeders for hay, forages, and grain need to be large enough to accommodate all cattle at once. This may require the producer to use more than one feeder. This alleviates the dominant, aggressive animals from keeping other cattle away.

Equipment to handle cattle is needed in all beef operations. A good corral makes it easier to work cattle. Corrals should also decrease labor and time, reduce stress on cattle, decrease chance of injury and weight loss, and increase safety for the producer.

Corral designs vary. In planning a new corral, allow for expansion in the future. Sharp turns and corners should be avoided. The corral should be built so cattle can be easily moved and trailers loaded.

7-37. A squeeze chute for working cattle. (Courtesy, University of Missouri Extension Service)

Working chutes and headgates are used to hold and restrain cattle for tagging and vaccinating. There are many models of chutes and headgates. Select the design that best allows efficient cattle handling.

SHOWING BEEF CATTLE

Showing beef cattle can be fun. Showing beef cattle allows producers to promote their cattle to prospective buyers and other breeders. This allows breeders to compare their beef cattle with other beef breeders' cattle. For 4-H, FFA and other junior members, showing beef cattle allows them to learn about the promotion of their individual breed of beef animal. Furthermore, showing beef cattle creates a sense of responsibility by properly caring for, grooming, and showing the animal.

The well-being of show animals is very important. Never abuse an animal. Always keep the animal in conditions that are favorable to its well-being.

Learn more about animal well-being and showing by attending local seminars or field days designed for people who show animals.

ANIMAL SELECTION

Show animals need to be carefully selected. Use a highly qualified show person to help in the selection process. Get the animal from a producer who consistently has quality cattle. Some cattle producers will hold special sales of calves that have the genetic potential to be good show cattle. A quality animal is needed to be competitive in the show ring.

ANIMAL CARE

An animal being raised for showing must receive proper care. Feed must be appropriate to the desired gain. Facilities should be used that provide for the well-being of the animal. Get the assistance of a person who is an authority on caring for show animals. Local feed dealers often have the proper feed.

HALTER BREAKING

For an animal to be shown, it must be halter broken to stand and lead. *Halter broken* refers to an animal that is comfortable with a halter on its

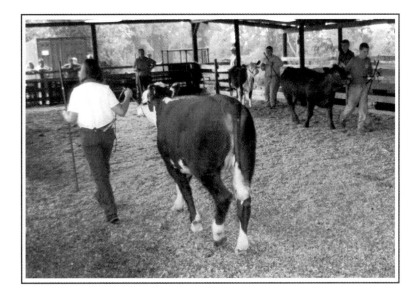

7-38. Leading a Hereford heifer in the show ring.

head and responds to the commands of the person who is leading it. Many hours may be needed to halter break an animal.

Begin halter breaking by tying the animal with a rope halter for short periods. Brushing the animal while it is tied up will help to calm the animal. It is important not to leave the animal unattended during this period to avoid injury to the animal. After the animal has been tied up several times, the next step is to lead the animal to water or to feed. This should be done in a secure area of a corral, so if the animal gets away, it can be caught again and tied. After the repetition of being led back and forth to feed and water, the animal will soon be accustomed to being led. Younger animals, in most cases, will be easier to break to lead.

GROOMING

Grooming, or fitting, the animal is a very important part of showing beef cattle. The initial step is to train the hair of the animal. This may be done by washing the animal at least once a day, then brushing and drying the hair on the animal. The next step is to clip the animal. This involves clipping the long hair off the animal to highlight its strengths. The areas of the animal that may be clipped are the head, neck, underline, body and tail. The hair may be left longer on the hind quarters and the body of the animal to make it look thicker and bolder. During the show, grooming products, which are similar to hair spray, may be used to hold the hair in place.

7-39. Using clippers to fit an Angus calf.

SHOWING

How an animal is managed in the show ring is important. When the animal is fitted and ready to go to the show ring, the rope halter should be replaced with a leather show halter. Upon entering the show ring, the exhibitor should pay attention to the judge of the animal. When asked to stop, position the animal's feet with the show stick. The animal should stand squarely and evenly distribute its weight on all four legs. Exhibitors should always be aware of where the judge is positioned and how their animals are positioned. Considerable practice is needed to develop skills in showing an animal.

7-40. Heifers competing in a show ring. (Courtesy, American Gelbvieh Association)

REVIEWING

MAIN IDEAS

Most beef cattle breeds originated in Europe or Great Britain. Beef cattle are produced by a series of systems: the cow-calf, backgrounding, and finishing. All of these systems are beneficial to the beef producer.

The producer has to follow sound management practices to be successful in producing beef cattle. These practices will increase profitability, decrease loss from disease, and reduce stress to the animals. Proper management practices will allow the beef producer to have higher-quality beef cattle. The promotion of these quality animals may be through showing the animals at local, district, and national level shows.

Proper facilities and equipment are necessary for a beef cattle operation. Properly planned facilities will decrease labor and make the beef operation more efficient. Designing facilities that will allow for expansion is important to facility design.

Equipment varies from one operation to another but is needed in all beef production. Corrals make it safer and easier to handle cattle, as well as reduce stress on the cattle. Watering equipment is important for most operations. Feeders and feed storage equipment will insure higher-quality feed.

Additional detailed information on beef cattle production is available in *Beef Cattle Science* available from Interstate Publishers, Inc.

QUESTIONS

Answer the following questions using correct spelling and complete sentences.

1. Distinguish between the *Bos taurus* and *Bos indicus* cattle.
2. List advantages and disadvantages of raising beef cattle.
3. What should be considered in selecting a breed of cattle?
4. What beef production systems are used?
5. Name and describe two important herd management practices.
6. What facilities and equipment may be needed in beef cattle production?
7. How are beef cattle identified?
8. Select any three breeds of cattle and provide a brief description of each breed.
9. What steps should be followed when preparing a beef animal for a show?

EVALUATING

Match the term with the correct definition. Write the letter by the term in the blank provided.

a. dual purpose
b. cutability
c. polled
d. anthelmintics
e. forager
f. halter broken
g. steer
h. Bos taurus
i. Bos indicus
j. backgrounding

___ 1. Chemical compounds used for deworming animals.

___ 2. Beef cattle that naturally do not have horns.

___ 3. Male calf castrated at a young age.

___ 4. Cattle used for both meat and milk production.

___ 5. Cattle that are from India and have long floppy ears and loose hide.

___ 6. The amount of saleable retail cuts obtained from a beef carcass.

___ 7. A system that takes the calf from weaning to the feedlot phase.

___ 8. Cattle common to the United States of European origin.

___ 9. An animal that is comfortable with a halter on its head.

___ 10. Makes good use of grasses and other plants for food.

EXPLORING

1. Prepare a computer-based presentation on a breed of beef cattle. Include pictures and descriptions of the breed. Each student in class should choose a different breed and provide a report to the class.

2. Construct a bulletin board that depicts the components of the beef industry. All components should be identified and pictures should be included.

3. Tour various beef operations in your area and prepare a report on your observations.

4. Make a field trip to a livestock show. Identify the breeds that you see. Describe the conformation of the animals. Note how they are fed, bedded, watered, and groomed.

5. Invite a beef cattle producer to serve as a resource person and cover important management practices with beef production.

8

SWINE PRODUCTION

What are they: pigs, hogs, or swine? All three refer to the same animal. The word, swine, is more appropriate to use. This is because of the casual ways we use "pig" and "hog." And, "dirty as a hog"—forget it. Hogs are clean!

Swine are popular because of the delicious meat, which is known as pork. They reproduce at a high rate, grow fast, require low amounts of labor, and give a fast return on investment. Pork is often referred to in advertisements as white meat. As pork quality increases, consumption increases as well.

A big change is underway in how swine are raised. Some small farms have been replaced with large corporate farms. These are sometimes known as factory farms. With the new approach, a large volume can be raised efficiently. This keeps the price of pork down and assures a quality food product.

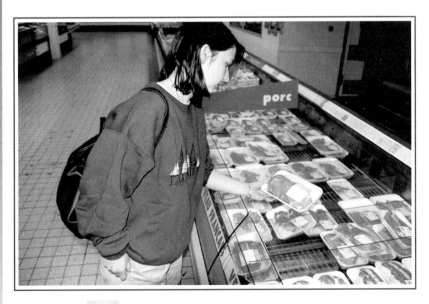

8-1. Pork is a popular meat product worldwide.

OBJECTIVES

This chapter has basic information on swine production. It has the following objectives:

1 Explain the importance of swine production

2 Describe swine as organisms

3 List and describe common breeds of swine

4 Explain the possibilities of pork production

5 Describe pork production systems

6 Explain important management practices in swine production

7 List nutritional requirements of swine

8 Explain additives and withdrawal time

9 Explain health management practices for swine

10 Describe facility and equipment needs for swine

TERMS

additive	piglet
barrow	porcine somatotropin
boar	porcine stress syndrome
contract production	probe
farrowing	prolific
feeder pig	sow
gilt	specific pathogen free
meat-type hog	tail docking
needle teeth	ultrasonics
pedigree	withdrawal time

IMPORTANCE OF SWINE PRODUCTION

Swine have been produced a long time. They were first domesticated in Asia about 9,000 BC. Swine were brought to America by Christopher Columbus in 1493. North America already had wild hogs, often known as boars. They had been used only as hunted game animals.

The swine brought to America were of European and Asian breeding. There have been many changes in these animals to provide desired products. Breeds were developed. Today, the breeds have given way to crossbred meat hogs.

8-2. Swine are raised in clean facilities and given care to assure their well-being.

SWINE INDUSTRY

The swine industry closely parallels the production of corn. Three-fourths of the hogs produced in the United States are produced in the Corn Belt. When corn yields are high, corn prices are low and hog production increases.

More than 60 percent of the hogs in the United States are grown in confinement. Many of these are produced under contract with a processing company. Hogs are the second largest livestock population in the United States, with nearly 70 million head. Iowa, Illinois, and Indiana are the leading states in swine production. North Carolina has sharply increased production in recent years.

Favorable Factors

Swine production has many factors that make it a favorable enterprise. Swine are efficient in converting feed to meat. Fewer pounds of feed are needed to produce a pound of pork than for a pound of beef. Less than 5 pounds of feed are needed for a pound of pork. Nine pounds of feed are needed for a pound of beef.

Swine are very prolific. *Prolific* means that they will produce a large number of young. Sows will usually farrow (give birth) 7 to 12 piglets twice a year. They also excel in dressing percentage. They will yield 65 to 80 percent of their live weight. Comparably, cattle only dress out around 50 to 60 percent, and sheep at 45 to 55 percent.

Labor requirements are lower because hogs are good at self-feeding. The capital investment (money) is low because they require little land or buildings, depending on the type of operation. The approaches used on factory farms require greater investments in buildings and waste disposal. However, the level of production is higher. The time to get a return on investment is relatively short. A return can be made within 10 months.

Unfavorable Factors

Several factors make the swine enterprise unfavorable. Hogs are very susceptible to disease and parasites. Since hogs have a simple stomach, they must have a large amount of concentrates and a minimum of forages. Hogs also require special attention at farrowing time. Because of the relationship between corn and hog prices, economic conditions can become unfavorable

8-3. A modern swine facility in Pennsylvania.

at times. The disposal of wastes is a major issue in hog production. Communities may refuse permits for hog farms and protest plans to build them.

CORPORATE SWINE PRODUCTION

Swine have typically been produced on small, family-operated farms. The farm might also grow grain crops, primarily corn, as feed for the swine. In recent years, changes have been taking place. These changes have become a major issue with some traditional pork producers.

More swine are being grown in large, factory farm systems. The producer may be under contract with a company that provides the piglets, monitors production, supplies feed, and markets the hogs. This approach is often known as vertical integration.

The swine are bred for the most efficient growth and yield. During growth, careful attention is given to nutrition and disease control. People are not allowed in facilities where the hogs are growing. Producers enter only after bathing and wearing carefully laundered clothing. These precautions allow faster growth to market weight for meat—about 220 pounds.

SWINE AS ORGANISMS

Swine have organ systems similar to most animals. They are mammals, with sows giving birth to litters of pigs. Swine differ from cattle in that they have a monogastric stomach. This means that the stomach has a single compartment. They must be fed feed that is more concentrated. Swine make poor use of forage.

Swine are classified as mammals with hoofs. The scientific name for the species produced in North America is *Sus scrofa domestica*. Other species include pygmy hogs (*Sus salvanis*), bearded pigs (*Sus barbatus*), and the warthog (*Phacochoerus aethiopicus*).

Swine used for food products are typically young hogs only a few months of age. They have grown rapidly and been protected from disease. Older hogs, especially boars, may develop a strong flavor in the meat.

CLASSIFICATION

Swine are classified by age and sex. A young swine is known as a *piglet* or baby pig. At a young age, male pigs are castrated, unless they are kept for

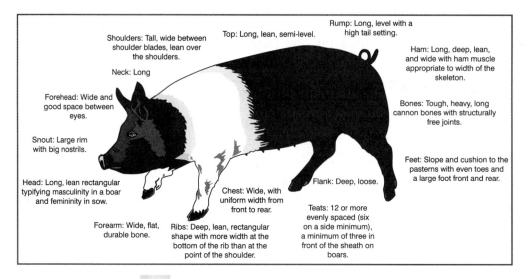

8-4. Major characteristics of a meat-type hog.

8-5. Location of major meat cuts. (The % symbol is the percentage that part of a carcass is of the entire carcass.)

breeding. The male castrated at a young age is a **barrow**. A young female that has not farrowed is a **gilt**. An older female is a **sow**. A male hog is a **boar**. Sometimes, male hogs are castrated after they reach sexual maturity. If so, they are known as stags.

The hogs used for high-quality meat are barrows and gilts. Sows and boars may be made into lower-quality meat, such as sausage or cooked food products. Remember, lower-quality meats are wholesome, but may not have the same flavor as the higher quality.

SWINE TYPE

Swine are produced for meat. People want lean meat without much fat. This reflects a change from years ago when fat was used to make lard. Consumer demand has resulted in a ***meat-type hog***. Lard-type and bacon-type hogs have declined in use. The meat-type gives the greatest amount of lean meat in high-value cuts, such as ham.

Producers need to know the desired features of a hog. These are useful in describing the desired conformation. Conformation is the type and shape of an animal.

Meat-type hogs need to grow fast and efficiently. They need to have plenty of muscle tissue that will be used for lean food products. In general, a meat-type hog is a long animal with deep, well-developed muscles in the hams. Breeding animals should have good bone structure to be able to carry their weight. A meat-type hog will not have thick layers of fat.

The major meat products from hogs are ham, bacon, loin cuts, and roasts. Other parts produce cuts that are of lower value. Some low-value cuts are ground into sausage.

BREEDS OF SWINE

Several breeds of swine are found in the United States. Swine breeds are very important to the purebred producer. Hogs grown for meat are typically crosses of the purebreds. The breeds of hogs include Duroc, Hampshire, Yorkshire, Hereford, Berkshire, Poland China, Landrace, Tamworth, and Chester White.

Some producers have detailed breeding systems to get the type wanted. For example, a Duroc boar may be bred to a Yorkshire female. A Hampshire boar may then be bred to female offspring from the Duroc-Yorkshire cross. A Yorkshire boar may then be bred to the females from the second cross.

Factory-type producers often use "hybrid" hogs. These result from crossing two or more breeds and selecting animals based on specific traits. Specific breeds are not important to many producers. The conformation of the animal is important.

UROC

The Duroc is light to dark red and has ears that droop over the eyes. It is a good breed because of prolificacy and hardiness. Durocs are known today as a good meat-type hog. Durocs should not have any white on their bodies. This is a popular breed of hog.

8-6. A Duroc boar. (Courtesy, *Seedstock, West Lafayette, Indiana*)

HAMPSHIRE

The Hampshire has the most striking appearance as a black hog with a white band or belt encircling the shoulders and front legs. The ears standout and do not cover the eyes. Hampshires cannot be registered if the band does not encircle the body and if there is too much white on the head and hind legs. Hampshires have been popular as show animals.

8-7. A Hampshire barrow. (Courtesy, *Seedstock, West Lafayette, Indiana*)

YORKSHIRE

The Yorkshire is a large, white hog. The females are known for their mothering ability. Yorkshires have long bodies that may tend toward a bacon-type of hog. They are sometimes faulted for small hams.

8-8. A Yorkshire sow and litter. (Courtesy, Land O Lakes, FT, Dodge, Iowa)

HEREFORD

The Hereford is a newer breed. The breed was founded by R. L. Webber of Missouri by inbreeding Chester Whites, Durocs, and hogs of unknown ori-

8-9. A Hereford gilt. (Courtesy, National Hereford Hog Record Association)

Connection

QUALITY ASSURANCE PROGRAM

The National Pork Producers Council (NPPC) has a producer education program. The program promotes good management to produce a quality pork product. It is known as the Pork Quality Assurance (PQA) program. Some pork packers have made participation in PQA mandatory for producers who sell to them.

The PQA has four goals:

■ To improve management practices in hog production.

■ To avoid violative drug residues in pork.

■ To decrease the costs of hog production.

■ To increase food safety awareness among hog producers.

Information on the PQA is available from the National Pork Producers Council, P. O. Box 10383, Des Moines, IA 50306. The World Wide Web site is www.nppc.org/

gin. The color markings are similar to those of Hereford cattle. They should be red with a white face and white underneath the body. At least ⅔ of the body must be red. The ears can be red or white. At least two of the four feet must be white. Herefords are popular in Indiana, Illinois, Wisconsin, and Ohio; though, they are found throughout the United States.

BERKSHIRE

The Berkshire is easily identified by its short, upturned nose. The hog is black with six white points: the feet, some white on the face, and white tail switch. Splashes of white may be in other places on the body. Berkshires are lean and typically have less fat than other hogs.

POLAND CHINA

The Poland China is black with white spots on the feet, the tip of the tail, and the nose. In the past, Poland Chinas were known as large, fat hogs. Pro-

8-10. A Berkshire boar. (Courtesy, American Berkshire Association)

8-11. A Poland China barrow. (Courtesy, *Poland China Record*, Peoria, Illinois)

ducers have selected breeding stock based on more lean and less fat. Selection efforts have resulted in the breed being more the meat-type.

PORK PRODUCTION POSSIBILITIES

Carefully select hogs to establish a herd. Selecting the wrong hogs can result in failure as a hog producer. Three important considerations are type and pedigree, health, and production testing.

TYPE AND PEDIGREE

Selection based on *type* means to select an animal that is as close to the ideal as possible. A *pedigree* is a record of an individual's heredity. In selecting for type, test for meatiness, porcine stress syndrome, and genetic defects.

Meatiness can be measured with a probe, lean meter, or ultrasonics. All show the amount of backfat present on an animal. The *probe* is a tool for measuring the thickness of backfat. The lean meter is based on the difference in conductivity of electricity in the fat and muscle. Muscle and blood are better conductors than fat. *Ultrasonics* use bursts of high frequency sound. It measures the reflection of the pulses and the time of reflection.

Porcine stress syndrome (PSS) is a nonpathological disorder in heavily muscled animals that results in sudden death loss. A couple of tests can be used to detect PSS, and they are 100 percent accurate.

Litter mates and parents should be noted for genetic defects to avoid. Commonly inherited genetic defects include scrotal hernia, cryptorchidism, hermaphrodites, and umbilical hernia.

8-12. Swine shows help producers to refine breeds for the desired type. (Courtesy, Hampshire Swine Registry)

HEALTH OF THE ANIMAL

Selected stock should be disease free. Specific swine diseases will be discussed later in this chapter.

Many producers purchase hogs that come from a *specific pathogen free* (SPF) herd. SPF pigs are free of disease at birth. They are raised in an aseptic environment. The diseases that SPF animals should be free of are brucellosis, leptospirosis, lice, mange, pneumonia lesions, pseudorabies, swine dysentery, turbinate citrophy, and snout distortion.

PRODUCTION TESTING

Several purebred swine registries, as well as producers, have a production registry and a meat certification program. The most important way to evaluate the productivity of potential animals is through production testing.

- Boar Selection—One boar should be selected for every 15 to 20 sows. Because they will be producing so many offspring, their evaluation is extremely important. Table 8-1 summarizes the major characteristics of a good boar.

   The boar should also be docile and have an easy temperament, yet be aggressive enough to breed the herd. The boar should be purchased at six to seven months of age. Use of boars in breeding may begin at eight months of age. Boars should be of good health and have a masculine appearance. The testicles should be well developed. With purebred boars, the standards of the breed should be met.

- Female Selection—Gilts and sows should be carefully selected. Only gilts from large litters should be selected. Their backfat should be less than 1.2 inches. The gilt or sow should be sound and free of any flaws or defects that would interfere with normal production. They should be the fastest growing and the leanest of the bunch. Gilts in purebred herds should meet the standards of the breed.

   Gilts are often selected from the animals that are being fed for meat production. When gilts reach 180 to 200 pounds (81.6 to 90.7 kg), they should be

Table 8-1. Characteristics of a Good Boar	
Trait	**Standard**
Litter Size	from litter with 10 or more farrowed, 8 or more weaned
Underline	12 or more well-placed, undeveloped nipples
Feet and Legs	medium to large boned; wide stance both in ront and in the rear; free in movement; good cushion to both front and rear feet; equal sized toes
Age at 230 lbs.	155 days or less
Feed/cwt gain	275 lb/cwt
Daily Gain	2.00 lb/day or higher
Backfat Probe	1.0 inches or less

8-13. A champion Hereford boar that has the characteristics of a good boar. (Courtesy, National Hereford Hog Record Association)

8-14. A Hampshire gilt that has the characteristics of a good gilt. (Courtesy, Hampshire Swine Registry)

evaluated for their reproductive potential. If selected as a replacement, they should be removed from the meat animals and fed separately.

PRODUCTION SYSTEMS AND CONTRACTS

A production system is the way the swine are raised. These are largely based on the age or weight at which pigs are obtained and marketed.

SYSTEMS

Four primary production systems are used:

■ Feeder Pig Production—A *feeder pig* is a pig weighing about 40 pounds (18.2 kg) and has just been weaned. This production system involves maintaining a herd of brood sows and having the necessary facilities for successfully breeding and farrowing litters of pigs. Boars may be maintained for natural breeding or for semen collection. Artificial insemination of the sows may be used. The feeder pigs are sold to farms that will feed them to market weight.

8-15. Feeder pigs of mixed breeding.

■ Finishing—Finishing is feeding feeder pigs until market size is reached. Most hogs are marketed for slaughter at about 220 pounds (100 kg). Some are lighter; others are heavier. Many packers may have specific weights. The price paid for pigs that vary from the weight more than a few pounds may be less. The goal of producers is to get the best possible weight gain on the minimum investment in feed.

8-16. Pigs being finished to slaughter weight.

■ Farrow-to-Finish—Pigs are farrowed and fed to slaughter weight on the same farm. The producer must have farrowing facilities as well as facilities for feeding out the pigs. This type of system is less specialized than the feeder pig or finishing systems.

■ Purebred—Purebred hog production is raising hogs that are subject to registration by the breed associations. All standards of the breed association must be met. If not, the pig cannot be registered. The goal is to produce breeding stock. The breeding stock is sold to other producers who are producing pigs.

CONTRACTS

An agreement between two or more people is a contract. It is usually in writing and signed by all parties. **Contract production** involves a contract between a producer and a buyer before hogs are raised. The contract specifies the kind of hog to be raised and other details. Contracts may be used for feeder pigs or market hogs. Feeder pig contracts are with growers who will provide feeder pigs to finishing operations. Market hog contracts are with finishing operations that may buy feeder pigs or farrow their own feeder pigs.

Contract production is increasingly popular. This is because of the high capital investment, the difficulty in financing an operation, and the willingness to forego large profits in return for security. The producer is assured of a market.

SWINE MANAGEMENT

Hogs are unique livestock. Except for a few breeds of sheep, they are the only livestock that naturally produce offspring twice a year. Because hogs reproduce at a younger age and more quickly than other livestock, genetic gains can be made quite rapidly. But, this requires good management. Good management is important because there is little difference between producing a litter of five pigs and a litter of ten pigs.

NORMAL BREEDING

Swine reach puberty from four to eight months of age. This wide variance is due to sex, environment, breed, and breeding lines. Gilts can generally be bred to farrow at 11 to 12 months of age. They should weigh at least 225 pounds when bred.

The heat period or estrus lasts from one to five days and averages two to three days. Signs of heat include restless activity, mounting of others, swelling and discharge from the vulva, increased urination, and loud grunting. The gestation period is 114 days.

CROSSBREEDING

Most meat hog producers use crossbreeding, which is the mating of different breeds. Crossbreeding allows for heterosis or hybrid vigor. Heterosis is a biological phenomenon, which causes crossbreds to outperform either of their parents.

Crossbreeding is widely used because it allows for increased production and profits. Roughly 90 percent of commercial hogs are crossbred.

THE BREEDING PROCEDURE

Sows can be bred naturally by a boar or artificially. Natural breeding results when the boar and female are placed together for breeding. This is sometimes known as hand mating, which is the most common method. A breeding crate is used when a heavy boar is to be mated with a young gilt. This reduces the risk of injury to both.

Artificial insemination is increasing in use with hogs. It has advantages and disadvantages. AI decreases the risk of disease and increases genetic improvement. However, the main problem with AI is that swine sperm cannot be frozen, thus decreasing the opportunities for use.

Thirty to 45 days after breeding, the sow or gilt should be pregnancy checked. The use of ultrasonics has increased detection up to 95 percent accuracy.

8-17. Preparing swine semen for microscopic examination prior to breeding.

CARE OF THE PREGNANT SOW

Many producers are using confinement options to care for pregnant sows. The sows are put in individual pens. The advantages of confinement are a decrease in labor requirements, the use of automatic feeding, freeing land for other purposes, a controlled environment, improved control over disease and parasites, and better management.

Some disadvantages to confining pregnant sows are the increased need for facilities, higher initial investment, possible delay in sexual maturity, lower conception rate, and the requirements of better management.

Sows should receive feed with more bulk shortly before parturition. A drug to induce labor may be given 111 to 113 days after breeding. The sow will usually farrow within 18 to 36 hours.

CARE OF THE BOAR

Boar care is important. A boar will be expected to service dozens of gilts and sows and produce healthy offspring. The boar should have plenty of room to exercise, which is usually in a pasture. The pasture should be well fenced. If kept in confinement, boars should have clean pens with slotted floors or concrete. Boars should be kept in individual pens.

Boars that are not being used should be kept separate and away from the rest of the herd. Also, never allow them to grow tusks as this drastically increases their dangerousness.

8-18. Boars are penned separately to prevent unwanted breeding.

CARE OF SOW AT FARROWING

Giving birth to pigs is *farrowing*. Care of the sow at farrowing is important to have a strong, healthy crop of pigs. Only 70 percent of the piglets born reach weaning, thus increasing the importance of good, sound management practices at farrowing time. This means that 30 percent die due to disease, mashing by the sow, bad weather, or other conditions. Increasing the survival rate makes a farm more profitable.

- Signs of Parturition—Signs of parturition include nervousness, uneasiness, enlarged vulva, and mucous discharge. Milk will be present in the teats and the sow will begin to make a nest for the young. She should be placed in a crate or pen by at least the 110th day of gestation. Farrowing crates are commonly used because they reduce the number of young that are crushed.

- Sanitation—Sanitation is essential. Before the sow is moved to the crate or pen, she should be scrubbed to remove dirt, manure, and any parasites. Bedding needs to be clean and fresh. Good sources of bedding are wheat, barley, rye, or oat straw; chopped hay; corncobs; and shavings. The farrowing unit should be scrubbed and cleaned with a disinfectant between uses and left empty for five to seven days.

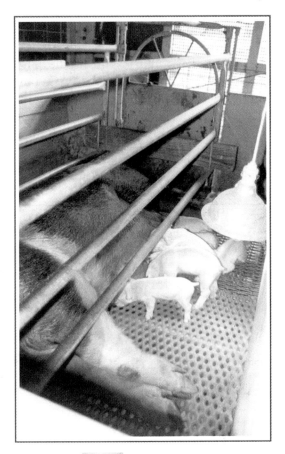

8-19. Sow with litter.

- Environmental Factors—Sows and piglets are sensitive to their environment. Good ventilation and protection from extreme weather are needed. Their quarters should be 60 to 70°F (15.5 to 21.1°C). Heat lamps or mats are placed in the farrowing crate or pen to keep piglets warm. An attendant on the sows frequently during and after the birth of the young. This will decrease loss and increase the health and survival of the pigs.

THE SOW AND LITTER

Getting the pig litter off to a good start is important for economic gains. Sanitation is necessary because swine suffer heavy losses due to disease and parasites. A diseased herd can be wiped out leaving the producer in debt and the facilities highly contaminated.

- Starting Piglets on Feed—Newly born pigs need to be observed to be sure they nurse within a few hours. If not, a pig may need to be placed near a nipple. Pigs will often begin learning to eat feed within 7 to 10 days. Use a manufactured baby pig feed. Feeds for pigs may have protein content of 20 percent. The feed should be in meal form. Using feed containing dried whey has shown positive signs in research trials.

- Clipping Needle Teeth—When pigs are born, they have eight *needle teeth*. Two are located on each side of the upper and lower jaws. They are sharp and long, much like a short needle. These teeth serve no purpose and will cause injury to the mother as they are nursing and to each other if allowed to grow. These teeth can be clipped off with pliers or forceps. Care should be taken to not injure the pig. The time of clipping needle teeth is good for treating the umbilical cord with an approved disinfectant solution.

- Tail Docking—*Tail docking* is clipping the tail from baby pigs. It is done at the time that needle teeth are snipped. The tail is cut about 1 inch or slightly

8-20. Clipping needle teeth prevents injury to the teats of the sow by the nursing pigs. (Courtesy, Mississippi State University)

less from the bone of the tail. Tail docking prevents tail biting that may happen in hogs as they grow in confinement.

■ Ear Notching—The most common method of identifying hogs is to use a special V-notcher to notch their ears. This is especially important in herds with thousands of hogs. This enables producers to exactly identify the animals, which is necessary when selecting breeding stock and replacements. Plastic ear tags are sometimes used. A disadvantage is that they will tear out easily. Branding and tattooing are also used to some extent, but are hard to read.

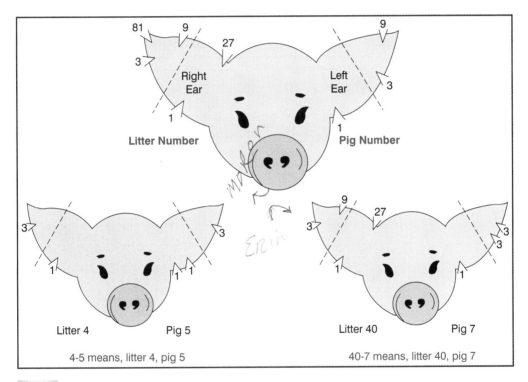

8-21. Ear notching is done on baby piglets as a permanent identification system. The right ear is the litter number. The left ear is the pig number within the litter. The position of each notch has numeric value.

■ Castration—All male pigs being raised for meat should be castrated (removing testicles) before they are weaned. A sanitary and confined environment promotes healing without infection. Males kept for breeding should not be castrated.

■ Vaccination—Vaccinations should be done before weaning. Pigs are vaccinated to prevent diseases that may occur. The common disease problems of hogs are discussed later in the chapter.

8-22. A barrow showing proper castration as a piglet. No scars are evident.

8-23. A piglet being vaccinated. (Courtesy, Mississippi State University)

■ Feeding the Sow—The sow should be fed liberally before parturition to stimulate milk production. During the first three days after giving birth, she should be fed minimal feed. Slowly increase the amount to full feed at two weeks. Use a manufactured feed designed for lactating sows.

GROWING FEEDER PIGS

Feeder pigs can be finished to market size in a few months with proper feeding and management. Feeder pigs are typically produced in confinement. They are given complete rations for their stage of growth. From 40 to 125 pounds, the feed is 16 to 18 percent protein. For pigs over 125 pounds, the feed is 14 to 16 percent protein. Main ingredients are typically yellow corn and soybean meal. Calcium, phosphorus, salt, and vitamins are added. A manufactured feed is typically used.

GENERAL NUTRITIONAL REQUIREMENTS OF SWINE

Proper nutrition is essential for survival and growth. Hogs raised in confinement must receive a complete feed. Costs need to be kept down to assure a profit. About 75 percent of the total cost of hog production is feed.

Each individual varies in the kinds and amounts of nutrients needed. These requirements are influenced by age, function, disease level, environment, and other factors.

- Carbohydrates and Fats—Carbohydrates and fats provide energy for the various functions of the body. Cereal grains are the biggest source of energy, mostly carbohydrates. Roughages are a good source of energy, but are not suitable for nonruminants, such as swine. Roughages are too bulky for the restricted size of the digestive system. Corn is the most widely fed concentrate supplying energy. Other sources of energy fed to hogs include barley, sorghum, wheat, and fats and oils.

- Protein—Protein is needed to build and repair tissue. Protein is broken down in the stomach and small intestine into amino acids. These amino acids are then recombined into proteins that the animal needs for muscle development and repair of other tissue. Hogs, as nonruminants, are unable to synthesize their own protein, thus increasing the importance of having a well-balanced diet. Soybean meal is the most widely fed protein. Because of its high palatability to hogs, it should be mixed with grain to prevent overeating. Young, growing hogs are fed feed higher in protein than older hogs. Protein supplements of 35 to 40 percent protein may be used with corn and other feedstuffs. Commercial feeds with 20 percent protein are used to start pigs at three weeks of age. Growing and finishing feed may have 12 to 16 percent protein. Pregnant sows are fed 12 percent protein feed. Lactating sows and boars are fed 13 to 16 percent protein feed.

- Minerals—Hogs are the most common farm animals to suffer from mineral deficiency. Those minerals needed in the largest quantities include salt, calcium, and phosphorus. Cobalt, copper, iodine, iron, manganese, selenium, and zinc are needed in smaller amounts, but are still essential.

  Mineral supplements are usually mixed with the feed ration. Newborn pigs are given iron shots shortly after birth.

■ Vitamins—Vitamins are needed in small amounts, yet are essential for performance of normal body functions. Hogs also suffer from more instances of vitamin deficiency than any other livestock. This is due to confinement and the restricted variety of feeds that they can consume. Fat-soluble vitamins A, D, and E and water-soluble vitamins biotin, niacin, pantothenic acid, riboflavin, and vitamin B-12 are the most likely to be deficient.

■ Water—Hogs will generally need $1/4$ to $1/3$ of a gallon of water for every pound of dry feed consumed. Ideally, they should have access to automatic waterers or be hand-watered at least twice a day.

8-24. Pigs will quickly learn to drink from automatic waterers.

ADDITIVES AND WITHDRAWAL

An **additive** is a substance added to feed to meet a particular need. Some additives are standard in hog rations. They do not provide nutrients, but provide other advantages. Anthelmintics may be added as wormers. Antibiotics are used to stimulate growth, improve feed efficiency, and control infestations. Piglets do not begin producing antibodies until they are five to six weeks old. Under ideal conditions, antibiotics can increase gain by 10 percent, while decreasing feed consumption by 5 percent.

Porcine Somatotropin (PST) is a growth hormone for swine. It increases protein synthesis and growth in most tissues. Research shows that PST can increase feed efficiency by 20 to 30 percent, average daily gain by 15 to 20 percent, and improve muscle mass by 10 to 15 percent.

Feeds with most additives should not be fed for a period before slaughter. **Withdrawal time** is the period before slaughter when an animal should not receive a particular medication or other product. The days of withdrawal allow the hog's system to purge itself of the medications. This assures pork that is free of medication residues above the amount that may be allowed.

Withdrawal time is for feed additives and medications given orally, by injection, or in other ways. The label on medicines and growth products gives withdrawal information. For example, feed containing sulfathiazole should not be used within seven days of slaughter. Other additives have similar withdrawal times, ranging from 2 to 70 days.

HEALTH MANAGEMENT PRACTICES

Healthy animals vary widely. Some animals may appear healthy, but may actually be suffering from pain or disease. Knowing the normal behavior of animals helps detect and correct any ill health. A producer or animal owner should learn to diagnose good and ill health.

PARASITES

Parasites live in and on hogs. The parasites feed on the blood, flesh, and other tissues of a hog. This reduces the rate of growth and overall health.

- Internal Parasites—Internal parasites live inside a hog. They may be taken in through the mouth when eating contaminated feed or when eating feed on a dirty floor. Most internal parasites are obtained through feces in the food. Pastures with very short grass also increase the likelihood of ingesting parasites.

8-25. Swine in drylots are at high risk of parasite infestation.

Hogs on pasture are more likely to have internal parasites. Minimize the amount of feed eaten on the ground. Roundworms, tapeworms, and other worms are parasites that can infect all animals. They are found in the intestines and the stomach. Tapeworms can grow several feet long. When they become this large, they use much of the feed that the animal has eaten. Animals with low infestations of small-sized worms may not show symptoms. As the number and size of worms in the digestive tract increases, the animal loses weight, becomes anemic, and may have diarrhea. Sanitation is a good control measure.

■ External Parasites—External parasites live on a hog's body. They get their food from the blood and tissue of the animal. External parasites can transfer contagious diseases or other parasites to other animals. Lice are small insects that suck their host's blood. Animals may become anemic. Lice cause animals to itch. This is why hogs with lice are often seen scratching or rubbing against trees or posts. Lice usually appear in the winter. Back rubbers and other means can be used to administer pesticides to control lice. Mange is a dermatitis (skin disease) caused by one of several types of mites. Mange results in intense itching around the head and neck and wrinkling of the skin. It spreads rapidly. If untreated, it can be debilitating and even fatal. Dipping or high-pressure sprays may be used for treatment.

NUTRITIONAL DISEASES

Most nutritional diseases are caused by an unbalanced diet. Animals require specific amounts of nutrients. Too high or too low amounts of any of these may result in a nutritional disease decreasing profitability. Mineral and vitamin deficiencies are quite common in swine.

COMMON DISEASES OF SWINE

Several infectious diseases of swine are included here. Some can be prevented. Others require the assistance of a veterinarian in administering treatments.

■ Bang's or Brucellosis—Bang's or Brucellosis affects the reproductive tract of the female causing young to be aborted. Prevention of Bang's is by vaccination. Sanitation, testing, and bringing only Bang's-free animals into a herd help to control this disease.

■ Hog Cholera—Hog cholera is a viral disease that is highly contagious. Hogs will have a high fever, lose their appetite, become weak, and drink much water. Considerable effort has been spent eradicating hog cholera in the United

States through federal government programs. There is no known treatment. Any hogs with hog cholera are destroyed.

- Leptospirosis—Leptospirosis is a bacterial disease. Symptoms include high fever, poor appetite, bloody urine, and females will abort their fetuses. Antibiotics are sometimes used to treat leptospirosis. Animals are often vaccinated against this disease.

- Pneumonia Lesions—Pneumonia lesions is a specific type of pneumonia that affects swine and can have a major economic impact. Pneumonia itself is not that significant, but the secondary infections that follow can be very costly. Symptoms are a chronic cough and lung lesions, thus the name pneumonia lesions. Lung lesions can result in increasing the incidence of lungworms and migrating ascarid larvae. Control is usually unsuccessful. Use of sulfas and antibiotics helps to decrease secondary infections. Depopulation and the use of SPF animals will lower the incidence of pneumonia lesions.

- Pseudorabies—Pseudorabies is a viral disease attacking swine. Adults may serve as unaffected carriers. Suckling pigs usually show clinical signs. Piglets will develop a fever, paralysis, coma, and death, in as little as 24 hours. It may cause adults to abort or to have stillborn pigs. It is spread through contact with nasal and oral secretions. A vaccine is available to prevent pseudorabies.

- Swine Dysentery—Swine dysentery or bloody scours is a type of diarrhea caused by *Treponema hyodysenteriae*. It is usually found in young pigs. It is spread by ingesting infected feces. Outbreaks are most common in late summer or early fall. Adding new pigs to the herd, especially ones of unknown origin, can bring in swine dysentery. Once this disease occurs, it will show up periodically, usually once a month. Symptoms include loss of appetite, soft feces, and a slight temperature. The diarrhea will become worse and will get bloody. Because carriers cannot be identified, control is difficult. Antibiotics can be somewhat successful.

FACILITY AND EQUIPMENT NEEDS

Hog production requires good facilities and equipment. The facilities must be properly located, constructed, and maintained.

GENERAL FACILITY TYPES

Production facilities determine how hogs are raised. In general, three types of approaches are used:

8-26. A drylot with shelter provided in a metal hoghouse.

■ Pastures and drylots—Hogs are placed in fenced pastures or dry lots. Some protection from the weather should be provided. Overall, the trend is away from pastures and drylots except for brood stock on some farms.

■ Buildings with concrete slab floors—The buildings may have roofs that cover half to all of the concrete area. The concrete has a gradual slope to allow easy cleaning. Bedding is typically not used. Automatic waterers and feeders are used.

■ Enclosed buildings with slotted floors—These facilities provide a completely controlled environment. Manure collects below the slotted floor and is washed out each day into a tank or lagoon. Automatic feeders and waterers are used. This type of facility works well in cold climates. Producers exercise control over the total production environment.

8-27. A concrete slab facility.

8-28. An enclosed facility.

LOCATION

A good site must be selected for hog production facilities. Location is more important as fewer and fewer people live on farms and understand what is involved. Neighbors may dislike the odors and noise coming from the facilities. Locating buildings downwind of housing areas would be beneficial. Protection from wind and snow and access to good roads are also things to consider in site selection.

BUILDINGS

Swine facilities usually include buildings for farrowing, growing, and finishing. The farrowing barn may contain a nursery. Pasture may be used with gilts and sows depending upon the climate, availability, and time of year. Attention should be given to three areas:

- Ventilation—Well-ventilated facilities help to remove extra moisture in the wintertime. During the summer, it helps to control temperature. Hogs are generally healthier when fresh air is present. Diseases are reduced.

- Manure Disposal—Hogs create much waste. Disposing of manure is a major problem for producers with large numbers of hogs. Waste management is needed to maintain good health, avoid air and water pollution, and comply with government regulations. Manure pits are usually located under confinement buildings for temporary storage.

- Cleaning—Facilities should be easily cleaned. Construction materials should resist damage from water and be durable.

8-29. A modern swine building with good ventilation.

FENCING

Some hogs are raised in fenced pastures and other areas. Sows and boars are sometimes put in pastures. Strong, woven wire fencing should be used to make the fence at least 3 feet high. Temporary fencing may be made of electric wire. Gates need to be made of wood or metal, be of sturdy construction, and have good hinges.

HANDLING EQUIPMENT

Swinging gates make holding and handling swine easier. A series of gates throughout the building allow for easier sorting. Loading chutes are best located at the end of the building.

FEEDING AND WATERING EQUIPMENT

Hogs are hard on feeding and watering equipment. The equipment must be sturdy and not easily damaged. The equipment should also be

8-30. Portable scales for weighing individual hogs.

easy to use and clean. Good automated feeders and waterers makes a more efficient hog operation.

REVIEWING

MAIN IDEAS

The swine industry is the second largest livestock enterprise in the United States. More than 60 percent of hogs are raised in confinement facilities.

Hogs are good to grow because they are prolific, labor requirements are low, the capital investment is low, and the return can be fast. Hogs are susceptible to numerous diseases and parasites. A farrowing enterprise requires more labor because of the attention to sows and piglets.

Four production systems are used—feeder pig, finishing, farrow-to-finish, and purebred. The goal of the first three is to produce high-quality pork. The goal of the latter is to produce quality breeding animals.

Good breeding management of swine is needed to attain high numbers of pigs twice a year. Keeping facilities clean is essential.

Because hogs have a simple stomach, they require concentrates and eat very little forages. Cereal grains, soybean meal, and mineral and vitamin supplements are used in a total mixed ration. Additives, such as antibiotics, sulfas, and porcine somatotropin, may be used to increase efficiency. Always follow withdrawal time guidelines when using additives and administering medicines in other ways.

Hogs require good facilities. Selecting the location for a hog facility is an important decision. Locate the facilities downwind of large populations. Building requirements vary based on the type of operation and the management system adopted.

QUESTIONS

Answer the following questions using correct spelling and complete sentences.

1. Why is the swine industry a good or favorable enterprise?
2. What is a meat-type hog?
3. Name five breeds of swine and distinguish between them.
4. List and explain the three different systems of the swine industry.
5. What are the characteristics of a good boar? Good sow?
6. How should the sow be managed at the time of parturition?
7. Describe management practices with newborn pigs.

8. Describe the nutritional requirements of swine.

9. Name three diseases of swine and describe how each affects hogs.

10. Describe the facility and equipment needs of swine.

EVALUATING

Match the term with the correct definition. Write the letter by the term in the blank provided.

a. prolific   e. specific pathogen free   i. needle teeth
b. tail docking   f. gilt   j. withdrawal time
c. pedigree   g. barrow
d. probe   h. feeder pig

_____ 1. A record of heredity.

_____ 2. Free of disease and raised in an aseptic environment.

_____ 3. Tool for measuring thickness of backfat.

_____ 4. Pig weighing 40 pounds.

_____ 5. Clipping or cutting the tails on newborn pigs.

_____ 6. Young female swine that have not had pigs.

_____ 7. Produce a large number of offspring.

_____ 8. Male swine castrated at a young age.

_____ 9. The period before marketing in which medications and other additives should not be given to hogs.

_____ 10. Sharp teeth in baby pigs that injure the teat of the sow.

EXPLORING

1. Visit a swine farm. Note the management of reproduction, nutrition, and health. Interview the manager or a worker about the farm. Determine the breed or cross being raised. Take a camera and make photographs to prepare a poster about the farm.

2. Write a report on the use of porcine somatotropin. Use reference material to help in preparing the report. Contact the animal science department at the land-grant university in your state for information. Your agriculture teacher can provide the name and address of the individual to contact.

3. Make a field trip to a livestock show. Study the breeds of hogs and the conformation of the body parts. Prepare a report on your observations.

9

SHEEP AND GOAT PRODUCTION

Sheep and goats look a lot alike. They are closely related and have been domesticated for a long time. Both are ruminants but eat different plants. People who know a little about them can tell them apart easily.

Here are a few distinctions: goats have a beard, sheep have foot glands, and male goats have a strong smell. Slight differences exist in the horns and skeletons. Goats are more intelligent, independent, and have a better ability to fight and protect themselves.

Goats were one of the first animals to be domesticated. They were once used to plant seed by trampling them into the ground. They were used as a source for food and fiber, and their skin was used for bottles. Sheep and lamb have also long been used for food and fiber.

9-1. Goat (left) and sheep (right).

235

OBJECTIVES

This chapter provides basic information on sheep and goat production. The following objectives are covered:

1   Explain the importance of the sheep and goat industry

2   Describe sheep and goats as organisms

3   List and describe the major breeds of sheep

4   List and explain the groupings of goats

5   Describe breeding management of sheep

6   Explain nutritional needs and feeding practices with sheep

7   Explain important health management practices used in sheep and goat production

8   Describe facility and equipment needs with sheep and goats

TERMS

billy	lamb
browse	lamb feeding
buck	lambing
chammy	mutton
confinement method	nanny
docking	orphaned lamb
drenching	purebred flock
ewe	ram
farm flock method	range band method
kid	wether
kidding	wool

THE SHEEP AND GOAT INDUSTRY

Sheep and goats are raised for food and clothing. They provide many important products. Goats were probably domesticated before sheep, with goats eing domesticated about 9,000 years ago and sheep more than 8,000 years ago.

Sheep originated in Asia and Europe. Goats came from the eastern Mediterranean area and Asia. The first were brought to America by the settlers some 400 years ago. The numbers of both grew until the mid 1900s in North America.

9-2. Sheep are often pastured on rangeland that is unsuited for other uses. (Courtesy, American Sheep Industry Association and *National Lamb and Wool Grower Magazine*)

WHERE SHEEP AND GOATS ARE RAISED

Australia and New Zealand are the leading producers of sheep. Several European countries produce large numbers of sheep. The per capita consumption of sheep products is greater in New Zealand at 60 pounds a year. The world per capita consumption is about 3 pounds a year.

China and India are the leading producers of goats. These two countries have more than 460 million head. North America has only about 2.5 million head of goats.

The number of sheep in North America decreased between 1940 and 1990. This decline was due to lower returns, higher risks than cattle or crops, increased loss due to predators, scarcity and high wages of good sheep

9-3. Lamb carcasses.

herders, and uncertainty in price. Increased interest in goats has occurred in the 1990s.

On a global basis, the United States ranks 27th in sheep numbers. Over half of the sheep are found in ten western range states and Texas. Farm flocks range from a few head to thousands.

PRODUCTS FROM SHEEP AND GOATS

Sheep and goats provide many products that are important to people. The major products are food and clothing.

Food

The food products from sheep and goats are meat and milk. The meat from sheep is lamb and mutton. **Lamb** is the meat from a young sheep that is less than one year old. Lamb includes both male and female sheep. Lamb is a delicate meat prepared in many delicious ways. **Mutton** is the meat from a sheep that is more than one year old. Mut-

9-4. Lamb chops in a supermarket display case.

ton has a stronger flavor than lamb and is less desirable as a meat in the United States. Goat is referred to as goat except for younger goat meat, which is known as kid. Many people enjoy barbecued goat meat.

Milk is primarily from goats. Goat's milk is easier for people to digest than cow's milk. Goat's milk contains more vitamin A than cow's milk and has smaller fat particles (known as globules) than cow's milk. Goat and sheep's milk are used to make cheese, such as Roquefort cheese from sheep's milk.

Clothing

Clothing is made from the hair or wool and hides of goats and sheep. **Wool** is the soft coat of sheep and is a major fiber used in making clothing. Mohair and cashmere are made from the fiber covering of specific breeds of goats. Sheep yield 5 to 15 pounds of wool each shearing.

Sheep and goat skins are used for leather products. **Chammy** is a soft, pliable leather made from sheep and goat skin. It is used in cleaning and polishing–often on automobiles.

9-5. Wool is shorn from a sheep.

9-6. Bags of wool loaded for hauling to market.

ADVANTAGES AND DISADVANTAGES OF SHEEP AND GOATS

Many factors make sheep and goat production favorable. Compared to other types of livestock, they are better suited to more arable land for grazing. They are excellent scavengers. Sheep are more efficient at converting feed to meat than cattle. They are dual-purpose animals producing wool and meat. Lambs are usually marketed in eight months yielding a fast return on investment.

Goats and sheep can be pastured together because they eat different plants. Goats eat *browse* (woody plants) and broadleaf plants. Sheep graze short grass and some broadleaf plants.

Several factors make sheep and goat production unfavorable. The price of wool is low and unstable. Competition from synthetic fibers has hurt the industry. Consumption of lamb is very low. Sheep are susceptible to disease and parasites with little resistance to keep themselves healthy. Sheep and goats are also susceptible to attack from predators. Management of sheep and goats is important because they are dual-purpose animals. About 80 percent of sheep production is lamb for meat purposes, while only 20 percent is for wool production.

Sheep and goats are often raised by youth. They are relatively small animals and require less elaborate facilities than cattle and do not create the waste disposal issue as with hogs. Lamb and goat shows have increased considerably in recent years.

9-7. Spanish goats thrive on brushy rangeland. (Courtesy, Texas Department of Agriculture)

SHEEP AND GOATS AS ORGANISMS

Sheep and goats are alike in many ways. The differences give advantages in different production situations.

Sheep and goats are mammals with ruminant digestive systems. They have cloven (divided) hoofs. Both are scientifically classified in the Bovidae family. Domesticated sheep are *Ovis aries*. Goats vary in scientific name, with the Spanish goat being *Capra pyrenaica* and other domestic goats being *Capra hircus*.

Sheep are far more important than goats in the United States, but goats are increasing. Greater emphasis in this chapter is on sheep.

DIFFERENCES BETWEEN SHEEP AND GOATS

Management of sheep and goats is basically the same. However, there are a few differences between them. For example, sheep herders work behind the herd. Goat herders work ahead of the animals to lead them.

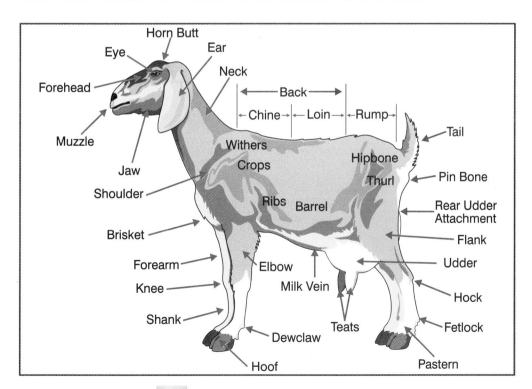

9-8. Major external parts of a dairy goat.

Connection

HOW TO MILK A GOAT

Hand milking a goat is easy with a few techniques. The goal is to have clean, usable milk. With practice, a beginner can rapidly advance to more speed in milking.

The goat must be tied or have her head in the milking stanchion. Brush any dirt off the udder, teats, and nearby area. Use a cloth or paper towel wet in a warm sanitizing solution to wash off the teats and udder. Use a massaging motion. Milk the right teat with the left hand and the left teat with the right hand. A beginner may want to milk only one teat at a time. Grasp the teat where it joins the udder and follow the procedure in the diagram. The container to catch the milk must be held or placed underneath. After milking, dip the teats in a mild disinfecting solution.

1. Thumb and finger **pinch off** milk.

2. Fingers **push** milk through streak canal and supply massage action.

3. Release teat to allow milk to return.

4. Repeat steps 1, 2, and 3.

Size

Goats and sheep are of similar size. Variations exist within the species as well as between.

Mature goats range from 20-pound (9 kg) dwarfs to 150 pounds (68 kg) or more for common domestic species. Heights range from 1.5 to over 4 feet (45 to 120 cm). Goats live eight to ten years.

Mature sheep range from 100 to 225 pounds (45 to 102 kg) or more. The wool on a large sheep may weigh 15 pounds (7 kg) or more. Sheep live 7 to 13 years.

Age and Sex Classification

The names for sheep and goats vary based on sexual classification and age. A lamb is a young lamb of either sex that is less than one year old. A *ewe* is a female sheep of any age. A *ram* is a male sheep kept for breeding purposes. A male sheep castrated before sexual maturity is a ***wether***. Its secondary sexual characteristics have not developed at the time it is castrated.

Goats have similar life cycles to sheep, but the terms applied to age and sex classifications are quite different. A female goat is a ***nanny*** or doe. A male goat is a ***buck*** or ***billy***. A young goat under a year of age is known as a ***kid***. Most all kids reach puberty by a year of age.

Reproductive Traits

The breeding season of goats is regulated by climate and weather. In hot climates, goats may mate year round. In climates away from the equator (including most of the United States), goats typically mate in the late summer and fall. Kids are born in the early spring the following year.

Does are bred to have their first kids at two years of age. The process of giving birth is known as ***kidding***. Most kids weigh about 5 pounds at birth, but weight depends on the breed of goat. Goats typically have two or three kids.

The gestation period of goats averages 151 days, whereas 148 days is common in sheep. The estrus period lasts 18 to 19 days, which is one to two days longer than sheep. Newborn kids require greater care than lambs. Kids

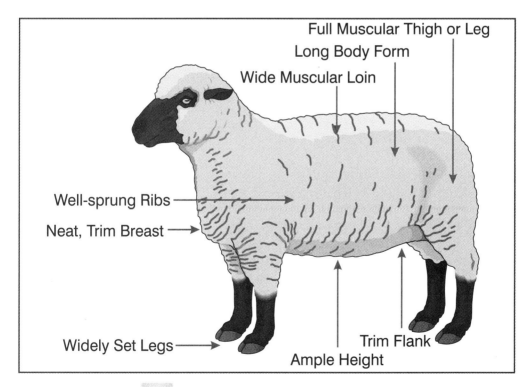

9-9. Desired conformation of a lamb for meat.

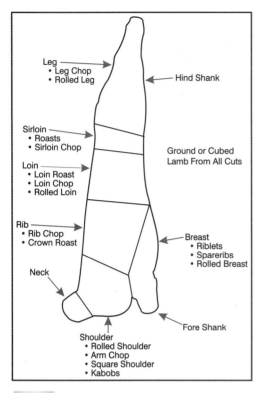

Leg
• Leg Chop
• Rolled Leg

Hind Shank

Sirloin
• Roasts
• Sirloin Chop

Ground or Cubed
Lamb From All Cuts

Loin
• Loin Roast
• Loin Chop
• Rolled Loin

Rib
• Rib Chop
• Crown Roast

Breast
• Riblets
• Spareribs
• Rolled Breast

Neck

Fore Shank

Shoulder
• Rolled Shoulder
• Arm Chop
• Square Shoulder
• Kabobs

9-10. Major wholesale and retail lamb cuts.

are either staked near a small A-shaped box or put in pens or corrals. Goats require little, if any, shelter. If they are well fed, they will not suffer in cold weather.

More information on sheep reproduction is presented later.

CONFORMATION AND TYPE

The desired conformation and type in sheep vary with the product to be produced. Sheep for meat should have large amounts of higher-valued cuts. Sheep for wool should have larger amounts of wool. Producers need to know the major external parts of a sheep and the locations of major wholesale and retail meat cuts.

BREEDS AND CLASSES OF SHEEP

Over 200 breeds of domestic sheep are in existence today. Three-fourths of the sheep raised belong to only six breeds. These top six include Suffolk, Dorset, Hampshire, Rambouillet, Polypay, and Columbia. Other breeds in North America include Oxford, Southdown, Corriedale, Montadale, Shropshire, Cheviot, and Katahdin.

Sheep may be classed by the type of wool produced or by their breeding use. Type of wool produced is broken down into fine-wool, medium-wool, long-wool, crossbred wool breeds, carpet-wool, and fur sheep breeds. Classification by breeding use breaks down into ewe, ram, dual-purpose, and other breeds.

SUFFOLK

The Suffolk is a medium-wool breed of sheep. It originated in England and is the most popular breed in North America. The head, ears, and legs are

9-11. Suffolk ewe.

black, without any wool on the head and ears. Suffolks are polled, though some males may have scurs.

DORSET

The Dorset is a medium-wool breed. The sheep may be polled or horned, with most being polled. The Dorset is entirely white and has no wool on its face. Ewes breed out of season and are known as good milkers. The Dorset breed originated in England.

HAMPSHIRE

The Hampshire originated in England and is a medium-wool breed. The face, ears, and legs are dark brown and often nearly black. Both males and females are hornless, though males may have scurs. The Hampshire is known as an early-maturing, large sheep.

9-12. Dorset ewe.

9-13. A Hampshire lamb groomed and ready for the show ring. (The way the lamb is standing is known as bracing. This enhances muscle development in the loin and rear legs.)

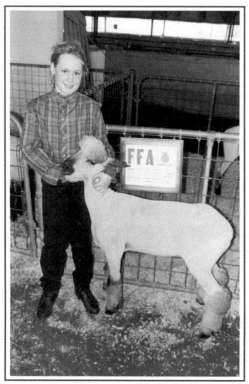

9-14. Oxford lamb fitted for showing.

OXFORD

The Oxford is a medium-wool breed that is popular with youth in lamb shows. Originating in England, the Oxford is a cross of Hampshire and Cotswold. The face, ears, and legs may have brownish or grayish coloring. It has a topknot of wool (wool cap) on its head. At maturity, the Oxford is among the largest breeds.

KATAHDIN

The Katahdin is a breed of sheep with hair rather than wool. Originating in the Virgin Islands, the Katahdin was first brought to Maine along with the English breeds of sheep. Colors vary from white to cream or red. The coloring may be spotted or solid. Most Katahdins are polled, though some males may have small horns or scurs.

9-15. Katahdin sheep.

The Katahdins are not tail-docked as are the wool breeds. Docking is not necessary because the hair does not collect feces, as does the wool on other sheep.

RAMBOUILLET

The Rambouillet is a descendant of Spanish Merino sheep. Half of all sheep in the United States have some Rambouillet blood. Rambouillets are

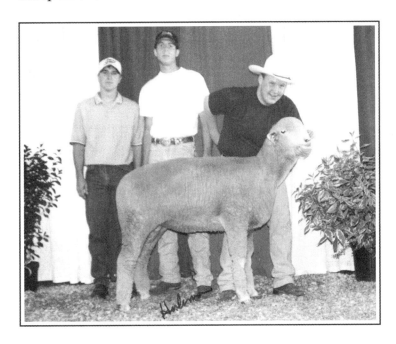

9-16. Ramboulllet lamb following a successful show. (Courtesy, The American Rambouillet Sheep Breeders Association)

good dual-purpose sheep. Rambouillets are white, have blocky bodies, and may be horned or polled. All ewes are polled.

GROUPINGS OF GOATS

Some 300 breeds of domestic goats are known. Wild goats are still found in many areas, such as the Rocky Mountains. Domestic goats are in five groups.

Select goats based on your goals. Goats should be healthy and disease free. Milk goats should have a well-developed udder. Color and the presence of horns may also be factors. A discussion of the groups of domestic goats follows.

ANGORA GOATS

There are more than a million Angora goats in the United States. However, few people really know what they look like. Texas is the leading producer of Angora goats. They are well adapted to living in areas where the grazing is not suitable for other livestock.

Angora goats are almost totally white. Occasionally a black one will be found. Kids may sometimes be red, but will usually shed their hair and the new growth will be white. Their coat consists of long locks of hair, known as mohair. This fleece covers the entire body, except the face. It is of very fine quality and has a high luster. The average Angora produces up to seven pounds of mohair each year, and is sheared twice a year.

DAIRY GOATS

Dairy goats produce 1.8 percent of the world's supply of milk. An average doe may produce 5 pounds of milk per day during a ten-month lactation.

Goat's milk has a sweeter taste and contains more minerals than cow's milk. During digestion, goat's milk forms into a soft curd. This makes it easier for children and the elderly to digest than cow's milk. The presence of bucks around does causes the milk to take on an unpleasant odor. The major breeds of milk goats are Nubian, Lamancha, Alpine, Toggenburg, Saanen, and Oberhasli. A description of three of the breeds follows.

- Nubian—The Nubian is considered a dual-purpose breed for milk and meat. It has larger, drooping ears than other breeds. The record milk production in a

9-17. Nubian doe. (Courtesy, American Dairy Goat Association)

year is 4,420 pounds. Though the amount of milk is less than some other breeds, the milk has a higher fat content.

■ Lamancha—The Lamancha breed originated in California. A major characteristic is the absence of external ears. The Lamancha has higher milk production than the Nubian.

■ Alpine—The Alpine is a good milker, with annual production up to 5,700 pounds a year. It has distinctive coloring, with white and darker colors forming unique patterns.

9-18. Lamancha doe. (Courtesy, American Dairy Goat Association)

9-19. Alpine doe. (Courtesy, American Dairy Goat Association)

MEAT GOATS

Meat goats are derived from the Mexican Criollo, thus, they are also called Spanish Goats. Most meat goats are found in Texas. The coloring varies from white to black with some being orange to sandy in coloring.

9-20. Spanish goat.

CASHMERE GOATS

Cashmere goats have a fine down undercoat. This is the finest animal fiber used. Cashmere produces clothing that is light and soft, yet warm. It also has three times the insulating value of wool.

PYGMY GOATS

Pygmy goats are miniature goats used for research and as pets. People find them enjoyable to watch.

9-21. Pygmy goat.

SHEEP SELECTION AND PRODUCTION

The management of sheep is more complex than the management of most animals. The complexity is the result of trying to get two products (wool and meat) from one animal, instead of one product, which is the case with most animals. Sheep meat makes up about 80 percent of the market and wool only 20 percent.

SELECTION

When selecting sheep, consider the production goals. Is the goal to raise purebred, crossbred, or grades? Breeds are selected for their characteristics. Also consider, the size of the flock or herd; the time of year to start; and the uniformity, health, age, soundness of the udder, and price of the animal.

Usually only experienced breeders need to buy purebred animals. Beginners should start with a crossbred ewe. Selection of breed or class should be based upon the purpose for raising them. Size of the flock should be based upon the space and facilities available. The animal itself should be healthy, sound, and uniform for its age and breed.

PRODUCTION SYSTEMS

Five types of sheep production systems are used. Success in any of these types depends upon managing a healthy and productive flock—that is, economically marketing lambs and wool.

The Farm Flock Method

The *farm flock method* is the most popular method used in the central, southern, and eastern United States. Production of market lambs is the focus of this method with the production of wool being secondary. Usually, the producer raises other livestock, which compete with the sheep for pasture.

9-22. Farm flock in good pasture with shade structures and good fences.

The Purebred Flock Method

Few farms raise a *purebred flock*. Their main purpose is the sale of rams and ewes. This requires a high level of management ability. The producer must raise sheep of ideal type. A full-time caretaker is usually employed.

The Range Band Method

With the *range band method*, each band of sheep has its own herder who moves them over a large area of land. The emphasis on lamb or wool depends upon the region's rainfall and vegetation. In arid regions, such as in the southwest, wool production is emphasized because there is not enough grass to finish a lamb.

The Confinement Method

The **confinement method** is becoming more popular. Raising animals in confinement means that they are raised totally indoors. Reasons for the popularity of this method include the success of raising other types of livestock in confinement, high land prices, fewer parasite problems, and the increase in young produced.

9-23. Lambs being fed concentrate. (Courtesy, *National Lamb and Wool Grower Magazine*)

The confinement method has several unfavorable factors for most producers. Equipment and building costs are higher. This type of system is more restrictive than the other methods. More skill and better management practices are required.

Lamb Feeding

Lamb feeding is a specialized method. After weaning, lambs are sold to feedlots. These feedlots are usually located near good, cheap food supplies. Feedlots feed out the lambs to a desired slaughter weight.

BREEDING MANAGEMENT PRACTICES WITH SHEEP

Commercial and purebred breeders alike strive to have a high lamb crop. The average lamb crop is 106 to 108 percent with the mortality as high as 25 percent from birth to weaning. Breeding and selecting ideal animals is important in meeting these objectives.

NORMAL BREEDING

Ewes reach puberty at 8 to 10 months of age, while rams reach puberty at 5 to 7 months. Ewes are usually bred to give birth (lamb) the first time at 24 months of age. Giving birth is known as *lambing*. The time of the year when ewes have lambs is known as the lambing season.

The heat period lasts an average of 30 hours. Ewes show no visible signs of heat other than the acceptance of the ram. Estrus occurs every 16 to 17 days. The gestation period of sheep lasts around 148 days. This varies among breeds.

RAM CARE AND MANAGEMENT

Rams are normally kept separate from ewes. A dry barn or lot with room to get plenty of exercise is adequate. Good hay and pasture are needed. Concentrate supplement may be needed in the winter and if the pasture is short. Rams do not require much, if any, grain. Many rams are pastured. Excess fat is harmful to a breeding ram.

EWE CARE AND MANAGEMENT

A pregnant ewe needs feed, water, shelter, and exercise. Quality hay and pasture have important roles in ewe rations. Supplemental concentrates and minerals may be used. A ewe that does not lamb is not profitable because lambs are more profitable than wool. Because of this, pregnancy testing is very important. Detection of one or multiple fetuses affects nutritional and health management practices.

9-24. Ewe with nursing lamb.

Care at Lambing

Most death loss of lambs occurs in the first few days after lambing. As parturition approaches, the ewe should be sheared around the udder, flank, and dock; placed in a dry, roomy pen; and the amount of grain should be reduced. Proper care of the ewe and her lamb(s) include use of lambing pens, helping lambs if they are being born in the wrong position, nursing chilled and weak lambs, and examining and treating any health problems.

Orphaned Lambs

If a ewe dies or is unable to nurse, its lambs are orphaned. An *orphaned lamb* needs a foster mother. A ewe who is very healthy may accept an additional lamb. A ewe who has lost her own young may accept an orphaned lamb, if the orphan is first rubbed with the dead lamb. A more effective method is to tie the skin of the dead lamb onto the orphaned lamb. The skin may then be removed a piece at a time after a few days. If the lamb is not accepted, cow's milk or milk replacer may be used to feed the lamb.

9-25. Orphaned lamb being bottle fed by California students.

MANAGEMENT FROM LAMBING TO WEANING

Between the time that lambs are born until they are weaned, they should be docked and castrated.

Docking

Docking is cutting off all or part of the tail. This is one of the first things done after lambing. It should be done when they are 3 to 10 days old. Docking may be done with a knife, hot docking iron, electric docker, or elastic band. The Katahdin lambs are not docked because they have hair rather than wool. Goats are not docked.

Castration

Ram lambs should be castrated when they are young. Many producers dock and castrate simultaneously. Castration may be done by similar methods as docking. Remember to be as sanitary as possible to prevent infection.

NUTRITION AND FEEDING OF SHEEP

Success in the sheep enterprise can be measured by the percentage of lambs raised and the pounds of lamb marketed. Nearly 100 percent of a sheep's diet is made up of roughages. Sheep have a ruminant digestive system, which allows for use of pasture and hay. Sheep are not fed grain except just before and immediately following lambing.

NUTRITION NEEDS

Sheep, like other animals require the six basic nutrients of carbohydrates, fats, proteins, minerals, vitamins, and water.

Carbohydrates and Fats (Energy)

Energy, which comes from carbohydrates and fats, is provided chiefly by pasture, hay, and silage. Corn, oats, wheat, barley, or grain sorghums may be added to the feed ration during times of drought, overgrazing, or snow-covered pastures. During the fall, they may have access to stalk or stubble fields.

Protein

Sheep need high levels of protein because they produce wool, which is made up of protein. Alfalfa, clover, soybeans, and green pasture are all good sources of protein. These crops are best when harvested before the protein quantity begins to decline.

When these are not available, a protein supplement may need to be added. Soybean, cottonseed, linseed, canola, peanut, and sunflower meals are good sources.

Range sheep may develop a protein deficiency. Protein blocks may be placed in the pasture to alleviate this problem.

Minerals

Sheep require the macrominerals salt (sodium chloride), calcium, phosphorus, magnesium, potassium, and sulfur. The microminerals that they require include cobalt, copper, fluorine, iodine, iron, manganese, molybdenum, selenium, and zinc.

Phosphorus deficiency may be experienced by range sheep. Self-feeding salt and mineral mixtures may be placed in pastures to alleviate any mineral deficiencies.

Vitamins

All of the fat-soluble vitamins (A, D, E, and K) are required by sheep.

9-26. Quality hay and concentrate promote good lamb growth.

Bacteria in the rumen produce adequate quantities of the water soluble (B) vitamins. Vitamin A is the most likely to be deficient. Vitamin mixtures may be fed free choice.

Water

The average sheep needs one gallon of water each day. Sheep obtain water not only from obvious sources, but also from feed, dew, and snow. When

9-27. Good drinking water is essential.

feeding on lush, green pasture, sheep may go for weeks without drinking water.

FEEDING

Sheep and goats benefit from pasture and other roughage. They are suited to rangeland where cattle would not thrive well.

Sheep do best on quality pasture and hay. They may also be given supplements with protein, minerals, and vitamins. Lambs may be creep fed while still nursing.

Goats tend to thrive in more rugged terrain. Milk goats are often kept in improved pastures and provided quality hay and concentrate supplements.

HEALTH MANAGEMENT

Keeping sheep and goats healthy means developing a preventive health program. When an animal gets sick, the expense of treating the animal usually outweighs the value of the animal. Therefore, prevention is clearly best.

GENERAL PRACTICES

Some recommendations for prevention of ill health of sheep and goats include monitoring the animals regularly checking for any signs of illness; providing a well-balanced ration; reducing stress whenever possible; controlling outside traffic from other animals and vehicles; rotating pastures to reduce parasite problems; and keeping facilities and equipment sanitary.

A good vaccination and checkup program should be established with the local veterinarian. If an animal becomes ill or has just been purchased, isolate the animal from the rest of the flock or herd. New animals should be isolated for at least 30 days.

DISEASES

Several diseases afflict sheep and goats. Prevention is especially important. Chapter 4 has general information on diseases. Examples of sheep and goats' diseases follow.

- Actinobacillosis (Lumpy Jaw)—Actinobacillosis causes economic loss to the producer. It affects the jaw and parts of the animal's head. Tumors or lumps of yellow pus appear on the jaw. The muscles and organs of the animal may also become infected, causing parts of the animal to be condemned upon slaughter. The animal will usually not die from the disease.

- Actinomycosis (Wooden Tongue)—Actinomycosis causes lesions on the animal's head and swelling of the lymph glands. The tongue may have lesions on it as it becomes hard and immobile. The animal finds it more difficult to eat and will lose weight. They will eventually die. Animals that have actinomycosis should be isolated from the rest of the herd or flock. Contaminated feed usually spreads this disease.

- Blue Tongue—Blue tongue is a viral disease that is spread by gnats. Blue tongue weakens the immune system of sheep and goats. Thus, most deaths occur due to secondary diseases. Signs of blue tongue include a high fever, loss of appetite, and sluggish behavior. The lips, muzzle, head, and ears will become swollen. The mouth will turn red or blue and will develop ulcers. The animal will have trouble eating. Blue tongue does not have any available treatments. Vaccination at shearing time is the best method of prevention.

- Enterotoxemia—Enterotoxemia is a bacterial disease that usually affects lambs and kids. They are commonly found dead with their heads being arched up. This occurs because of the convulsions that they experience shortly before death. There is no treatment for enterotoxemia. Vaccination, good management, and proper feeding are keys to preventing this disease. Ewes should be vaccinated two and six weeks before giving birth. Lambs should also be vaccinated. Increasing the amount of roughage and using chlortetracycline may be necessary.

- Foot Abscess—Foot abscess affects the foot's soft tissue. It is usually found when conditions are wet and muddy. Bacteria enter the foot causing abscesses or pockets of pus. The joints and tendons may also be affected. Put the animals on clean, soft bedding, drain the abscesses, and use antibiotics.

- Foot Rot—Foot rot is caused by bacteria, which are different from the one that infects cattle. It is extremely contagious and occurs when the animals are in wet conditions. Foot rot causes animals to lose weight and requires added labor to treat these animals. Animals will become lame and will have an unpleasant odor about the feet. Keeping their quarters dry and clean and isolating in-

9-28. Healthy lambs being prepared for showing.

fected animals will help reduce losses from foot rot. A vaccination is available.

■ Johne's Disease—Johne's disease causes the intestinal walls to become thicker. Animals will appear to have diarrhea, lose weight, and eventually die. There is no effective treatment of Johne's disease. Purchased animals should be checked for a history of this disease.

■ Lamb Dysentery—Lamb dysentery is caused by bacteria and affects lambs when they are only one to five days of age. They will lose their appetites, become depressed, have diarrhea, and will suddenly die. Death loss can be very high. Sanitation and good management practices are the best methods of prevention.

PARASITES

External parasites cause losses to sheep and goat producers because of the decrease in the quality of wool, mohair, meat, and milk. Controlling of external parasites is done through good sanitation and proper use of pesticides. Common external parasites of sheep and goats include blowflies, bot flies, lice, mange, and sheep ked or tick.

The biggest problem of raising sheep and goats is the loss caused by internal parasites. Infested animals lose weight and lose milk and meat production. They have poor-quality wool and decreased reproductive efficiency. Signs of internal parasites include weight loss, rough hair coat, loss of appetite, diarrhea, and anemia. Animals should be wormed regularly and rotated between pastures frequently. Common internal parasites of sheep and goats include coccidia, liver fluke, lungworms, and stomach and intestinal worms.

Drenching is giving a liquid medication by oral means to control internal parasites. A dose syringe or gun is used in the mouth. Proper dosage must be used. Only use approved products to control parasites.

FACILITY AND EQUIPMENT REQUIREMENTS

Sheep do not require expensive shelter. They are relatively hardy animals. Some producers raise sheep in confinement facilities using automatic feeders and waterers. This increases costs, but decreases labor requirements. The average farm or ranch does not need confinement facilities.

9-29. Ewe and lamb in warm facility during the winter. Note the heat lamp is located to allow the lamb to enter but keep the ewe away.

HOUSING

Barns or sheds for sheep usually open to the south. This provides protection from winter weather. The barn should be well bedded, dry, and well ventilated. Where lambing is to take place, there should be electricity to hook up heat lamps. Troughs, feeders, and waterers should be placed in an easily accessed location.

Milking does are commonly kept in free stalls. Loose housing may be used for kids and yearlings. Both require extra bedding and care during the cold winter months. Goats kept in loose housing or tie stalls must be dehorned.

9-30. Rams used for semen collection are housed in individual pens. (Courtesy, Elite Genetics, Waukon, Iowa)

FENCING

Fencing is needed to keep predators out, more than keeping sheep and goats in. Sheep need a 60-inch, or higher, fence. It could be made of woven wire or barbed wire with only 4 to 5 inches between each strand.

EQUIPMENT

Loading chutes, corrals, portable shelters, weigh crates, and pregnancy testing cradles may also be needed, depending upon the operation.

9-31. A specially designed corral for working sheep.

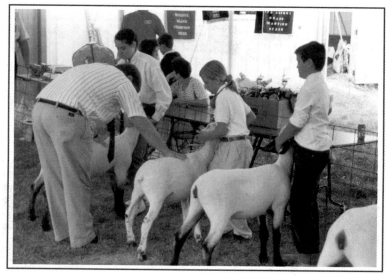

9-32. Judging is underway in a local lamb show for youth.

REVIEWING

*M*AIN IDEAS

Sheep and goats are closely related. They have relatively minor differences in physical characteristics and management practices. The number of sheep in the United States has been declining since the 1940s.

Sheep are used for their meat and for their wool. Meat production is more profitable. Goats are raised for their meat, milk, mohair, and cashmere, or as pets.

Selection of sheep and goats should be based on many things. They should be healthy and without defects. Sheep and goats should fit into your facilities and management abilities.

Sheep and goats may be raised by various means. They may be part of a farm flock, a purebred flock, a range band, a confinement system, or they may be fed out in a feedlot.

With the mortality of lambs as high as 25 percent, proper breeding and health management are essential. Unbred ewes that do not have lambs are unprofitable. Lambs should be docked and castrated when young. They consume mostly roughages and very little concentrates. Drenching is a common practice used to prevent internal parasites.

Minimum facilities and equipment are required for raising sheep and goats. However, a good fence is needed to keep out predators.

(Additional information on sheep and goat production is available in *Sheep and Goat Science*, available from Interstate Publishers, Inc.)

*Q*UESTIONS

Answer the following questions using correct spelling and complete sentences.

1. How are goats and sheep similar in appearance and in management practices?

2. Describe how sheep are classified.

3. Describe how goats are grouped.

4. What should be considered when selecting a sheep or goat?

5. What are the five types of sheep production systems? Describe each.

6. Which type of production system would best suit the area in which you live? Explain.

7. How should a ewe be managed in terms of reproduction, nutrition, and health?

8. Describe a good health management system.

9. Describe facilities needed on an average sheep or goat farm/ranch.

EVALUATING

Match the term with the correct definition. Write the letter by the term in the blank provided.

a. ram   d. wether   g. lamb
b. nanny or doe   e. mutton   h. ewe
c. kidding   f. chammy

_____ 1. Soft, pliable leather made from a sheep skin.

_____ 2. Sheep less than one year old.

_____ 3. Male sheep castrated before sexual maturity.

_____ 4. Male sheep.

_____ 5. Female sheep.

_____ 6. Female goat.

_____ 7. When a nanny gives birth.

_____ 8. Meat from a sheep that is more than a year old.

EXPLORING

1. Visit a local sheep farm. Note the management of reproduction, nutrition, and health. Take a camera and record major observations. Prepare a poster or bulletin board that depicts what you have learned.

2. Visit a local goat farm. Note the management of reproduction, nutrition, and health.

3. Fit a lamb for showing. Use your own lamb or assist someone who owns a lamb. Determine the different procedures in fitting and why these are followed. Write a report on your experiences. Give an oral report to the class.

10

DAIRY PRODUCTION

Milk is secreted by the mammary glands of female mammals. Its purpose is to feed her babies. As the primary food for baby animals, milk is a good source of human food.

Milk and the foods made from milk are "nutrient dense." This means that they contain large amounts of essential nutrients compared with calories. This has made milk and dairy foods popular with consumers.

Humans use milk from several animal species. Cows are most important in the United States. Goats, camels, llamas, reindeer, sheep, and water buffalo are also sources of milk.

10-1. A Holstein dairy calf that is receiving good care.

OBJECTIVES

This chapter provides general information on dairy cattle and dairying. It has the following objectives:

1   Describe the dairy industry

2   Describe dairy cattle as organisms

3   Explain dairy conformation and type

4   Identify common breeds of dairy cattle

5   Explain important management practices in dairy production

6   Explain reproduction in dairy cattle

7   Describe dairy feeding and nutrition

8   Describe how the environment is modified for dairy cattle

9   Explain health management practices with dairy cattle

10 Describe facility and equipment needs with dairy cattle

TERMS

alveoli
animal model
cold housing
culling
Dairy Herd Improvement
   Program (DHI)
dry cow
functional type
homogenization
immunoglobulin
ketosis
linear evaluation
management intensive
   grazing

metabolic disorder
milk fever
milking parlor
nutrient dense
pasteurization
predicted transmitting
   ability
progeny
selection
total mixed ration (TMR)
type production index (TPI)
udder
warm housing

THE DAIRY INDUSTRY

The dairy industry provides milk and other dairy foods to consumers in North America and many foreign markets. Its success can be attributed in part to the ability of the dairy cow to efficiently convert feed to milk. As a ruminant, the cow converts plant nutrients into forms humans will eat. The cow converts plant nutrients into the nutrient-dense product called milk. *Nutrient dense* means that the food has large amounts of nutrients relative to calories.

10-2. A modern milking facility.

BEGINNINGS

Dairy cows were first brought to the Jamestown Colony in 1611. Early farms had only one or two dairy cows for their own use. Lack of refrigeration made it more difficult for people living in large cities to obtain milk.

Pasteurization, refrigeration, and bottled milk were developed in the second half of the 19th century. This allowed milk to be stored and transported to population centers.

Pasteurization is a process that destroys bacteria and other tiny organisms. It involves heating the milk to 161°F for 15 seconds. This keeps people

10-3. A dairy facility in California.

from getting disease, such as undulant fever, from drinking milk from diseased cows.

Small family dairy herds began to develop in response to improved storage and handling technology. The dairy industry continued to evolve

Table 10-1. Major Events in the Dairy Industry	
Year	**Event**
1611	first cows arrived at Jamestown
1841	unrefrigerated milk shipped to New York by rail
1851	first cheese factory in Oneida, New York
1890	Babcok Butterfat Test introduced
1895	pasteurization available on a commercial basis
1919	homogenized milk marketed
1932	vitamin D fortification of milk introduced
1936	artificial insemination successfully used with cattle
1948	ultra-high temperature pasteurization first used
1975	embryo transfer began to gain acceptance
1994	bovine somatotropin (bST) used on commercial basis

throughout the 20th century. Compulsory pasteurization, homogenization, and addition of Vitamin D to milk became accepted marketing practices in the United States before World War II.

Homogenization is a process that is used to keep the fat and milk liquid from separating. Cream rises to the top of milk that is not homogenized. In homogenization, the fat droplets are broken into very small particles so they stay in suspension.

More changes occurred in the mid 1900s. Dairy farmers started using artificial insemination on their cows. Commercial dairy farms became market oriented. This pushed production, per cow, upward.

TODAY

There are 9.3 million dairy cows on farms in the United States. These cows annually produce nearly 150 billion pounds of milk. Each dairy cow, on the average, produces nearly 16,000 pounds of milk a year. This is three times the production in 1950. One dairy cow, today, provides nearly 25 people with their dairy product needs.

The number of farms with dairy cows has declined. There are 95 percent fewer dairy farms today than in 1950. Each dairy farm is larger today. Most herds today range from 35 to 500 cows. Some herds have as many as 10,000 cows.

Dairy farms use high-tech milking equipment. The hand labor of milking has been replaced with milking machines.

Biotechnology has contributed to the changing dairy industry. The use of rbST (bovine somatotropin) began in 1994 and has steadily increased. Bovine somatotropin is a growth hormone naturally found in cows, but at a higher level in high-producing cows. Giving additional bovine somatotropin to cows causes them to milk at higher levels for a longer period of time in well-managed operations.

Dairy production is concentrated near large cities. California leads all states in total milk production. Wisconsin is second, followed by New York, Pennsylvania, and Minnesota. All states have some dairy cattle.

Dairy cattle also contribute to the beef supply. Calves are used for veal. Older cows are used for processed meat products, such as bologna and hamburger.

DAIRY CATTLE AS ORGANISMS

Dairy cattle belong to the family bovidae, which includes ruminants with hollow horns. Members of this family also chew their cuds. The following shows the scientific classification of the dairy cow:

Kingdom Animalia: The animal kingdom

Phylum Chordata: Animals with either a backbone (in the vertebrates) or the rudiment of a backbone (in the chorda).

Class Mammalia: Mammals or warm-blooded, hairy animals whose off-spring are fed with milk produced by the mammary glands.

Order Artiodactyla: Even-toed, cloven-hoofed mammals.

Family Bovidae: Ruminants have numerous placental attachments and hollow, nondeciduous, upbranched horns.

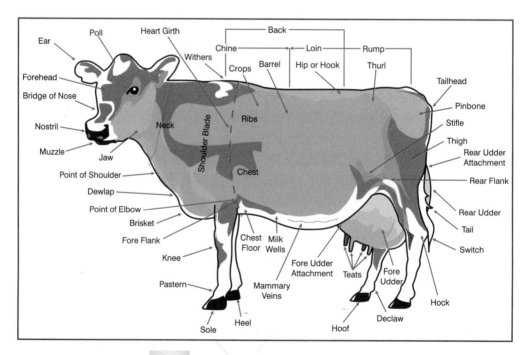

10-4. Major external parts of a dairy cow.

Genus Bos: Four-footed ruminants, including wild and domestic cattle, distinguished by a stout body and hollow, curved horns standing out laterally from the skull.

Species: *Bos taurus* and *Bos indicus*

DAIRY CATTLE CONFORMATION AND TYPE

Dairy cattle are milk producers. The body structure should contribute to high milk production.

CONFORMATION

Dairy cows appear angular and often lack muscle development in high-value meat cuts, as with beef cattle. In cows, the udder should be well attached and have the capacity to hold 50 to 70 pounds of milk. Four teats should be shaped and spaced uniformly for machine milking.

Since much feed is needed for the cow to produce milk, the body must be capable of eating and digesting feed. A cow needs a good appetite and the strength to compete with other cows for feed. They also need good body capacity to hold feed for digestion.

Dairy cows should regularly reproduce. Milk production is associated with the reproductive cycle. Lactation is the secretion of milk by the mammary glands. It begins at parturition and ends when the cow is dried up.

A *dry cow* is one that has stopped producing milk. Most dairy cows produce milk for 305 days. They are bred to calve each year. With a gestation period of 283 days, some rest from milk production is needed. Milking is stopped and they are dried-up 50 to 60 days before the next calving.

UDDER STRUCTURE

The *udder* is the bag-like structure that contains the mammary glands and to which teats are attached. Cows have four mammary glands or quarters. Each quarter has a teat. A canal in each teat allows removal of milk.

The udder should be large and strongly attached to the body. Udders sometimes break away from the cow. This shows problems in producing a

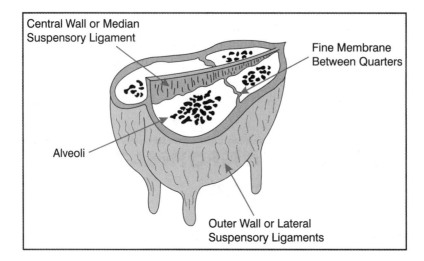

10-5. Structure of a cow's udder.

large amount of milk. The milk is produced in the *alveoli*. These are tiny structures that remove nutrients from the blood and convert the nutrients into milk.

Milk is 87 percent water. Fat and protein make up a little less than 4 percent each. It has a little more than 4 percent lactose (milk sugar). Cows must drink much water to carry on life functions and secrete milk.

DAIRY CATTLE BREEDS

Seven major breeds of dairy cattle are used. Differences in breeds include size, color, milk yield, milk fat, and milk protein.

Variation exists within each breed. Holsteins produce the most milk by volume. They are followed by Red and White, Brown Swiss, Ayrshire, Milking Shorthorn, Guernsey, and Jersey. The breeds ranked in order based on percent of milk fat and protein are Jersey, Guernsey, Brown Swiss, Ayrshire, Red and White, Holstein, and Milking Shorthorn.

JERSEY

The Jersey is from the Island of Jersey and was brought to North America in 1850. Jerseys are grayish to fawn and near white in color. They are smaller than other breeds. Jersey milk has the highest fat and protein content. The amount of milk produced is less than other breeds. Jersey numbers have been increasing in recent years.

10-6. A Jersey cow. (Courtesy, Pete's Photo, Wykoff, Minnesota)

HOLSTEIN-FRIESIAN

The Holstein-Friesian (commonly shortened to Holstein) breed came to North America from the Netherlands in 1621. It is the most popular breed of dairy cattle. Most Holsteins have distinctive black and white color. They produce the largest amount of milk though low in butterfat and protein. Holsteins are large, with cows weighing up to 1,500 pounds, and bulls, 2,200 pounds.

10-7. A Holstein-Friesian cow. (Courtesy, Pete's Photo, Wykoff, Minnesota)

Connection

MILK AND YOUR HEALTH

Many of our favorite foods are dairy products! In addition, these foods are good sources of important nutrients. What are your favorite dairy foods? Many people like ice cream, yogurt, and cheese.

Besides dairy foods, milk and milk products are used in many other foods. Cheese is used in pizza, cheeseburgers, and macaroni and cheese. What would these foods be like without cheese?

Milk is good in the human diet. Calcium, protein, and other nutrients are in milk. Strong bones and good health are promoted with dairy products. As this photo shows, ice cream is popular with young people.

BROWN SWISS

Brown Swiss were brought to North America in 1869 from the Alps of Switzerland. The color is solid, light to dark brown. The nose and tongue are

10-8. A Brown Swiss cow. (Courtesy, Pete's Photo, Wykoff, Minnesota)

black. The Brown Swiss has a calm disposition. The milk has a favorable protein to fat ratio for the present milk market.

GUERNSEY

The Guernsey was introduced into North America in 1831 from the Island of Guernsey. The color is fawn with white markings. Guernseys are quiet cattle and easy to work with. Their numbers have been declining. They are sometimes referred to as "golden Guernseys" because of the slight golden color in their milk.

10-9. A Guernsey cow. (Courtesy, Pete's Photo, Wykoff, Minnesota)

AYRSHIRE

The Ayrshire was brought to North America from Scotland in 1822. Color varies from light to dark red, brown, and white. The Ayrshire is known for well-attached udders and grazing ability.

MILKING SHORTHORN

The Milking Shorthorn breed was designated as a dairy breed in 1968. The coloring is similar to the Shorthorn, with red, white, and combinations of these. The breed is adaptable to a variety of situations.

10-10. An Ayrshire cow. (Courtesy, Pete's Photo, Wykoff, Minnesota)

10-11. A Milking Shorthorn cow. (Courtesy, Pete's Photo, Wykoff, Minnesota)

RED AND WHITE

The Red and White evolved from the Holstein. It is similar to other Holsteins in milk production, size, and disposition. As the name suggests, the color is red and white.

10-12. A Red and White Holstein cow. (Courtesy, Pete's Photo, Wykoff, Minnesota)

MANAGEMENT OF DAIRY HERDS

Good management is essential to success in dairy production. Dairying is a daily venture. Cows must be milked at least twice each every day. Some cows in a herd are producing every day of the year.

STARTING A DAIRY ENTERPRISE

Some people like dairy farming. The decision to start a dairy enterprise is a long-term one. Once started, it is difficult to change without serious consequences. As a result, the final decision should be carefully thought out.

Dairying has been one of the most stable agricultural enterprises. It has returned a reasonable profit and allowed long-term financial growth. Personal pride and satisfaction of ownership are additional rewards for most dairy farmers.

The rewards of dairying also come with risks and challenges. The capital investment is sizeable and carries with it risks just as any investment does.

The final decision is to choose the type of operation best suited to resources and personal goals. This can include choosing between a commercial (grade) and purebred operation, a family-size (100 or less) or large herd, a to-

10-13. This Wisconsin dairy farm required a large financial investment.

tal herd (cows and replacement heifers) or a milking herd, and whether the feed will be raised or purchased.

Selection and Culling

A key to success in dairying is the ability to select and cull cows and heifers. **Selection** is choosing or picking the animals for the herd. **Culling** is choosing the cows or heifers to remove from the herd. Criteria in selection and culling are production factors, genetic makeup, and type.

Dairy Herd Improvement

Milk production can be assessed by the **Dairy Herd Improvement (DHI) Program.** DHI is a national dairy production testing and record-keeping program. USDA personnel work with dairy producers to get information and compare a herd with others in the area, state, and nation.

The state and local Dairy Herd Improvement Associations (DHIA) have the most common official testing plans. All cows must be properly identified and all DHI rules enforced. Cows are tested every 15 to 45 days. The results are published and used by the USDA to evaluate sires (bulls).

The Dairy Herd Improvement Registry (DHIR) includes the standard DHIA requirements plus the added requirements of breed associations. Additional testing is needed when an individual cow's milk production exceeds specified standards. The records for registered cows in herds enrolled in DHIR are sent to the respective breed registries.

These rules apply to both the DHIA and DHIR programs:

1. All cows in the herd must be entered into the official testing program.

2. All animals in the herd must be permanently identified.

3. Copies of pedigrees of all registered cows must be made available for DHIR.

4. Testing is done each month with not less than 15 days nor more than 45 days allowed between test periods.

5. Testing is conducted over a 24-hour period.

6. An independent supervisor must be present for supervising the weighing of the milk and the sampling of the milk for milk fat and other determinations.

7. Milk or milk fat records that are above values established by the breed association require retesting of the cow to assure that an error was not made. An owner may request retesting if he or she feels the test does not properly reflect the production of the cows.

8. Surprise tests may be made if the supervisor suspects that they are needed to verify previous tests.

9. Any practice that is intended to or does create an inaccurate record of production is considered a fraudulent act and is not allowed.

Several unofficial DHI testing programs also exist. These programs often involve the owner doing the sampling rather than a neutral supervisor. They may only involve the recording of milk weights. Unofficial records cannot be published or used to promote the sale of cattle.

Progeny Information

Progeny information is useful in selecting dairy animals. ***Progeny*** refers to the offspring of animals. Information about the offspring is important. Calf identification is the cornerstone to successful progeny data. Progeny information identifies the genetic worth of sires and dams.

The dairy industry has benefitted from progeny information. The USDA gathers DHI information in a large, computerized data bank. Sire and dam predicted transmitting abilities are calculated using the animal model.

Predicted transmitting ability is an estimate of the traits an animal will transmit to its offspring. It includes both genetic superiority and inferiority. In other words, both good and bad traits are transmitted.

The ***animal model*** is the genetic evaluation for dairy cattle production used by the USDA. It uses pedigree data and production-related factors.

Linear Evaluation

The type or physical appearance of a dairy cow may be used in culling or selection. Purebred producers want outstanding type. This improves the market value and pedigree of an animal. Commercial herds are more interested in functional type.

Functional type refers to the traits of a cow that are good enough to allow her to complete a useful life in a herd. ***Linear evaluation*** is the coding of 15 primary traits of dairy cows. It was started in 1978 and has been well received. Nearly one-half of the herds that use artificial insemination use linear evaluation.

With linear evaluation, 15 primary traits of cows between biological extremes are coded. Several numerical scales have been used, but the most common is a 50-point scale with 25 being the midpoint. The most limiting biological extreme is usually given a score of "1."

Most experts agree on the use of linear evaluation. A dairy farmer should select a group of sires that meets specified production standards (goals) for their operation. The selected group of bulls should then be mated to each cow in the herd to maximize type improvement.

Breed associations also calculate predicted transmitting abilities for type (PTAT). A higher PTAT indicates to a dairy farmer that it has a potential to

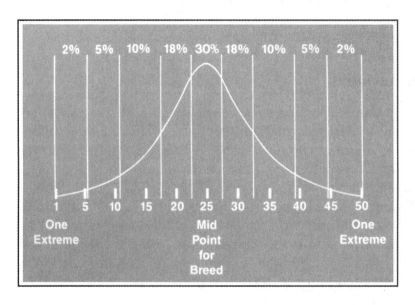

10-14. Linear evaluation uses 15 traits of dairy cows between biological extremes. This shows typical frequencies on a 50-point scale. (Courtesy, Lake-Plains Education Materials, Crookston, Minnesota)

10-15. Udder extremes: separation from body (left) and well-attached (right). (Courtesy, Lake-Plains Educational Materials, Crrokston, Minnesota)

improve the type classification of a particular mating. Dairy farmers may use PTAT together with linear scores. PTAT is also included in the ***type production index*** (TPI), which combines several type and production factors into a single figure. The type production index ranks cows based on overall performance. Higher type production indexes are a popular sire selection tool.

The ability to judge dairy cattle is often practical with both commercial and purebred herds. Judging is comparing animals as a whole based on their body type. Four animals are in a class. The animals are compared and ranked. Judging competition is popular throughout the United States.

10-16. Dairy judging at the World Dairy Exposition in Madison, Wisconsin.

10-17. Dairy Cow Unified Judging Scorecard.

DAIRY CATTLE REPRODUCTION

A dairy cow must reproduce annually while under the stress of high milk production. A combination of proper nutrition and good health are necessary. (The basics of reproduction are covered in chapter 5.)

REPRODUCTION

Reproductive goals include (1) a 12 to 13.5 month calving interval, (2) less than 2 units of semen per pregnancy, (3) heifers first calve at 23 to 25 months, (4) a live and healthy calf crop of 95 to 98 percent, and (5) a minimum of reproductive tract disorders. Embryo transfer is used. Some consideration is being given to cloning.

When to Inseminate Based on Signs of Heat				
Coming Into Heat	Standing Heat - 12-18 Hours		Going Out of Heat	
Bunts other cows. Bellows. Attempts to ride other cows, but will not stand. Non-elastic clear mucus discharge from vulva.	Stands to be ridden. Attempts to mount the front end of other cows. Elastic clear mucus discharge. Very restless.		Smells other cows. Attempts to ride other cows, but will not stand. May still have clear mucus discharge from vulva. Ovum Released ✳	Metestrus Bleeding ✳
Too early to inseminate	Can be inseminated	Best time to inseminate	Can be inseminated	Too late to inseminate

10-18. When to artificially inseminate a dairy cow based on signs of heat. (Courtesy, Gary Stegman, Crookston, Minnesota)

Achieving the reproductive goals requires a good knowledge of animal health. Many diseases that may cause reproductive problems, such as bovine virus diarrhea (BVD), brucellosis (Bang's disease), and leptospirosis, must be included in a herd vaccination program. Other problems, such as metritis and retained placentas, should be analyzed and treated by a veterinarian.

The danger of many genital diseases, such as trichomoniasis, vaginitis, and vibriosis, can be virtually eliminated with artificial insemination. (Some diseases are presented in more detail elsewhere in the book.)

Heat detection in cows and keeping good reproductive records are important. Synchronization methods may be used to bring cows in heat or to group the heat cycles of many animals. Most dairy cows are artificially inseminated.

SELECTING SIRES

Selection of quality sires is important. Dairy farmers have only five or six cow generations to achieve the production goals that have been set. Breeding artificially with bulls that are in the top 20 percent of the breed could be the difference in making or not making a profit.

Semen from five to eight bulls should be adequate for small herds of 60 cows or less. Intermediate size herds (61 to 200 cows) may use 7 to 10 bulls in their breeding program. Large herds (over 200 cows) may use 10 to 15 bulls. Semen is available from many sources. Bulls that meet the standards established by a given herd can easily be selected using predicted transmitting ability data.

FEEDING A DAIRY HERD

Dairy animals have inherited genetic potential for performance. How well that potential is realized depends upon the environment of the animal. Nutrition is the most important environmental factor.

A great need for energy and protein is created by lactation. A cow that weighs 1400 pounds and gives 80 pounds of milk per day needs 2.5 times more energy for milk production than for maintenance.

Modern dairy technology is used by dairy farmers to feed many times per day. Computerized feeders, mechanized track feeders, or auger-type delivery

10-19. Scientist assessing the energy value of grain. (Courtesy, Agricultural Research Service, USDA)

systems all allow for multiple feedings. Total mixed rations are increasing in use. A *total mixed ration (TMR)* is one in which all feedstuffs are mixed together. They provide all feed ingredients in each mouthful a cow eats.

RATIONS

Rations for dairy cattle begin with high-quality roughages. Roughages are usually the lowest-cost sources of nutrients. High-quality roughages can be bought or grown. Proper soil fertility and harvesting will be needed. The common roughages for dairy cattle are hay, silage, green chop, and pasture. Roughage can be restricted, fed free choice, or as a part of a total mixed ration.

High-producing dairy cows cannot get an adequate supply of nutrients from roughages alone. While 50 to 60 percent of a ration may be roughages, the other 40 to 50 percent must be obtained from concentrates. Concentrates refer to high-energy feeds, such as grains or protein supplements.

A higher percent of concentrates is usually fed to a dairy cow during the first four months of lactation. The nutrient demand is greatest at this time. Lower-producing cows or cows in late lactation are fed a reduced percent of concentrates to reduce feed cost.

NUTRIENT REQUIREMENTS

The nutrient requirements of dairy animals can be summarized easily. Maintenance needs are related to body size, but imply no weight gain or loss.

10-20. Dairy cattle often graze improved pasture.

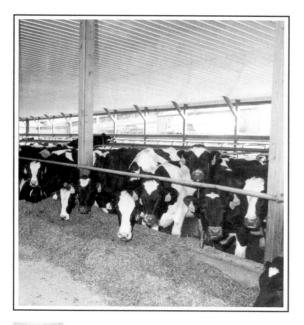

10-21. Dairy cattle on a ration that includes silage. (Courtesy, University of Minnesota, Crookston)

Most dairy animals under five years of age also require nutrients for growth. Lactation requires large amounts of nutrients.

Growth rates and nutrients needed for growth vary according to the stage of maturation. Bred heifers and cows require nutrients for reproduction, including both gamete production and fetal growth. The greatest nutrient requirement is for a pregnant lactating cow.

Body condition scores (1 = thin, 3 = average, 5 = fat) are often used to monitor proper nutrition. Ideally, a cow should spend her entire life with a body condition score that fluctuates from 2.0 to 4.0. Peak lactation (60 to 90 days into lactation) requires a cow to use her body reserve to supplement her ration. Her body condition may drop to 2.0 or 2.5 during that phase of lactation. A cow should replenish her body reserve during the last half of lactation, entering the dry period with a body condition score ranging from 3.5 to 4.0.

Feeding Calves

Heifer calves must be healthy and well-nourished. These are vital for the calf to become a high-producing dairy cow. Colostrum fed to a calf ensures that it can use the immunioglobulins. *Immunoglobulins* are antibodies in the milk to help the baby calf develop passive immunity. Colostrum is the first milk given by a female after giving birth. One-half gallon of colostrum should be fed within the first hour of birth and a full 6 percent of body weight within the first six hours of life. An additional 6 percent of body weight should be fed between the sixth hour and one day of age, for a total of 12 percent of body weight during the first 24 hours.

Colostrum is usually fed to a calf from its mother, but there are times when colostrum from other older cows in the herd should be fed. This is necessary when cows have been purchased shortly before calving or bred heifers have been raised separately from the older cows. Isolated heifers do not develop immunioglobulins to the diseases in the herd.

10-22. A growing dairy calf receiving proper nutrition.

Preferably, a calf should be fed whole milk for about five days. From five days until weaning, it can be fed whole milk or a high-quality milk replacer. A high-quality milk replacer should contain 20 percent fat, 20 percent protein, and less than .5 percent fiber. In addition to milk or a milk replacer, a calf should be fed a calf starter as soon as it will begin to eat. At four to six weeks of age, a calf should be eating 1.5 pounds of calf starter per day.

Forages should not be fed until after weaning. If hay is fed to a preweaned calf, it should be a leafy, high-quality alfalfa or alfalfa-grass hay fed only after four weeks of age.

FEEDING YOUNG STOCK

Feeding postweaned calves becomes easier due to a functional rumen. Some basic feeding guidelines are still necessary.

Forage can be fed free choice, with concentrates being supplemented at a rate up to 6 pounds per head. This depends on forage quality. Pasture can be the only forage used. Pasture quality should be carefully monitored so an appropriate supplemental grain mix can be fed.

Just as with lactating cows, body condition can be a useful tool in monitoring nutritional needs. If the nutritional level for young stock appears correct, but the body condition level and growth rate are less than desired, look for related problems. A common problem is parasites, which should be controlled.

FEEDING DRY COWS

Concentrates are not fed when a cow is dried off. For two to three weeks, average to good roughage may be the only feed. Concentrates, as a special dry cow formulation, are often reintroduced to a dry cow about midway through a standard 50 to 60 day dry period at a rate of 4 to 7 pounds per day. Toward the end of the dry period, some producers challenge or lead feed. Lead feeding refers to feeding concentrates that more closely resemble the lactating ration (7 to 14 pounds per day).

A few basic rules should be followed for a dry cow ration: (1) the calcium:phosphorus ratio should fall in the 1:1 to 1.5:1 range to reduce the incidence of milk fever; (2) a maximum of 35 pounds per day of corn silage should be fed; (3) alfalfa hay or haylage is not a recommended dry cow forage, but if it must be fed, it should never make up more than 50 percent of the forage dry matter; and (4) salt should be reduced to .25 percent of the ration (dry matter basis).

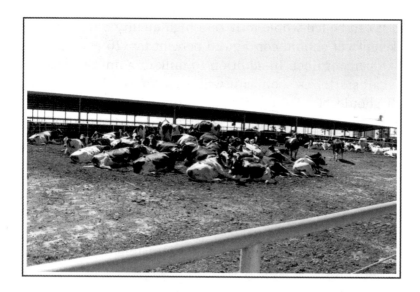

10-23. Cows in a drylot at a California dairy.

FEEDING THE LACTATING COW

The most challenging ration to develop is that for a high-producing cow. Each cow is unique and should be fed that way when possible. At a minimum, a cow should be fed a ration that is being fed to a group of cows that closely resembles her needs.

Several cow factors must be considered when developing lactating rations. These are age, size, stage of lactation, milk composition, and labor ca-

10-24. Feed-stuffs stored in bulk on the farm are becoming more common to reduce feed costs. (Courtesy, University of Minnesota, Crookston)

pability. Feed quality, availability, cost, and palatability to the cow must be considered.

High-producing cows typically require 1 pound of concentrate (grain) for each 2.4 to 3 pounds of milk produced. Concentrates rarely make up more than 60 percent of the total dry matter (a 50:50 ratio of concentrates to roughages is more desirable). Forages make up the balance of the ration and usually are fed at a rate of 1.5 to 2.0 percent of body weight. Feed protein in a ration should be 19 percent in early lactation and decreased slightly in later lactation. A higher level of bypass protein (undegradeable) is necessary in early lactation.

The fiber level in a ration should be 19 to 21 percent acid detergent fiber (ADF) and 25 to 28 percent neutral detergent fiber (NDF). Acid detergent fiber is the residue of the fiber content of feed. Neutral detergent fiber is cell walls and other feed content that are not readily digested. A maximum of 5 to 6 percent fat level in the ration is suggested. Salt should be fed at .5 percent of the total ration, a calcium/phosphorus mineral source at 1 to 2 percent of the grain mix, and vitamins as recommended.

ENVIRONMENTAL MODIFICATION

Modification of a dairy cow's environment began with domestication. Most environmental change is to provide for the well-being of the cow. It includes people who handle the cow, the climate in which a cow lives, facilities,

10-25. A dairy facility with an open ridge and curtains allow for environmental control at Windgate Dairy in Minnesota. (Courtesy, University of Minnesota, Crookston)

and economic considerations. Modifications are to improve a cow's environment. Conditions should provide for the well-being of the cow.

ANIMAL/HUMAN INTERACTION

The major environmental modification for most domesticated animals is increased interaction with humans. This is definitely true with dairy cattle.

A good dairy herdsperson is patient and understands animal behavior. Dairy producers should understand normal behavior to detect and treat animals that are acting abnormally. Dairy animals bond to kindness.

STRESS MODIFICATION

Many people feel that stress should be eliminated from both human and animal lives. A better viewpoint is that stress should be modified to a desirable level. It should be at a level that prevents boredom, but does not create unnecessary fear. Finding the right level of stress requires knowledge of the past experiences of an animal. Feeding time is an example. Dairy animals accustomed to eating at a specific time are stressed by a one-hour delay.

Some potential stressors are under human control. These include regular care, space allocation, transportation, and presence of strangers. These stressors can be easily managed with proper planning.

Other potential stressors cannot be controlled by humans. However, some can be modified. Examples these stressors include weather changes, social-order fights among animals, and physiological cycles, such as heat or calving.

Keeping dairy animal stress at an appropriate level requires knowledge of animal behavior, animal history, and coming changes in the life of an animal. A bred heifer raised on a large pasture with little human interaction will require a longer acclimation period to the dairy cow housing facility. A bred heifer raised with close human interaction in a facility will usually adapt quickly.

CLIMATE MODIFICATION

Four climate factors affect dairy animals: (1) temperature, (2) humidity or precipitation, (3) wind, and (4) radiation. While dairy cattle have a fairly wide tolerance range to individual factors, combinations of extremes can quickly have an adverse effect on milk production, conception rates, and growth.

A combination of heat and humidity with little wind is especially difficult for dairy animals to cope with. Shade and misting are used to reduce the negative effects. While dairy animals are more tolerant of cold, protection against severe cold, wind, and snowstorm conditions is beneficial.

Understanding an animal as it reacts to the weather is useful. For example, cows eat more before storms and less during hot, humid weather.

Dairy animals prefer to travel in groups or herds to water while grazing if the water is further than 500 to 700 feet away, while they travel individually at closer distances. Hot weather requires greater water intake and disrupts grazing if the water is at a distance that encourages herd travel.

DAIRY CATTLE HEALTH

Good health is essential for milk productivity. A good manager can use practices that promote health.

The primary objective of a health program is to prevent and control disease. Knowing the normal and abnormal dairy animal behavior is the first step in management. Noticing a cow in heat, a calf with dull eyes, or a cow whose body condition is declining makes it easy to respond.

Several diseases are important in dairying. Prevention and fast action when problems arise help keep a healthy herd. Many of the diseases of beef cattle may also afflict dairy cattle. Chapter 4 has basic information on disease.

10-26. Checking for rumen movement in a dairy cow. A lack of rumen movements may indicate a displaced abomasum, which is often referred to as a "twisted stomach." (Courtesy, University of Minnesota, Crookston)

METABOLIC HEALTH DISORDERS

Nutrition overlaps most areas of dairy cow health and management. Similarly, several metabolic disorders are directly or indirectly related to nutrition.

A *metabolic disorder* is a chemical transformation of energy by the body. Most occur at or shortly after calving. This is due to the stress associated with high milk production. The three most common metabolic disorders are milk fever, ketosis, and fat cow syndrome. Other metabolic disorders, such as bloat, retained placentas, and grass tetany, are not as specific to dairy cows.

Milk fever is a metabolic disorder characterized by low blood calcium and paralysis. It is caused by overfeeding calcium during late lactation or the dry period. Milk fever most often occurs near calving time, when the need for calcium increases rapidly. A low calcium to phosphorus ratio and an adequate amount of vitamin D in the ration reduce milk fever problems.

Ketosis is characterized by a poor appetite and dullness. Blood sugar is low and should be treated with propylene glycol. While ketosis can be a primary problem, it is more often a secondary problem. It results when a cow is struggling with a health problem early in lactation.

Fat cow syndrome develops when cows become overfat in late lactation or the dry period. This leads to other health problems, such as ketosis. Avoid fat cow syndrome by proper feeding.

MASTITIS

Mastitis is inflammation of the mammary gland. Mastitis costs $200 or more per cow annually. Mastitis is sometimes visible to the human eye. It is

10-27. Mastitis is reduced by removing the inside dewclaws of heifers. (Courtesy, University of Minnesota, Crookston)

often not visible. Laboratory tests are needed to measure somatic cells. A cowside California Mastitis Test (CMT) may be used.

Controlling mastitis requires clean yards, barns, and milking equipment. Spread of mastitis is reduced with good sanitary milking practices. Common practices include using individual towels for each cow, pre and post teat dipping, drying of teats, and elimination of overmilking. Cows with mastitis that do not react to treatment during lactation should be dry treated. If a cow is a problem after reasonable treatment, it may be wise to sell her.

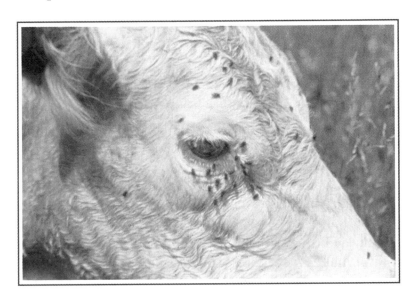

10-28. Flies may be pests on dairy cows and should be controlled. (Courtesy, Mississippi State University)

FACILITIES AND EQUIPMENT

Facilities and equipment help make efficient dairying possible. These also consider the well-being of the cows.

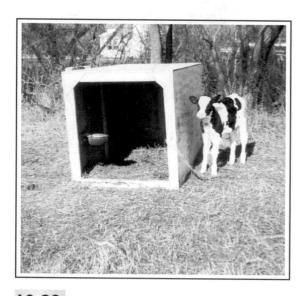

10-29. Calf hutches provide a healthy environment for young calves.

Housing

Housing protects dairy cattle and the humans who work with them. The goal is a system that does not allow conditions to get outside the "acceptable" range. Each dairy operation has a level at which additional expenditures will not increase profits.

Cold housing is an unheated building kept cold during the winter. Natural air movement keeps these buildings cold and removes moisture. *Warm housing* refers to buildings kept warm in the winter. Body heat from the animals provides the heat source while insulation helps retain it in the building.

Cold housing for cows usually takes the form of loose housing. Loose housing allows individual cows freedom, while the herd is handled on a group basis. Loose housing may be an open area with a manure pack or a free-stall system in which cows can enter individual stalls to rest and ruminate. Free stalls often reduce bedding requirements by 75 percent.

Warm housing for cows can also include a free-stall barn with insulation. Northeastern and midwestern United States dairy farms commonly use stall barns. Stall barns usually consist of two rows of cows confined in tie stalls, comfort stalls, or stanchions. Each cow has an area (usually 4 feet by 6 feet for large breeds) where it spends most of its life. Most often, cows are milked in their stalls.

Calves, young stock, and dry cows are commonly housed in a natural, cold environment. Calf hutches for 0- to 6-week-old calves and super hutches or open-sided sheds provide shelter for older calves, young stock, and dry cows. Other sheds are occasionally used, but frequently they lack proper ventilation and temperature control to provide the calf with a healthy environment.

Cattle on open lots in the southern United States or on pasture should have shade and a clean, readily available water supply. Northern pastures may require wind protection or shelter during the early and late grazing season.

MILKING SYSTEMS

Many cows are still milked within their stall in a flat milking barn system. This requires a pipeline to carry the milk from the stalls to the bulk tank.

Milking parlors are more common in the southern and western United States. They are rapidly gaining acceptance in the northern and eastern dairy areas. A ***milking parlor*** is a concrete platform raised above the parlor floor (pit). Cows enter the parlor in groups or individually and stand on the platform for milking. Milking parlors improve labor efficiency, working conditions, and cleanliness. A wide range of milking parlor designs are used.

10-30. Modern milking facilities are clean and efficient. (Courtesy, Mississippi Cooperative Extension Service)

FEEDING SYSTEMS

Nearly twenty tons of feed are fed to a cow in one year. Labor efficiency, ability to individualize to each cow or groups of cows, and cost must be considered in selecting a feeding system. Storing and feeding forages can require different equipment from feeding concentrates or can overlap in the case of total mixed rations (TMR).

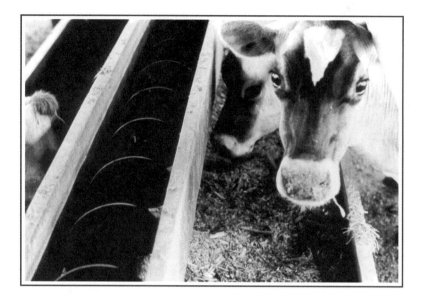

10-31. Automated systems can be used to feed forage. (Courtesy, Mississippi State University)

Storage is needed for silage, haylage, dry hay, and straw. Silages can be stored in traditional upright silos, bunker silos, oxygen-limiting silos, or silage bags. Dry hay and straw can be baled in small square bales, large square bales, or round bales and can be pelleted or put up in loose stacks.

Fresh or ensiled forages are usually fed with mechanized delivery systems or a wide range of silage/green chop wagons. Dry forages are moved mechanically with a tractor or manually in flat stall barns. Mechanical chopping of hay is required in certain feeding systems.

Management intensive grazing is a system that allows the cows to harvest their own forage. Cows are placed in a pasture that can be eaten down to the desirable height in a 24- to 48-hour period. The cows are then moved to another pasture. This concept encourages proper nutrition by forcing the cow to eat both leaves and stems. It also maximizes the number of cows that a pasture can support. Management intensive grazing has the disadvantage of requiring more labor as cattle and waterers must be moved each day.

Concentrates are stored in grain storage bins. A wide range of mechanical feeding equipment is available. Simple push carts, mechanized feeding carts, augers, cable or track feeders, computerized feeders, and lever-operated storage boxes are all available to meet a wide range of housing/feeding arrangements.

Total mixed rations are gaining acceptance as it provides rumen microorganisms with a stable food supply. Chopped hay and silage are weighed and mixed with concentrates in a mixer wagon. Cows can be fed in groups according to production level.

MANURE HANDLING SYSTEMS

Dairy animals produce urine and feces at a rate of about 8 percent of body weight daily. This is about 15 to 20 tons a year for a mature dairy cow. Manure from one cow has nitrogen, phosphorus, and potassium worth over $100.

A well-designed manure disposal/handling system serves three functions: (1) keeps dairy animals clean through frequent removal of manure/waste; (2) provides efficient and economical collection of manure/waste; and (3) disposes of manure/waste in an environmentally sound manner.

Solid manure systems usually require the lowest investment in equipment and facilities. Manure can be hauled on a daily basis or stored and hauled periodically. Daily disposal reduces odor and fly problems. All solid manure systems are labor intensive. Some require daily labor and others require intensive labor for a few days.

Liquid manure handling requires a larger investment in equipment. These systems meet most environmental standards and preserve nutrients. Conventional liquid systems may have above or below ground storage tanks/basins. Water is frequently added to the waste material to allow it to be handled by pumps. A minimum of four to five months storage capacity is needed. Adequate land for manure application must be available to avoid leaching of the nutrients to the underground water supply.

10-32. A dairy farm lagoon being agitated prior to removal of manure. (Courtesy, University of Minnesota, Crookston)

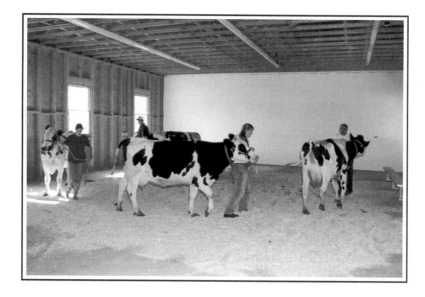

10-33. Students at a New Hampshire school practice showing dairy cows.

An alternate liquid handling system uses lagoons (a type of pond). Lagoon systems can be aerobic (require oxygen), anaerobic (do not need oxygen), or a combination. Aerobic systems are relatively odor-free, but require larger surface area. Anaerobic systems can handle more waste and require less land, but can cause a bad odor problem.

Waste disposal is a concern. Manure should be viewed as an asset and considered as such when designing a dairy operation.

REVIEWING

*M*AIN IDEAS

Dairying is concerned with providing wholesome milk. Producers follow many practices to keep costs down and earn a profit.

Milk production is associated with the reproductive cycle. Cows are milked about 305 days following parturition. They are allowed to be dry 50 to 60 days before the next calving. Cows normally have a calf each year and complete the reproductive cycle.

Dairy cows are selected based on their milking potential. Some differences exist among the breeds. The Holstein is the most popular breed and gives the largest amount of milk. The Jersey gives the milk with the highest butterfat and protein content.

Selecting and culling dairy cows may be based on Dairy Herd Improvement records, linear evaluation, and progeny information. Cows that do not produce or are diseased should be sold for slaughter.

Providing a good environment improves efficiency. The environment should meet the needs of the cows and make high production possible. A ration with high-quality feed is needed. An abundance of good water is essential. Preventing and controlling disease requires continual observation of the herd. Reproductive management and a herd health program are needed. Feed is the greatest cost in dairy production.

QUESTIONS

Answer the following questions using correct spelling and complete sentences.

1. How has dairying changed since dairy cows were first brought to the United States?

2. What are the important breeds of dairy cattle? Distinguish between the breeds.

3. What should be considered before starting a dairy farm?

4. How are cows selected and culled?

5. How does linear evaluation differ from judging?

6. Describe the ration for a lactating dairy cow.

7. How would you react to the statement that "stress is bad for animals"?

8. What are four climatological factors that affect the performance of dairy animals?

9. Identify three types of dairy cattle housing. How does each modify the weather?

10. What is a milking parlor?

11. What is a total mixed ration (TMR)? What are its advantages?

12. What are the functions of a manure disposal/handling system?

13. What is the overall objective of a health program?

14. Why is dairy cattle reproduction important?

EVALUATING

Match the term with the correct definition. Write the letter by the term in the blank provided.

a. immunioglobulin
b. alveoli
c. dry cow
d. homogenization

e. pasteurization
f. progeny
g. animal model
h. Dairy Herd Improvement Program

_____ 1. A cow that is not lactating.

_____ 2. The genetic evaluation for dairy cattle production.

_____ 3. Antibodies in colostrum that develop passive immunity in a calf.

_____ 4. Breaking the fat globules in milk into smaller pieces so they stay suspended in milk.

_____ 5. Offspring.

_____ 6. Heating milk to destroy microorganisms.

_____ 7. Industry-wide dairy production testing and record-keeping program.

_____ 8. Structures in the mammary gland that convert blood nutrients to milk.

EXPLORING

1. Tour a dairy farm. Discuss with the dairy manager how the environment is modified for various age dairy animals.

2. Ask a breed association classifier or an AI organization cow evaluator to discuss the linear traits and how they relate to functional type. Refer to the Activity Manual for a linear mating problem. Pay close attention to the goals of the dairy family as you complete this problem.

3. Collect several forage samples and send them to an infrared laboratory for analysis. Discuss the analysis with your instructor.

4. Contact a local veterinarian about preventive health visits. How do these visits differ from emergency calls?

11

POULTRY PRODUCTION

Thanksgiving without turkey on the menu would hardly be Thanksgiving! Roast turkey for Thanksgiving dinner is an American tradition. In many homes, the special day would not be complete without turkey. And there are many other poultry products!

Poultry consumption has been zooming upward! People are eating more poultry than ever. Of course, chicken is the most popular poultry. The average person eats more than 70 pounds of chicken a year. Poultry and poultry products are economical and healthy food sources.

Poultry provide many products. Eggs are important in many dishes. Feathers are used in making pillows, clothing, and fishing lures. Eggs are used in laboratories to make human medicines, including vaccines. Some poultry are kept for ornamental uses and as a hobby. Much of the increased use of poultry is due to changes in how they are raised and marketed.

11-1. New approaches in restaurants have promoted chicken consumption.

OBJECTIVES

This chapter provides an overview of poultry production. It has the following objectives:

1 List and describe the major kinds of poultry

2 Explain the poultry industry

3 Describe poultry as organisms

4 Explain production systems for raising poultry

5 Describe sanitation and disease control programs

6 List facility and equipment needs for poultry production

TERMS

albumen
broiler
candling
capon
cock
cockerel
debeaking
down
drake
duckling
egg injection
gaggle
gander
gizzard
gosling

incinerator
layer
litter
molting
peacock
peahen
poult
poultry science
pullet
ratite
roaster
spent hen
tom
vertical integration
yolk

KINDS OF POULTRY

Poultry are domesticated birds raised primarily for meat, eggs, and feathers. Poultry products are used primarily by humans. Some products, often byproducts, are used as pet food and for other purposes. Eggs are used in making vaccines for humans and animals. Poultry are also called birds or fowl. Chickens, turkeys, ducks, geese, quail, peafowls, swans, ratites, pigeons, and several other species are classified as poultry. A few game birds, such as pheasants and quail, are sometimes included.

11-2. House filled with broilers that are near market size.

CHICKENS

Chickens are the most important poultry species. They are raised for meat and eggs. The type raised depends on the product wanted. A few other uses include game and fancy show chickens. The latter are used mainly for hobbies. Groups of gamecocks may be seen tethered to small individual houses on a range. They account for only a tiny part of the poultry industry.

The meat from a chicken is based on its age and sex. Most chicken is from broilers. A ***broiler*** is a young chicken six to seven weeks of age that weighs about 4 pounds (2.8 kg). They are tender and easy to cook. Chickens of either sex are used as broilers, which are sometimes known as fryers. About seven billion broilers are raised each year in the United States.

11-3. Cocks tethered to A-frame houses represent recreational poultry production.

Other chickens used for meat include roasters, capons, and spent hens. A *roaster* is a young chicken that is older and slightly larger than a broiler. A *capon* is a male chicken that has been neutered (castrated). Most capons are five to seven months of age and weigh about six pounds. A *spent hen* is a hen that is no longer laying. Spent hens go into processed foods, such as soup, or are baked. Older male chickens have tough flesh and are not used much in cooking.

Eggs are produced by mature female chickens, known as hens. A *layer* is a mature female chicken kept to produce eggs. Most hens in laying flocks pro-

11-4. White Leghorn cock and hens. The White Leghorn is primarily for egg production.

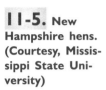

11-5. New Hampshire hens. (Courtesy, Mississippi State University)

duce almost 300 eggs in the year after they mature. A *pullet* is a young female chicken that is being raised for laying. Often, layers are force-molted so they can be recycled to lay for another year.

A mature male chicken is a rooster or *cock*. Young male chickens are known as cockerels. A *cockerel* is less than one year of age.

Breeds of chickens are not important in commercial poultry production. Varieties of chickens have been developed by poultry integrators to meet particular needs.

Examples of breeds of chickens are White Leghorn (used for egg production, white egg shell); Barred Plymouth Rocks (meat and eggs, brown egg shell); New Hampshire (meat and eggs, brown egg shell), and the White Rock (meat and eggs, brown egg shell). The White Leghorn is the smallest, with mature cocks weighing 6 pounds (2.7 kg) and hens weighing 4 pounds (2 kg). The White Rock is among the largest, with cocks weighing 9 pounds and hens weighing 7 pounds (3.4 kg).

TURKEYS

Turkeys are raised primarily for meat. Consumers want birds that have a high proportion of white breast meat. Nearly 300 million turkeys are raised each year in the United States. Turkeys are being raised on fewer farms but the size of the farms has increased. Many commercial producers raise more than 100,000 birds each year on a single farm.

11-6. White turkeys being raised on a range. (Courtesy, Agricultural Research Service, USDA)

A young turkey is a ***poult***. It has not grown to the point where its sex is easily determined. A mature male turkey is a ***tom*** or gobbler. A female is a hen. Younger and smaller turkeys are roasted whole. Larger turkeys are often made into boneless breasts and other further processed products.

Most turkeys are raised in confinement under contract with a processing plant. Many turkeys are marketed after 20 weeks of age. Toms weigh considerably more than hens, with the market for further processing requiring heavier and heavier turkeys. Good producers can have feed conversion ratios of 2.2 pounds (1 kg) of feed to 1.1 pounds (0.5 kg) of gain. Turkeys have been bred to be larger and have broad breasts. Colors vary from white to bronze to mottled shades of black. The broad-breasted white turkey is about 100 percent of those in commercial production.

Ducks

Ducks are raised for meat, eggs, and down and feathers. (***Down*** is the soft feathery covering that grows under the feathers.) Some are kept as hobby or ornamental ducks.

Young ducks are covered with down. Feathers develop as they grow. A young duck that still has down rather than feathers is a ***duckling***. A mature male duck is a ***drake***. Mature female ducks are known as hens.

Duck and duck products are not nearly as widely used as chickens. Some 15 million ducks are raised in the United States each year. White Pekin

11-7. Pekin ducks. (Courtesy, Mississippi State University)

ducks are most widely found. Ducks grow faster and heavier than chickens. They can also swim, which is not the case with chickens and turkeys. Most commercial ducks are now raised indoors.

GEESE

About one million geese are raised in the United States each year. Geese are raised for meat, eggs, and feathers and down. Many of these are kept for ornamental purposes. Geese like to eat tender grass in pastures and parks. Some are kept to control weeds and grass, and are known as "weeder geese." They are hardy birds and resist many poultry diseases. A baby goose of either sex is a *gosling*. A mature male goose is a *gander*. A mature female is a hen or goose. A flock or group of geese that are not flying is known as a *gaggle*. Wild Canada geese have been encouraged around lakes and resorts. In some places, they have become pests.

11-8. A Canada goose.

PEAFOWL

Peafowl are raised for large, beautiful feathers. A male is known as a **peacock** and a female as a **peahen**. Sometimes, either sex is known as a peacock. Male peafowls may spread their feathers, known as a train. The feathers may be five times the length of the body. Most peafowls are raised for ornamental purposes. The hens lay a few eggs each year.

SWANS

Swans are similar to ducks and geese in their preference for water. Several kinds of swans are found, with colors ranging from white to black. Most

11-9. A peacock.

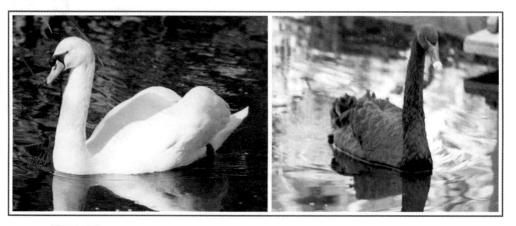

11-10. A white swan (left) and an ornamental black Australian swan (right).

Connection

QUICK CHICKEN

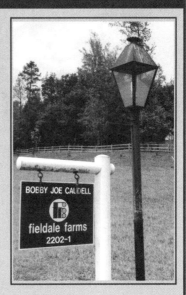

"Quick chicken" is the way it is done by producer Bobby Joe Caudell. He does it the Fieldale Farms way. Here are the events in quickly raising top-quality broilers:

The four modern, computer monitored and controlled houses:

- hold a total of 80,000 genetically improved birds
- grow day-old chicks to market-size (5 lb.) broilers in 49 days
- have 97$^+$ percent of the chicks live to market size
- have nearly 80,000 birds weigh 400,000 lbs. at market
- have birds eat 800,000 lbs. of feed (2:1 feed conversion ratio)
- use chicks that are *in vivo* vaccinated on day 18 of incubation
- re-vaccinate chicks again at 8 and 21 days of age in drinking water
- have lighting for 24 hours a day the first week; cut back to 6 hours a day the second week; and increase it 4 hours a day each week thereafter until 22 hours a day
- produce six batches a year

Amazing! Every production day counts. Similar schedules are repeated by poultry producers across the nation. Few do it better than Caudell!

swans are kept for ornamental purposes. They are frequently used with pools of water at resorts and similar locations.

RATITES

No other birds have created more excitement in recent years than a group known as ratites. A *ratite* is a group of flightless birds that have some commercial uses. Ratites include the ostrich, emu, rhea, cassowary, and kiwi.

The largest and best-known ratite is the ostrich. An ostrich can weigh as much as 350 pounds and stand 10 feet tall. Ostriches have the longest life span of most birds, living up to 70 years. These birds are raised for feathers, meat, skin, and oil products. The feathers are known as plumes and are used in decorations.

11-11. The ostrich is the best-known ratite.

Interest in ratites was high in the late 1980s but declined by 2000. Many growers paid high prices for breeding animals but were unable to sell offspring at a profit. The novelty of ratites has worn off.

GUINEA FOWL

Guineas are raised for food, as novelty birds, and to stock game preserves. They are hardy and often roam free. Their eggs are smaller than chicken eggs and have a thicker shell. Because the eggs are not as easily broken, they are often dyed and decorated in various ways.

11-12. A guinea fowl.

THE POULTRY INDUSTRY

The poultry industry has changed from a family farm with a few birds for home use to large commercial producers that raise hundreds of thousands of broilers each year.

*E*ARLY POULTRY PRODUCTION

The first poultry was raised more than 5,000 years ago in Asia. This was followed by Egypt about 3,500 years ago. Poultry were brought to North America by the early settlers. A few people in Jamestown had small flocks of chickens as early as 1607.

In colonial times, many families kept small flocks that ran about and often nested in brush and thickets. They ate whatever they could find – insects, seed, and tender leaves. Fox and other animals often preyed on them. Most poultry were kept for eggs, with young males and older hens used for meat. People often raised only what they needed, with extra eggs or chickens traded to neighbors for a ham or vegetables.

The turkey is native to the Americas. The Aztecs domesticated the wild turkey. Native Americans used the turkey for food and its feathers for decoration. Turkeys were exported to Europe before chickens were brought to Jamestown!

Dramatic changes began to occur in poultry production in the mid-1900s. These changes led to the modern poultry industry.

*M*ODERN POULTRY INDUSTRY

Poultry production is a large commercial industry. Thousands of birds may be raised in confinement in one house. Science and technology are used to assure their well-being.

The number of small flocks has declined considerably since 1950. Most people could not raise poultry, if they wanted to, because of where they live. Urban life is not well suited to having a small flock of poultry. People like to go to the supermarket and buy poultry products that have been carefully inspected and are ready to cook.

Providing an abundance of poultry products at low costs required major changes in how poultry were raised. Good information was needed. A system to market products had to be developed for consumers to have a year-round

11-13. Modern poultry houses have controlled environments and automated feeding systems. (Courtesy, McCarty Farms)

and uniform supply. Much has happened in the poultry industry since Harland Sanders opened the first KFC restaurant in 1956!

The leading states in broiler production are Georgia, Arkansas, Alabama, North Carolina, and Mississippi. California is the largest producer of eggs. North Carolina, Minnesota, and California are leading states in turkey production.

Poultry Science

Poultry science is the study and use of areas of science in raising poultry. It includes breeding, incubation, rearing, housing, feeding, sanitation, marketing, and other areas. The goal of poultry science is to provide consumers wholesome poultry products at a reasonable price.

Poultry science has resulted in improved production. A hen now lays almost 300 eggs a year (compared to 100 per year in 1900). Automation allows one worker to care for many birds. A single broiler house may have 40,000 birds. One worker can care for more than 100,000 broilers or 40,000 laying hens in cages. Most of the information used in poultry science has been developed in the last 50 years. It is based on biology and related areas of science, including environmental science.

Vertical Integration

Raising, processing, and distributing poultry has developed into one continuous chain. *Vertical integration* is one agribusiness being involved in more than one step in providing poultry products. The steps are linked together to assure a continual supply of poultry. The poultry industry has more vertical integration than other areas of agriculture. All areas of poultry are vertically integrated.

The best example of vertical integration is in broiler production. One company vertically integrates by carrying out all of the steps in the process of poultry production. A poultry company may sign agreements with growers, provide the chicks and feed, supervise the grower, process the broilers, and distribute the chicken to buyers. The grower (farmer) provides a poultry house and looks after the chicks. The company may pay the grower a fee for each bird raised. The amount is based on how well the birds grow or other conditions.

11-14. Vertical integrators process the birds. (Courtesy, Mississippi State University)

POULTRY BIOLOGY

Poultry are terrestrial animals with feathers, a backbone, and wings that have limited use in flying. Modern poultry do not need to fly, nor do they have the ability to do so. Some species can swim, such as ducks and swans. Chickens and turkeys cannot swim. Poultry are classified in the Aves class of vertebrata of animals. Each species has a scientific name. The scientific names of common poultry species are shown in Table 11-1.

LIFE PROCESSES

Poultry carry out the same life processes as other animals. The structures for doing so vary. Two major areas of difference are food digestion and reproduction.

Table 11-1. Scientific Names of Common Poultry Species			
Kind of Bird	**Scientific Name**	**Kind of Bird**	**Scientific Name**
Chicken	*Gallus domesticus*	Ostrich	*Struthio camelus*
Duck	*Anas domestica*	Pigeon	*Columba domestica*
Goose	*Anser domesticas*	Swan (black)	*Cygnus atratus*
Guinea fowl	*Numida meleagris*	Turkey	*Meleagris gallopavo*

Digestion

Birds take in food with their beak. Beaks are strong mouth parts made of material similar to a person's fingernails. It can be used for pecking and breaking foods apart. In large poultry houses, birds may be debeaked to reduce damage they can inflict on each other. *Debeaking* is removing the tip of the beak so they cannot peck other birds.

Poultry do not have teeth and rely on a *gizzard* to grind food. The gizzard is a strong, muscular organ that may contain grit that the bird has eaten. Grit (e.g., grains of sand) helps in the food grinding process.

Reproduction

To reproduce, a female bird must mate with a male for eggs to be fertile. After laying, eggs are incubated. The length of time for the embryo to develop varies by species: chickens need 21 days; geese, 29 to 31 days; turkeys, 27 to 28 days; and ducks, depending on their species, need 28 to 35 days. The incubation period is also influenced by the temperature and humidity conditions. Eggs go through distinct changes in incubation.

Most eggs sold in supermarkets are infertile. Hens do not need to mate to produce eggs. There is usually no difference to the consumer between fertile and infertile eggs, if they are fresh. A fertile egg that was not refrigerated quickly and stored in a warm environment may show signs of embryo development.

Artificial insemination is used in the poultry industry. Some birds, particularly turkeys, are unable to mate because their mass of flesh presents a mechanical barrier. Natural fertilization is still predominant with chickens.

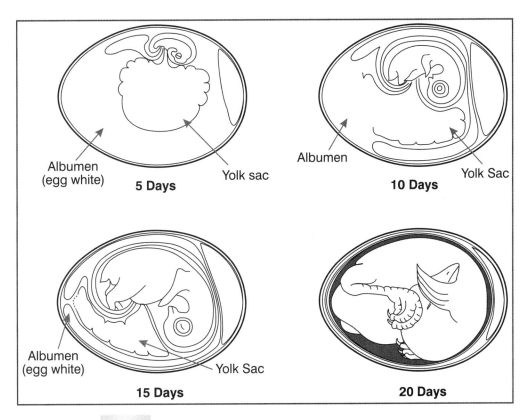

Albumen
(egg white) **5 Days** Yolk sac

Albumen **10 Days** Yolk Sac

Albumen
(egg white) Yolk Sac
15 Days

20 Days

11-15. Development of a chick embryo in an incubating egg.

APPEARANCE

The external parts of most birds are similar. Some, such as the game chicken, have long spurs near their feet. Chapter 2 provides additional information on biology.

The health and condition of a bird are evident in its exterior appearance. For example, some breeds have yellow shanks; others have white shanks. The disappearance of the yellow pigment in those with yellow shanks is a sign that the female has been laying eggs. A large, red comb is a sign that the bird is in good health. Ragged feathers may indicate poor condition.

It is fairly easy to distinguish between most types of poultry. Some, such as ducks and geese, live in and around water. Chickens and turkeys cannot swim. Turkeys have beards and other fleshing structures from their heads. Some chickens have serrated combs and wattles. Geese and ducks have rather plain heads. Looking at the heads of poultry is a good way to tell them

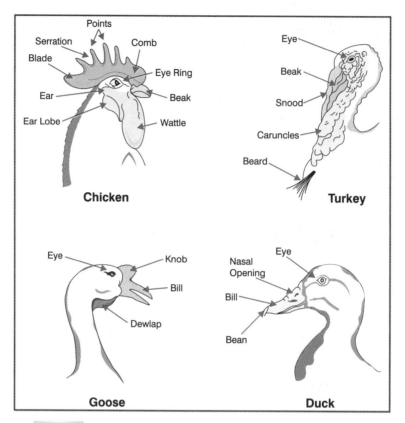

11-16. Head and neck features help identify common poultry.

apart. The heads also vary with the sex of the poultry. Males tend to have larger head features, such as the comb.

POULTRY PRODUCTION SYSTEMS

Chickens are the dominant form of poultry in the United States. Most chickens are produced on large poultry farms that specialize in birds for a particular use. With vertical integration, producers usually have little investment in the chickens, and large investments in the land for the farm, housing and equipment, electricity and water, and labor. The poultry company provides the chicks, feed, medications, and field services in an effort to assure that the birds receive proper care.

The chicken industry has four areas: broiler production, egg production, pullets for egg production, and pullets and cockerels for broiler egg production.

BROILER PRODUCTION

The goal of broiler production is to produce the most meat as quickly and efficiently as possible. Growers take extra care to keep chicks healthy and growing. Mortality (death) is kept to a minimum—no more than 5 percent while the broilers are growing.

Most broilers are raised in large poultry houses. Six to seven weeks are needed to raise a broiler to market size, or about 4 to 4.5 pounds (2 kg). The time required is declining as genetic stocks and diets are continually improved. Housing is designed to meet the needs of young chickens.

Growers attempt to provide an environment that is ideal for broiler production. A day or so in the time required to raise a broiler makes a big difference in feed cost. The goal is to get broilers to grow as fast and efficiently as possible. Most growers get a pound of growth on slightly less than two pounds of feed. Young chicks are fed a finely ground meal feed that has 18 to 20 percent protein. As they get older, the percentage of protein is reduced to 12 to 15 percent. Protein in feed is typically from soybean meal and fish and other meat-meal ingredients. The feed also contains ground yellow corn, other grains, minerals, vitamins, and, in some cases, medication.

A small niche market exists for broilers raised on a range. They are outside and allowed to run about inside fenced areas. The efficiency of feed conversion and control over product quality is not as good as that in a modern house.

11-17. Five-day-old chicks are learning to eat from a pan.

Housing

Houses for broilers are designed to meet the needs of newly-hatched chicks through market-size birds. A day-old chick needs between one-quarter to one-third of a square foot (0.02 to 0.03 m²) of floor space. Space requirements expand as the chickens grow. A four-week-old chicken needs about three-quarters of a square foot (0.06 to 0.09 m²) of floor space. Older, larger birds would need more space. Growers often use a portable fence to isolate baby chicks to a small part of a house. The fence is removed as they grow.

Poultry houses have a controlled environment. The controls provide for the well-being of the chickens. Environmental control characteristics:

- Litter—*Litter* is the floor covering in a poultry house. Litter is often made of wood shavings. The litter absorbs moisture from the manure. Litter is normally changed between batches of chickens.

11-18. The floor of this house is covered with clean wood shavings used as litter.

- Lighting—Modern poultry houses rely almost exclusively on electric lights. No natural light may be allowed in some houses. The lighting is timed to give chickens the light needed to eat and grow. Shorter days are made longer by artificial lighting. Most poultry houses use a specially designed bulb that provides adequate light on a low amount of electricity.

- Temperature—Baby chickens are especially sensitive to temperature changes. Many growers keep house temperatures in the 85 to 95°F (29 to 35°C) range for small chicks. Large hood-type brooders with electric or gas heat are used. By six weeks of age, a temperature of 70°F (21°C) is adequate. It is lowered as

11-19. Economical lighting is needed to promote efficient growth.

11-20. A thermostat controls the temperature of this gas-powered brooder. The large, round, reflective hood assures chicks of adequate heat.

the chickens grow. In warm weather, cooler outside air may be pulled into the house for ventilation.

■ Humidity—Humidity control is used to provide the ideal level of water vapor in the air of chicken houses. Heat tends to dry the vapor from the air. Water vapor may be added to the air with mist systems. The humidity level should be 50 to 75 percent in broiler houses.

11-21. Humidity and ventilation are regulated by large fans that automatically turn on when needed.

11-22. Sensors above the level of the chickens in a poultry house help control the environment.

■ Ventilation—Ventilation involves using large fans to bring fresh air into a chicken house. The fans are often on controls to help regulate temperature. Ventilation also removes excess moisture from a house.

■ Alarm systems—Modern poultry houses should have systems to alert people to problems. Failures of feeders, waterers, ventilation systems, and other areas are monitored.

■ Standby electrical generators—Poultry farms need standby electrical generators in case of power failure. Chickens cannot live long without feed, water, and environmental control.

11-23. An emergency generator is on standby if the electricity fails.

Feeding and Watering

Automated feeding and watering equipment are used. Young chicks may require feed in trays, or pans, and waterers that are easy to use; however, they soon learn to eat from automated feeders. After a few days, most producers rely only on automated systems.

Feeder systems may consist of tubes or troughs located throughout the house. Feed bins and weighing equipment are located outside at the middle of a long house. A conveyor chain moves the feed throughout the house. A large flock will eat a lot of feed. Plan ahead to be sure there will be plenty. Adequate feed storage must be provided. Feed is usually stored in bins designed

11-24. Automated feeding systems carry feed throughout a house.

11-25. Feed is stored, weighed, and metered into the feeding system by these automated bins.

11-26. Two types of watering systems are shown here: watering nipple (left) and watering trough (right).

to keep the feed from damage and contamination. Bins should be easy to use with access for delivery by trucks.

Watering systems may involve troughs and nipples attached to tubes. The water is carefully metered. The amount consumed is an indication of the health of the flock. Water consumption drops when chickens are sick or in less than optimum environment. Medications may be added to water. Waterers need to be cleaned each day.

Most producers use water from wells or municipal systems. Many have standby sources in case one source fails.

EGG PRODUCTION

The goal of egg production is to produce high-quality eggs for human use. Hens are given an ideal environment to produce. Producers strive to get one egg each day from hens. The eggs used for food are not fertile. They will not hatch if incubated.

Eggs are graded and placed in cartons designed to protect them from damage. Eggs are graded by size, such as small, medium, and large. The eggs are put in a carton with the small end downward. The larger end should be up to help maintain the quality of the egg. The large end has the egg's air cell. Dirty, misshaped, or cracked eggs are removed.

Managing hens for egg production requires facilities where their well-being is met and the eggs are of high quality. The facilities must keep the eggs clean and prevent breakage. Regularly collecting the eggs or using automated egg collecting systems is needed. Most layers are kept in specially designed cages. Other layers are in houses with slatted or litter floors. A few layers are on a range, and the eggs they produce are for niche markets. Nests must be

11-27. Cage layers. (Courtesy, Mississippi State University)

provided for range hens to use in laying eggs. Range hens will lay in tall grass, leaves, or other places.

Feed and Water

Laying hens need proper feed. The formulation of a complete commercial feed should vary according to the hen's age, stock, and time of laying. Diets are supplemented with calcium because egg shells contain substantial calcium. A sound supply of water is critical because laying hens will drink 2 to 3 pounds (1 to 1.5 kg) of water for each pound of feed.

An average chicken egg weighs 2.2 ounces (0.062 kg). It is made of 66 percent water, 10 percent fat, 13 percent protein, and 11 percent other material, primarily calcium in the shell. The fat is mostly in the egg yolk. Hens that do not get feed with the needed nutrients cannot be productive.

Egg Quality and Color

Egg quality is determined by external appearance and interior condition. External appearance includes size, shape, color, cracks, and blemishes or dirt.

Egg color includes both the shell and internal parts of the egg. The shell is a protective layer of material, primarily calcium. Egg shell color varies. Nearly 95 percent of the market eggs in North America are white. In recent years, markets for brown eggs have increased in areas other than the New England states, which have long had a great demand for brown eggs.

The inside of the egg has two major parts: albumen and yolk. The *albumen* is the white part that surrounds the yolk. It should be firm and cling near the yolk when broken out. Albumen that spreads out indicates deterio-

ALBUMEN

Outer Thin
Firm
Inner Thin
Chalaziferous
Chalazae

SHELL

Cuticle
Spongy (Calcareous) Layer
Mammillary Layer

YOLK

Germinal Disc
Latebra
Light Yolk Layer
Dark Yolk Layer
Yolk Membrane

MEMBRANE

Air Cell
Outer Shell Membrane
Inner Shell Membrane

11-28. Major parts of a chicken egg.

ration and aging. The *yolk* is the center part of an egg that varies in degree of yellow color. Hens fed diets high in carotene lay eggs with yellower yolks.

Inside structure indicates egg quality. The presence of meat or blood spots is objectionable. An air cell is located inside at the large end of an egg. As an egg ages, the air cell gets bigger. Fresh eggs have small air cells. *Candling* is shining light through an egg to see the inside features. Most of an egg is fairly transparent. Dark spots are signs of problems.

Cracked shells lower the quality of eggs. In most cases, cracked eggs are unacceptable to consumers. Breaking plants use cracked eggs. The eggs are frozen and used in food products.

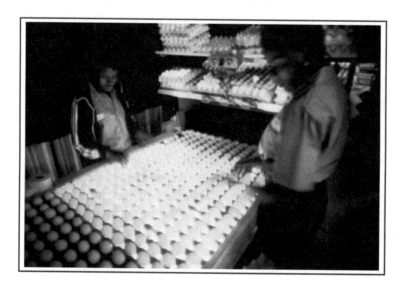

11-29. Eggs being candled. (Courtesy, Mississippi State University)

11-30. Eggs are packaged in cartons for easy use by consumers.

Molting

Molting is the process of shedding and renewing feathers. Hens usually stop laying during molting. If not, the laying rate is greatly reduced. Some growers use forced molting or recycling. This gives the hen a rest before returning to production. Forced molting is becoming popular.

A hen may deplete the calcium in her bones during egg production. The rest allows her to restore the calcium and begin high production again.

In the wild, birds molt in the fall so they have a good protective layer of feathers for the winter. Molting is associated with age. Chickens often molt at about one year of age. Molting takes about four months.

Molting can be induced by various procedures, with most involving a decrease in the length of the day. Since most poultry houses have artificial lighting, the light timer is set to make the dark time longer. Hens will molt with eight hours of light. Molting stops and egg production begins in about two months when the light is increased to 14 to 16 hours per day.

Pullets

Hens for egg production are raised by pullet producers. The grower takes specially bred, day-old chicks and raises them similar to broiler chicks. The chicks are sexed, so only females are raised. The sexing is done prior to delivery to the grower. The male chicks are raised as broilers for meat. The birds are bred for maximum egg production. The pullets are known as replacement hens.

Pullets are raised to 20 weeks of age by the pullet producer. They then go to laying farms as starter pullets. Pullets will begin egg production a few weeks after delivery. Pullets start laying regularly at 24 weeks of age and reach maximum production at 30 to 35 weeks.

BREEDER BIRDS

Some growers specialize in raising pullets and cockerels for fertile egg production. The fertile eggs are hatched to become broilers. Chickens are carefully selected for their potential in producing broilers with large amounts of meat. Many producers raise pullets and cockerels for breeding under a contract with a poultry company.

The pullets and cockerels are raised separately. A typical ratio is to raise one cockerel for each eight to ten pullets. At maturity, the cockerels and pullets are placed in a combined flock so mating can occur and the hens will lay fertile eggs.

The initial practices with breeder chicks are different from those with broilers in many ways. Growing conditions are used to prepare pullets for high egg production, and feed intake of both males and females is carefully controlled. In the breeder house, separate feeders for males and females are common, in order to control body weight of each sex.

SANITATION AND DISEASE CONTROL

Sanitation and disease control are important in successful poultry production. A disease outbreak can quickly wipe out a flock. It is far better to prevent disease than it is to try to treat an outbreak after it occurs.

SANITATION

Practicing good sanitation is important in preventing the spread of disease. Many diseases are spread by diseased birds, insects, people and vehicles that travel from one farm to another. Carcasses of dead birds, contaminated feed, water, and litter also spread disease. Keeping a poultry farm clean is a major part of sanitation.

Removing all litter from a poultry house and spraying the building and equipment with a disinfectant helps prevent disease. This is done after a

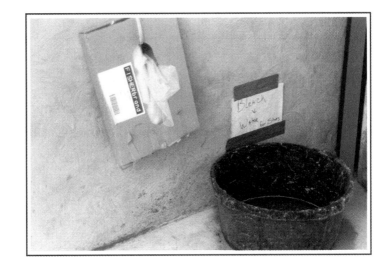

11-31. Disposable boots and shoe dip are both used at the entrance to this poultry facility.

batch of chickens has been moved to processing or other farms, and prior to bringing a new batch of chickens to a farm.

Many poultry farms restrict the access of vehicles and people. Tires, shoes, clothing, and equipment can transport a disease from one farm to another. Tires of a vehicle must be washed with a disinfectant solution before going onto a poultry farm. People dip their shoes in disinfectant solution or wear disposable boots.

VACCINATION

Vaccination is used to help poultry develop immunity to disease. The common diseases vaccinated against include Newcastle, Marek's disease, and infectious bronchitis.

Recent advances have allowed for vaccination *in vivo*. This is into the embryo of the incubating egg by **egg injection**. Egg injection is used on the eighteenth day of incubation to help the chick be immune to certain diseases at hatching. An egg injection system is used in the hatchery at a high rate of speed. A specially designed needle penetrates the egg shell with a minimum of damage to the shell or embryo. The dosage is delivered inside the egg. One machine and two operators can inject 20 to 30 thousand eggs an hour. Vaccinating during incubation results in the chicks hatching with immunity. Handling and stress are reduced. The chicks are said to be ready for market about two days earlier, if vaccinated before hatching.

If chicks are not vaccinated *in vivo*, they are vaccinated in the debeaking process. The vaccine is placed in the nose, eye, or wing. Medications are also placed in drinking water.

11-32. Day-old chicks being vaccinated and debeaked.

11-33. An automated in vivo system to vaccinate embryos on day 18 of incubation. (Courtesy, Embrex, Inc.)

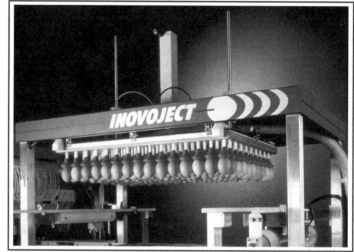

DEAD BIRD DISPOSAL

Any birds that die should be promptly disposed of to prevent the spread of disease to others in the flock. Dead birds are usually handled in one of three different ways: incineration, burying, and composting.

An *incinerator* is equipment that burns chicken carcasses so only ashes remain. Incinerators are made of steel and heated with propane, natural gas, or fuel oil.

Composting is a low-cost, self-contained method of layering the carcasses with straw or other litter, and leaving them to decompose naturally. A special compost bin is used. The nutrient-rich remains, or compost, can be used as fertilizer on crops.

DISEASES AND PARASITES

Poultry are affected by some 33 pathogenic diseases and 10 parasites. In addition, nutritional diseases and deficiencies affect poultry. These can cause huge losses. In some cases, the birds die. In others, the birds are less efficient. Either way, the owner has a loss of money from the disease. Further, who wants to eat a diseased chicken?

Diseases

Pathogenic diseases are caused by bacteria, viruses, or other living agents. Some examples follow.

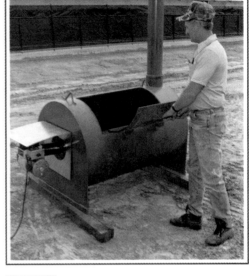

11-34. An approved incinerator for a poultry farm.

- Marek's disease—Marek's disease is also known as range paralysis. It is caused by a herpes virus. Symptoms include diarrhea, paralysis of legs or wings, loss of weight, and death. No treatment is available. Prevent Marek's disease by vaccination of day-old chicks or embryo vaccination. Some genetic resistance has been developed.

- Newcastle disease—Newcastle disease is caused by a virus. Affected birds gasp, wheeze, twist necks around, may be paralyzed, and lay soft-shell eggs, if any are laid. No treatment is available. Vaccination prevents Newcastle disease.

- Infectious bronchitis—Infectious bronchitis only affects chickens. Young birds wheeze and gasp and have nasal discharge. Older birds stop laying. Sanitation and isolation arc cffcctivc in prevention. A vaccine is available. Infectious bronchitis is caused by a virus.

- Fowl cholera—Fowl cholera affects many different birds. Caused by bacteria, birds get a fever, purple-colored heads, yellowish droppings, and sudden death may result. Some growers have success in treating fowl cholera with antibiotics and sulfonamides. Vaccination can be used to prevent fowl cholera.

Parasites

Sanitation and control of pests, such as flies and wild birds, help in preventing parasites in poultry. All poultry may be affected. Some examples of parasites:

11-35. Incubation equipment is kept clean to prevent disease in newly-hatched chicks.

■ Coccidiosis—Coccidiosis is an internal protozoan parasite disease. It is transmitted by the droppings of affected birds, including wild birds. The droppings may be bloody, and affected birds may be sleepy, pale, and listless. Anticoccidials may be put in feed or water.

■ Large roundworms—Birds with large roundworms are droopy, emaciated, and may have diarrhea. The worms may reach lengths of 3 inches (7.62 cm) in the intestines. Sanitation and rotation of range and yards help control large roundworms. Dewormers can be used to treat affected poultry.

■ Mites—Mites are external parasites that cause birds to be droopy, pale, and listless. Approved miticides can be used to treat mites. Mites are prevented by sanitation and controlling access of wild birds to the flock.

■ Tapeworms—Tapeworms are internal parasites that use snails, earthworms, beetles, and flies as intermediate hosts. Pale head and legs and poor body flesh indicate tapeworm infestation. Dewormers are available.

11-36. This poultry grower lost it all when a thunderstorm destroyed the house filled with market-ready broilers. Most of the 25,000 broilers escaped, died of exposure, or were captured by predators, such as fox.

REVIEWING

MAIN IDEAS

Poultry are domesticated birds raised for meat, eggs, and other products. Chicken is by far the most important poultry. Turkey is second in importance.

Broilers are the major source of chicken meat. A broiler is a young chicken 6 to 8 weeks old. Most commercial broiler producers can grow broilers to the weight of 4 pounds or more in six to seven weeks. Egg production, pullet production, and breeder pullet and cockerel production are also specialized areas of the chicken industry.

Vertical integration is used to link the poultry industry. Most producers specialize in one kind of bird, such as broilers or layers. They get the baby chicks or pullets from specialized producers.

Housing and equipment are used to provide a good environment for chickens. The well-being of birds is a major part of the environment. Birds are managed to efficiently grow and produce.

Sanitation is an important part of reducing disease problems. A method for properly disposing of dead birds is essential on every poultry farm. Many farms use incinerators.

Chicks are vaccinated for disease. *In vivo* vaccination is now widely used. It involves egg injection of the embryo on the eighteenth day of incubation. This reduces the handling stress newly-hatched chicks often experience.

QUESTIONS

Answer the following questions using correct spelling and complete sentences.

1. What are the major kinds of poultry? Briefly describe the two most important.
2. Explain how the poultry industry has changed chicken production.
3. How does the digestive system of poultry vary from other animals?
4. What is the reproductive process in poultry?
5. What are the distinguishing features of the heads of chickens, ducks, turkeys, and geese?
6. What are the four production systems with chickens? How do these relate to vertical integration?
7. What are the housing and equipment needs in raising broilers?
8. How does the feed vary from young chicks to older broilers?
9. How are layers managed for production?
10. What are the major parts of an egg?

11. What is molting? Forced molting?

12. Why is sanitation important in controlling disease?

EVALUATING

Match the term with the correct definition. Write the letter by the term in the blank provided.

a. egg injection   e. debeaking   i. layer
b. incinerator   f. gosling   j. capon
c. candling   g. tom
d. litter   h. cockerel

_____ 1. A mature female chicken kept for egg production.

_____ 2. A castrated male chicken.

_____ 3. A mature male turkey.

_____ 4. A young male chicken.

_____ 5. The floor material in a chicken house.

_____ 6. Removing the tip of the beak.

_____ 7. Vaccinating chicks as embryos through the egg shell before hatching.

_____ 8. Equipment to burn the carcasses of dead poultry.

_____ 9. A young goose of either sex.

_____ 10. Shining a light through an egg to grade quality.

EXPLORING

1. Invite a poultry technician or other qualified person to serve as a resource person in your class. Have them explain the nature of the poultry industry, particularly their role in providing poultry products.

2. Incubate a small batch of eggs. Use a laboratory educational incubator. Get the fertile eggs from a local producer or person with a small farm flock. Carefully monitor temperature, humidity, and other conditions. Follow instructions with the incubator to be sure the eggs are properly incubated. Keep records on all processes and activities.

3. Use a whole broiler from a local supermarket. Carefully cut the broiler into different pieces based on the natural joints of the bird. Study muscling and bone structure. CAUTION: Be careful with sharp knives. Practice proper sanitation. Wash your hands and all equipment thoroughly. Bacteria, especially salmonella, may grow in water or blood from the chicken.

12

AQUATIC ANIMALS

Fish are being farmed in ways never heard of just a few years ago. And, it isn't limited to just fish! More animals are included, such as shrimp, oysters, and abalone. Aquatic farming is science and technology in action!

Why is aquatic farming zooming in importance? People like the products. They want improved quality on a year-round basis. People also want foods that promote good health. This has resulted in greater demand for many aquatic animals.

Streams, lakes, and oceans cannot provide enough wild fish. Some natural waters have been over-harvested, resulting in no more wild catch. People have tried aquaculture for hundreds of years. It has become a major food source in the United States since 1970.

12-1. Students at a Montana high school study the tilapia growing in a school lab tank.

OBJECTIVES

This chapter provides background information on aquaculture. It has the following objectives:

1   Define aquaculture and list requirements of aquacrops

2   Distinguish between water on the basis of salinity and temperature

3   Describe fish, mollusks, and crustaceans as organisms

4   Identify fish production systems

5   Describe management practices in the fish production cycle

6   List the nutritional and feeding requirements of fish

7   Explain important health and predator management practices

8   Describe harvesting procedures with fish

TERMS

aquacrop	fry	production intensity
aquaculture	harvesting	raceway
bivalve mollusk	hatchery	sac fry
brackish water	mariculture	saltwater
broodfish	mollusk	smolt
cage	operculum	spawning nest
crustacean	parts per million	spring
dissolved oxygen	parts per thousand	surface runoff
expressing	pen	swim bladder
fingerling	plankton	water facility
fish	pond	water quality
food fish production	predator	well water
freshwater	production cycle	

AQUACULTURE

Aquaculture is the production of aquatic plants, animals, and other species, such as algae. Aquatic plants and animals are species that grow naturally in water. Most aquaculture in North America is the production of aquatic animals, primarily fish, a few shell fish, and crustaceans, such as shrimp and crawfish.

Important skills are used in aquaculture. Each species is different. Species are reproduced and raised to grow-out size. Appropriate water must be provided and managed to keep it in good condition. Feeding fish and controlling disease are important. The crop must be harvested and marketed for a profit.

AQUACROPS

An *aquacrop* is a commercially produced aquatic plant, animal, and other species. "Commercial" means that the aquacrop is to be marketed for income to the producer. This is different from a pond used for sport fishing where no money is gained.

Aquacrops are produced for several purposes. The most important purpose is for human food. Other uses of aquacrops include bait for sport fishing; recreation sport fishing; ornamental fish or pets; and feed, which is using a fish, such as goldfish, as feed for other fish crops.

12-2. The scallop is a bivalve mollusk that is increasingly being cultured in saltwater.

12-3. Shrimp are cultured in some saltwater locations.

The kinds of aquacrops vary with location. Climate, water, and consumer demand are important factors. A few examples:

- Oysters are grown in protected saltwater areas, such as Wilapa Bay off the coast of Washington.

- Catfish are raised in "pure" freshwater in warmer areas, such as the delta area along the Mississippi River.

- Trout are cultivated in cool freshwater in areas with mountain streams or in the north, such as Pennsylvania and Idaho.

- Tilapia, an imported species, is being raised in warm tank systems.

Important Species

Different aquatic animals are raised, with some more important than others. Species chosen are adapted to the available water and market. Some species have more demand than others. Species with high demand include trout, shrimp, oysters, salmon, and catfish. Examples are in Table 12-1.

General Requirements

Aquacrops differ from terrestrial crops. A major difference is the water environment. Farmers who are very good in producing terrestrial crops will find that water crops require new skills.

Six requirements must be met:

Table 12-1. Important Cultured Aquatic Species in North America

Species (scientific name)	Climate/Culture	Use
abalone (red) *Haliotis rufescens*	cool, saltwater	food, decorative shell
alligator *Alligator mississippiensis*	warm, freshwater	food, leather
bullfrog *Rana catesbiana*	warm, freshwater	food, sport
catfish (channel) *Ictalurus punctatus*	warm, freshwater	food
coho salmon *Oncorhynchus kisutch*	cool or cold, saltwater/freshwater	food
crawfish (red swamp) *Procambarus clarkii*	warm, freshwater	food
prawn (freshwater) *Macrabrachium rosengbergii*	warm, freshwater	food
minnow (fathead) *Pimephales promelas*	cool or mild, freshwater	food
oyster (American) *Crassostrea virginica*	cool to warm, saltwater or brackish	food, pearl
red drum (red fish) *Sciaenops ocellatus*	warm, saltwater	food, sport
salmon (chinook) *Oncorhynchus tshawytscha*	cool or cold, saltwater/freshwater	food
shrimp (western white) *Penaeus vannamei*	mild, saltwater	food
hybrid striped bass *Morone chrysops x M. saxatilis*	mild, freshwater	food, sport
sea scallop *Patinopecten yessoensis*	cool, saltwater	food
tilapia (blue) *Tilapia aurea*	warm, freshwater	food
trout (rainbow) *Oncorhynchus mykiss*	cool or cold, freshwater	food, sport

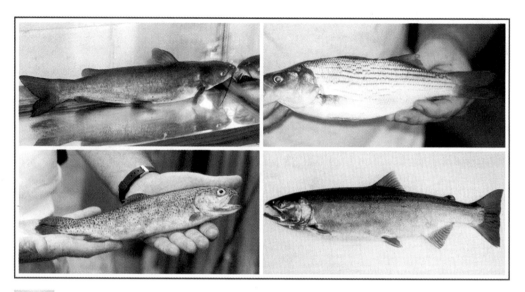

12-4. Four species of fin fish cultured in North America are (clockwise from top left) channel catfish, hybrid striped bass, rainbow trout, and coho salmon. (Courtesy (salmon), American Fisheries Society)

1. A suitable species must be selected. (The species should be adapted to the water and the available growing conditions.)

2. Water appropriate for the species must be available. (Without the "right" water, production will be difficult or impossible.)

3. The nutritional needs of the aquacrop must be met.

4. Practices must be followed to keep diseases under control.

5. The water must be properly managed for the aquacrop to grow.

6. A market must be available. (Large crops require mechanical harvesting equipment and automated processing plants. A way must exist to get products to consumers.)

Production Intensity

A stream, lake, or ocean naturally supports a low density of production. Foods for aquatic species grow in most bodies of water. These food sources include tiny plants and animals, known as *plankton*, and larger plants and animals. Producers want a higher production level than exists naturally. Ponds will naturally support the growth of 100 to 200 pounds of fish a year. This is far below the amount needed in fish farming.

12-5. Large aquaculture ponds may require aeration to assure adequate dissolved oxygen in the water.

Production intensity is the number or weight of aquatic species in a volume of water. Scientists call intensity biomass. Ponds may be stocked at 6,000 to 8,000 or more fish per acre in intensive production. This level of stocking may produce 8,000 pounds of fish per acre in a year. Tanks may have several fish per gallon, depending on the species and system. Carefully manage the water, feed, and control disease.

WATER ENVIRONMENTS

Aquatic animals vary in the water environment they need. Water is the limiting factor with most aquatic animals. Two features of water are its salt content and temperature.

Water Salt Content

Aquaculture is practiced in three different water environments based on salt (salinity) content. Species must be carefully selected to match the available water.

■ Freshwater—*Freshwater* is water with little or no salt content. Salt is measured as *parts per thousand* (ppt) or the number of parts of salt in a thousand parts of water and salt solution. Freshwater typically has less than 3.0 ppt salt. The water from most wells, streams, and surface runoff is freshwater. Species that are adapted to freshwater will not thrive, and may die, when put

12-6. Collecting a freshwater sample from a tank for testing.

into water that is not freshwater. Most aquaculture in North America has been with freshwater species, such as trout, catfish, and hybrid striped bass.

■ Brackish Water—*Brackish water* is a mixture of freshwater and saltwater. It is found where freshwater streams run into saltwater oceans and seas. The salt content is higher than freshwater, but lower than saltwater. Only a few species are cultured in brackish water. Two examples are mullet and crab.

■ Saltwater—Much of Earth's surface is covered with saltwater in oceans, seas, and lakes. Water that has more than 16.5 ppt salt is *saltwater*. The salt content of oceans and seas is typically 33 to 37 ppt.

   Many different species grow in saltwater. Salmon, oysters, shrimp, and red fish are a few examples. Some of these are produced in large cages that confine them. Producing aquacrops in saltwater is known as *mariculture*.

Water Temperature

Some species will live only in water that has a temperature within a suitable range. Species can be classified as warm water, cool water, and cold water. Most cultured species grow in either warm or cool water.

Water temperature is important in the metabolism of fish. For example, if the temperature is too cool or too warm, the fish will stop eating and fail to grow. The species selected must be appropriate for the water temperature.

■ Cool Water—The predominant cool water species cultured in the United States are trout and salmon. These species grow best at a water temperature of 50 to 68°F (10 to 20°C). The rate of growth slows below and above this temperature range. The species will not survive long in water that is warmer than

12-7. Water temperature can be measured with a thermometer.

80°F (26°C). Some research has been done with arctic char–a species that survives in cooler water.

■ Warm Water—Catfish, tilapia, crawfish, and hybrid striped bass are the predominant species of cultured warm water fish in the United States. Catfish grow best in water that is 75 to 85°F (24 to 29°C), but will survive in water below or above the temperature range. Tilapia prefer a similar temperature and will not survive in water below 50°F (10°C). Crawfish prefer water that is 65 to 85°F (18 to 29°C). Hybrid striped bass prefer 77 to 88°F (25° to 31°C), but will survive in water as low as 40°F (4°C).

Other Water Factors

Besides salinity and temperature, water must be suitable in other ways. Water quality is affected by pollution, pH, mineral content, gases, and the presence of other living plants and animals.

AQUATIC ORGANISMS

Aquatic organisms carry out the same life processes as other organisms. They do so in water rather than on land.

FISH

Fish are vertebrates, or animals with backbones. Fish are classified in the phylum of *Chordata*, meaning that they have a spinal cord. As organisms,

12-8. Tilapia are increasing in popularity as a cultured species.

they typically have bony skeletons, a skull, teeth (though not always well developed), nasal openings, swim bladder, and gills. A major distinction from other animals is that fish do not have lungs. Fish use gills to get oxygen from the water.

External Anatomy

Most fish have a streamlined body shape. The color of the fish matches their natural surroundings. Fins help fish move and keep their balance in the water. Some fins are sharp and can inflict painful wounds in human skin.

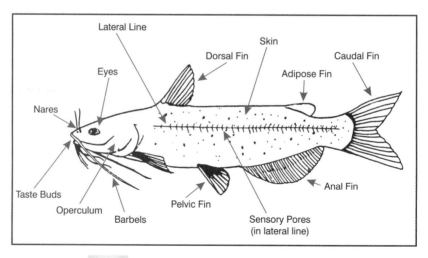

12-9. Major external parts of a typical fin fish.

The integumentary system consists of skin and scales. Some species have scales, such as trout and tilapia. Other species do not have scales, such as catfish. All species have skin.

The body of a fish is divided into three parts: head, trunk, and tail. The head extends to the back of the **operculum** (gill cover). The trunk begins at the back of the operculum and extends to the anus. The tail extends from the anus to the tip of the caudal fin.

Internal Anatomy and Systems

Fish have internal anatomy and systems similar to most other vertebrates. The major systems:

- Nervous system—Consisting of the brain and spinal cord and branching spinal nerves, the nervous system sends impulses or commands throughout the body.

- Circulatory system—The circulatory system consists of the heart and blood vessels. Blood contains red and white cells, similar to other animals.

- Sensory system—The sensory system collects information from the fish's environment. Skin and eyes are important to fish. Some fish also have barbels, or feelers, about the mouth to help in sensing danger, food, and movement.

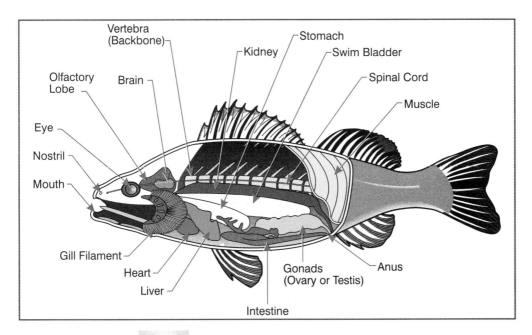

12-10. Major internal parts of a typical fin fish.

- Skeletal system—The skeletal system consists of bone and cartilage. The skeleton gives the body shape and protects the internal organs.

- Muscular system—The muscular system is the largest system in the body of fish. Muscles make it possible for fish to swim and move in other ways. The large muscles along the trunk and tail are highly desired as food.

- Respiratory system—Fish have gills that remove dissolved oxygen from the water and release carbon dioxide into the water. The blood flows through the gills on its way to other parts of the body.

- Digestive system—The digestive system prepares food for use by the body. Fish take in food through their mouths. Digestion is in the stomach and intestines. Wastes pass from the body through the anus.

- Reproductive system—Most fish spawn. Spawning (fish reproduction) is the process under which new fish are created. The female fish lays eggs that are fertilized by sperm from the male. After a few days of incubation, the eggs hatch as tiny young fry. A few species give birth to live young.

- Swim bladder—The *swim bladder* is a small balloon-like structure that helps a fish with buoyancy. Buoyancy allows a fish to stay at a particular depth in the water. The deeper a fish swims, the smaller its air bladder becomes.

CRUSTACEANS

A *crustacean* is any species in the *Arthropoda phylum*. Arthropods have exoskeletons, segmented bodies, and jointed legs. The species used for human food include shrimp, prawns, crawfish, crabs, and lobsters. Crawfish may also be known as crayfish or crawdads. Shrimp and crawfish are more

12-11. A freshwater prawn.

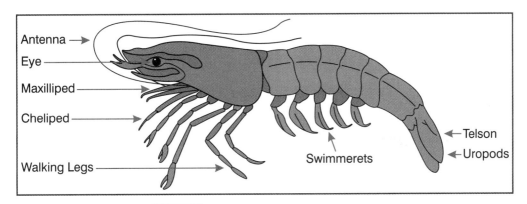

12-12. Major external parts of a shrimp.

- Antenna
- Eye
- Maxilliped
- Cheliped
- Walking Legs
- Swimmerets
- Telson
- Uropods

common as aquacrops. New methods have been developed for freshwater prawn culture. Lobsters are primarily harvested from the wild though some young are cultured in Maine and released into the ocean to grow. The crustaceans included here are decapod crustaceans, or decapods, because they have five pairs of limbs.

Crustaceans have exoskeletons that cover their bodies. The material is much like cartilage or the human fingernail. Crustaceans are in the same scientific order as insects. They have many other body features similar to insects, including three major body parts: head, carapace (similar to the thorax in insects), and abdomen. Decapods have five pairs of limbs, some of which are used for walking. A few decapods form one pair of strong pincers, such as the prawn and crawfish. Crustaceans have body systems similar to fish, but the systems are not as complex.

Mollusks

A *mollusk* is any species that has a thick, hard shell. They are in the *Mollusca phy-*

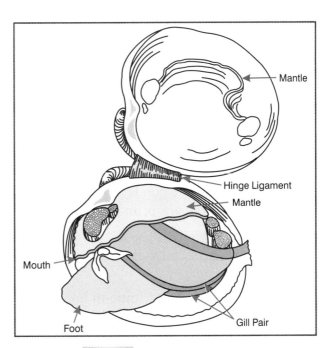

- Mantle
- Hinge Ligament
- Mantle
- Mouth
- Foot
- Gill Pair

12-13. Anatomy of an oyster.

12-14. A mussel has a smooth bivalve shell.

lum. Some shells are hinged and easily opened for the organism to capture food.

A mollusk with a hinged shell is a *bivalve mollusk*. Oysters, mussels, and clams are examples. The major muscle in a bivalve is the abductor muscle, which opens and closes the shell. A mollusk with a one-part shell is a gastropod or univalve mollusk. The abalone and snail are examples of gastropods.

Mollusks have simple organs and systems. They have gills for respiration. A heart forces blood over the gills and throughout the body. Most mollusks are filter feeders, meaning that they remove food particles from the water and use them for food. Quality water is essential; mollusks will eat pollution if it is in the water!

Pearls are a valued crop from mollusks. Formed naturally by oysters, pearls are cultured by implanting a tiny grain of sand or other irritant in the oyster.

PRODUCTION SYSTEMS

Production systems vary with the water facility. A *water facility* is the structure in which aquacrops are grown. Several kinds of water facilities are used; they include ponds, raceways, tanks, vats, aquaria, and pens and cages. Pens and cages are primarily used in natural bodies of water.

PONDS

A *pond* is a water impoundment made by building earthen dams or levees. Ponds may range from less than an acre to more than 50 acres in size. Many producers prefer ponds that are about 20 acres in size. The water in ponds does not flow and must be managed to keep fish growing.

Ponds should be designed for easy management and harvesting. Trees, stumps, and other obstacles are removed from ponds. The bottoms of ponds must be smooth, which makes harvesting easy. Levees should be wide enough for vehicles to travel on them and free of trees and utility poles. Ponds are typically rectangular and have water that is 3 to 5 feet (1 to 1.7 m) deep.

Careful management is needed to assure that a pond has a good environment for the fish.

RACEWAYS

A *raceway* is a water structure that uses flowing water. Raceways are typically long and narrow. The water may be 3 to 4 feet (1 meter) deep. More fish can be stocked in flowing water than in still water. Flowing water has more oxygen and carries waste away. Raceways vary considerably, with some built of earth, and others, of metal or concrete. Using pumps to keep the water flowing is expensive because of the cost of electricity to run the pump motors. Raceways also require more water and a method for disposing of used water.

12-15. A raceway used for trout production.

TANKS, VATS, AND AQUARIA

Tanks are made of many different materials, such as fiberglass, concrete, and metal. Large round tanks are often used for intensive fish culture. Vats,

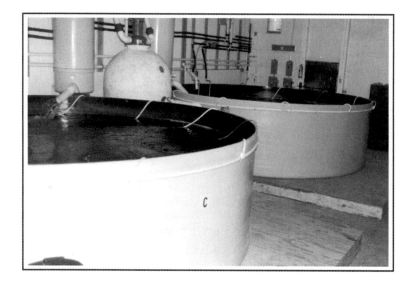

12-16. A tank system for tilapia.

often made of concrete, are similar to tanks and are often known as tanks. Aquaria are smaller tanks made primarily of glass.

The water in tanks may flow, much like a raceway, or it may be managed, as a pond. Intense production requires flowing water that has plenty of oxygen.

PENS AND CAGES

Pens and cages confine aquacrops in small areas in large bodies of water. A *pen* is a confinement structure made so it is joined with the earth in the

12-17. A round net pen for raising salmon.

12-18. A square cage for raising salmon.

pond, lake, or stream. Posts that support the pen may be imbedded in the earth. Pens are often used with broodfish where the nesting container rests on the mud at the bottom of the water. Water where pens are located is often only 2 to 3 feet (0.6 to 0.9 m) deep.

A *cage* is a confinement structure that floats on the surface of the water. Cages do not touch the earth at the bottom. They have frames covered with net-type material, such as wire or plastic. Cages can be in water that is deep and open, such as a large lake. Many other species may be growing in the water. Cages confine the aquacrop so it can be managed and harvested.

MANAGING FISH PRODUCTION

The water environment of fish must be carefully managed. This is more challenging than first believed. Water environments can be complex.

PRODUCTION CYCLE

A *production cycle* is the complete production of a crop. It begins with seedstock and ends with the crop reaching market size. In nature, these cycles often follow the seasons of the year. Reproduction may be in the spring, with the young growing all summer. A couple of years, or more, may be needed for the young to reach adequate size.

Connection

LOBSTER BABIES

Lobsters have an under-shell covering that provides a good place for eggs to incubate. One female produces many thousands of eggs. The fertile eggs are attached under the curve of her tail for nearly a year before they hatch.

A baby lobster appears little more than a piece of trash in the water. They are about ⅓-inch long. Other aquatic animals prey on the babies. For the first four weeks, the babies drift and swim about. Afterward, they sink to the bottom of the ocean where they live about 15 years unless captured or they die from disease.

Artificially hatching lobsters is limited. The Oceanarium at Bar Harbor, Maine, has had success with its lobster hatchery. This shows a female in the hatchery with many thousands of attached incubating eggs.

The production cycle described here is primarily for freshwater fish. It has four phases: broodfish and spawning, hatching and raising fry, producing fingerlings, and growing fish to food size.

Broodfish and Spawning

Young fish are used to stock food fish growing facilities. **Broodfish** are the sexually mature fish kept for reproduction purposes. Female fish produce eggs and the males produce sperm that fertilize the eggs.

Broodfish grown in warm water, such as catfish, are often kept in separate ponds from young, growing fish. Sometimes, broodfish will eat young fish if they are together. Broodfish are usually larger than food fish and three or more years old. At spawning time, pairs of broodfish may be isolated in pens or left in small ponds that contain spawning nests. A **spawning nest** is a container large enough for the pair of fish and provides some protection for the spawn (egg mass).

The size of eggs varies with the species of fish. Some eggs may be the size of a mustard seed or small pea. The red drum eggs are about the size of the head of a straight pen. Other species, particularly the smaller bait fish, may have tiny eggs.

Female fish may produce thousands of eggs at one spawning. For example, a female channel catfish may produce 2,000 or more eggs per pound (0.45 kg) of body weight. This means that a fish weighing 5 pounds (2.27 kg) might produce 10,000 or more eggs at one spawning.

Spawning varies among the species. Catfish spawn when the weather begins to warm in the spring. Trout spawn a little later in the spring or in the early summer. Some species spawn only once and die, such as the coho salmon, while others can be used in repeated years, such as the tilapia.

How fish respond to spawning also varies. In natural spawning, some females leave the eggs in a nest where the male tends them for 5 to 7 days. Other fish, such as some tilapia, incubate the eggs in their mouths. This is why they are called mouth brooders. Producers of mouth-brooding fish check the mouths of females for eggs and remove them for artificial hatching.

Spawning can be encouraged by regulating the environment of fish, such as water temperature, and by injecting the fish with a hormone. Of course, these are only practical in laboratories.

Eggs may be artificially removed from some female fish. With fish that die after spawning, the abdomen may be cut open and the eggs removed. With some species, particularly trout, the abdomen of the female is carefully squeezed to force the eggs out into a container. This procedure is known as *expressing*. Females that have eggs forced from them by gentle pressure on their abdomens have been expressed. Sperm, also known as milt, may be ex-

12-19. Eggs are being removed from a sacrificed female fish (top left), mixed with sperm expressed from a male (bottom left), and placed in a hatching jar for artificial hatching (right).

pressed similarly from the male. The milt is placed on and mixed with the eggs in a container so fertilization can occur.

Hatcheries and Fry Rearing

A *hatchery* is a place where eggs are artificially incubated and hatched. Incubation is the time between spawning and when the young fish are hatched from the eggs. A favorable environment must be maintained for the eggs during incubation. The water must be within an appropriate temperature range and the water around the eggs must typically be moving.

12-20. Paddle agitation is provided egg masses in a catfish hatchery.

12-21. Thousands of tiny catfish fry are being removed from a hatching trough.

12-22. An incubator tray system is used in trout hatcheries.

Hatcheries must have the proper equipment. Eggs are incubated in jars, trays, or troughs, depending on the species. The conditions provided must be ideal for maximum hatch and survival of the young fish. Red drum eggs hatch in about 30 hours at a water temperature of 72°F (22°C). Channel catfish eggs hatch in 5 to 9 days at water temperatures of 70 to 85°F (21 to 29°C).

Young fish are *fry* or larvae when they hatch. Fry are very small and delicate. Conditions must be provided so fry survive and grow free of disease. At first, a fry may have the yolk sac from the egg still attached for a few hours or a day. These are *sac fry*. (Newly hatched salmon are known as alevin.)

Feeding fry is essential. Since they are very small, they cannot eat feeds made for larger fish. Some

12-23. A basket of coho salmon eggs in a hatching trough. Incubation is about 60 days. White eggs are infertile (not developing a fish) and should be removed.

12-24. Eyed salmon eggs that will soon hatch.

12-25. Channel catfish fingerlings.

feed manufacturers produce a finely ground meal for fry. Fry may be fed very finely ground organ meat (heart or kidney), yeast, or natural foods. Plankton is sometimes grown for fry. The yolks of chicken eggs may be made into a slurry and used as fry feed with some species. Once the fry have grown sufficiently, they are moved to other growing facilities. With food fish, fry are about 1 inch (2.54 cm) long when moved to a fingerling area.

Fingerling Production

A *fingerling* is an immature fish. They are between the size of a fry and the size of a fish for stocking in a food fish growing facility. A salmon this size may be known as a *smolt*. A smolt has other characteristics, such as moving from the freshwater where it hatched to the saltwater where it will grow to maturity. A smolt's color changes from dark to light as it grows.

Fingerlings may be grown in tanks, small ponds, or other facilities. The rate of stocking is high because they will not grow to maturity in the facility. Fingerlings are fed a nutritionally complete feed. Since fingerlings are young and rapidly growing, their feed must have a higher percentage of protein (about 36 percent). A specially formulated commercial fingerling feed is convenient. Feed particle size is small so the young fish can·eat it. Disease control is needed for fingerlings to survive.

A channel catfish fingerling is from 1 to 8 or 10 inches (2.54–20 or 25 cm) in length. One growing season is needed for fingerlings to reach adequate size. Fingerlings should be at least 5 inches (12.7 cm) when stocked in growing ponds.

Food Fish Production

Food fish production is raising fingerlings to food fish size. Fingerlings are stocked in growing ponds, raceways, or tanks. Appropriate feed is needed at the rate of about 3 percent of the weight of the fish each day. The feed may have a protein content of 28 to 32 percent. Of course, the intensity of production has a big impact on the rate of feeding and other management details.

Fish may reach harvest size in six months or more depending on the climate and other conditions. Fish are considered food size when they are three-fourths pound or larger. Most processing plants are designed for fish within a certain size range. The fish that are too large may not process well with the equipment in the processing plant. With catfish, the processing plants most efficiently handle fish that are 1 to 2 pounds in size.

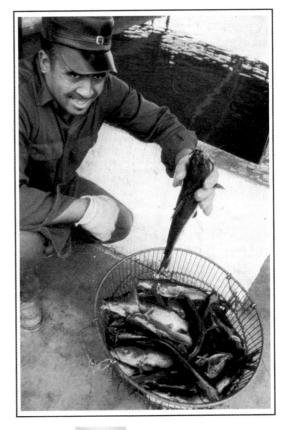

12-26. Food-size catfish.

All areas that affect the growth of a fish must be considered. This includes feeding, water quality, and disease and predator control.

WATER MANAGEMENT

Water must be well managed for an aquacrop to live and grow. This involves having a good source of the water, maintaining oxygen in the water, and disposing of used water.

Sources of Water

Water from several different sources may be used. This depends on the climate and requirements of the species.

Surface runoff is excess water from precipitation. It may collect behind dams or in streams, lakes, or oceans. Runoff may be okay if it is free of pollution and dependable in supply. Water must be available throughout the growing season for an aquacrop. Runoff tends to be seasonal and available only during certain times of the year. Legal regulations may restrict pumping water from a stream.

Well water is water pumped from aquifers deep in the earth. The water is often of high quality, but is expensive to get. Deep wells, pumps, and electricity to operate the pumps are costly. Occasionally, the water may be too cool for pumping directly into fish facilities. It may need to be pumped and stored in open ponds for a few days so the sun can warm it. Permits are needed to drill water wells for aquaculture.

A *spring* is a natural opening in the earth that provides water similar to well water. The water may be of good quality for aquacrops. Some

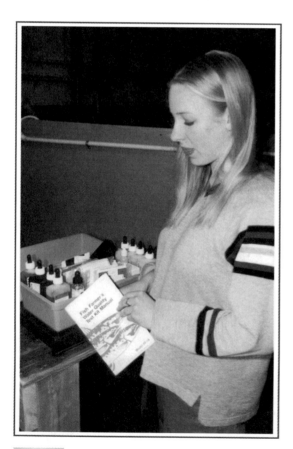

12-27. Preparing to use a kit to test water in a tank facility.

springs dry up at certain times of the year and are not dependable sources of water for aquaculture.

Other sources of water are industrial waste water and municipal systems. Waste water must be free of dangerous substances and have the right temperature. Municipal water systems may treat the water with chemicals that are harmful to fish. This source of water is often too expensive for large-scale aquaculture but can be used for small aquaria.

Water Quality

Water quality is the suitability of water for a particular use. It must have sufficient dissolved oxygen and be free of pollutants. In addition, acidity, alkalinity, hardness, nitrogen compounds, and carbon dioxide are important. The major emphasis in this chapter is on dissolved oxygen.

Dissolved oxygen (DO) is oxygen in the water that is available for living organisms to use. It is in a free, gaseous form and not the oxygen in a molecule of water. DO is measured as *parts per million* (ppm) in water. For example, 5.0 ppm DO in water means that there are 5.0 parts DO in 999,995.0 parts water, or a total of 1 million combined parts.

Cooler water holds more DO than warmer water. Fish are more active in warm water and produce more wastes. Water must be carefully monitored in warm weather. In ponds, DO is highest about noon on a bright, sunny day and lowest just before sunrise. This is because tiny plants in the water, known as phytoplankton carry on photosynthesis and produce oxygen. At night, when there is no sun, the phytoplankton does not produce much oxygen. Water facilities inside buildings respond differently.

Most aquacrops need 5.0 ppm DO or above to grow, though some will survive below that level. DO is frequently measured with an oxygen meter. If DO is low, the water is oxygenated by splashing it into the air, injecting pure

12-28. A DO meter for measuring oxygen levels.

oxygen into it, or in other ways. Fish can die quickly if the DO level drops too low.

NUTRITION AND FEEDING

Aquatic animals must have appropriate nutrition to grow. Deficiencies may result in disease outbreaks. Most aquacrops have nutrient needs similar to other animals.

NUTRITION

Fish need a ration (daily feed amount) that meets nutritional needs. The nutrient requirements of fish vary, but are very similar. Fish need the following:

- Protein—Research has given information about the protein needs of fish. Without sufficient protein, fish do not grow. Most commercial feeds have the needed protein. Some of the protein must come from animal sources, such as fish, shrimp, beef, or chicken scraps. These scraps must be properly prepared in the feed. Feed with 28 percent protein is needed for growing fish.

- Carbohydrates—Carbohydrates provide energy. Most commercial fish feed contains grains, such as corn and soybeans. These are good sources of carbohydrates. The amount of carbohydrate in the diet of a fish is about 10 percent.

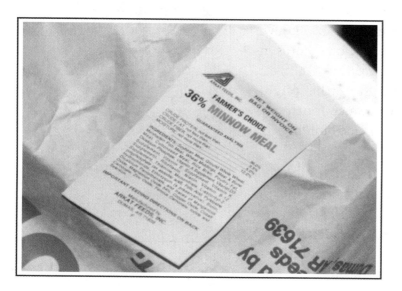

12-29. Bag with a label showing minnow feed analysis.

■ Fats—In fish rations, fats usually come from animal wastes, such as by-products from a processing plant. The fat content of feed is 4 to 15 percent.

■ Minerals—Fish need about the same minerals as other animals. These include calcium, iron, silicon, manganese, sodium, and seven others.

■ Vitamins—Fish need vitamins or they will develop severe deficiency symptoms. Eleven vitamins are essential for fish.

Feeds and Feeding

The intensity of production is a big factor in the amount of feed. If only a few fish are in a pond, adequate natural food may be available. This is not true with aquaculture, however. High stocking rates require appropriate feed.

Natural Foods

Plankton naturally grows in ponds, streams, and other water impoundments. It serves as food for fish and other aquacrops. Depending on the stocking rate, plankton can provide some of the needed food. The use of natural foods also depends on the species of aquacrop.

Crawfish eat vegetation in the water. Most crawfish growers plant rice in the crawfish ponds. The rice (stem, leaves, and grain) serves as food for the crawfish.

Manufactured Feed

Commercial mills produce feed for aquacrops. These products are often specially designed for a species, such as trout or catfish.

Feeds are manufactured in several forms:

■ Meal—A meal is a finely ground feed used with small fish.

■ Pellets—A pellet is a feed product made by binding meal into larger particles. Pellet sizes vary for different fish. Pellets are made to float, sink, or remain neutral in the water. Many food-fish farmers prefer floating pellets because the fish have to come to the surface of the water to get food. This helps see fish that otherwise may be unseen.

■ Other forms—Crumbles, blocks, and other forms may be used with fish feed. These forms are for specific species of aquacrop.

12-30. Floating pellets (left) and meal fish feed (right).

Frequency and Rate of Feeding

Frequency of feeding ranges from feeding every hour to once or twice daily or once a week. How often fish are fed depends on the size of the fish, climate, and rate of growth.

Fry are fed several times during the day, often each hour. Fingerlings may be fed less often. Food fish are fed once or twice a day. Broodfish may be fed

12-31. A mechanical feeder that blows feed out into a pond.

once a day or two or three times a week. In the winter, broodfish are fed less often.

Automatic feeders are sometimes used. Some provide feed on demand, such as when the fish touch a "trigger" that releases feed. Most automated feeders are on a timer and feed is provided at regular intervals.

The rate of feeding depends on fish size and water temperature. Small fish are fed at a higher rate than larger fish. A general rule is to feed fish 3 percent of their body weight each day. This depends on the water temperature and other factors. Most fish eat more in water as it warms. Too much feed is wasteful. The fish do not eat it. Over feeding fouls the water. Uneaten food particles settle to the bottom. Feed only the amount the fish will eat in 15 minutes.

HEALTH AND PREDATOR CONTROL

Diseases and predators can rapidly wipe out a fish population. Good management promotes health and well-being. Steps can be taken to control predators.

HEALTH

Health is the absence of disease. Fish that get adequate feed and have quality water are less likely to have health problems. Prevention of disease is the best control. Fish are difficult to treat once an outbreak of disease occurs.

Kinds of Disease

Diseases are infectious and noninfectious. Infectious diseases are caused by germs or pathogens. Bacteria, fungi, viral, and parasitic diseases may occur. Parasites include worms, flukes, and protozoa. These diseases attack the tissues of aquacrops and cause loss of growth and, sometimes, death.

Noninfectious diseases are from a variety of causes. Some are caused by organ systems that do not work properly. Others are caused by the environment in which an aquacrop lives and the food it gets. Sometimes the organs may not operate properly. This may be related to water quality, such as a sudden change in pH may cause blood problems. Nutritional disease results when the aquacrop has an inadequate diet. Environmental diseases result when the water contains gases, chemical residues, or other materials that are damaging to fish.

Disease Control

The best disease control is to keep fish healthy. Only disease-free fish should be brought on a farm or put in a tank. One fish with a disease can infect an entire population!

A few chemical treatments are available for fish. Some involve putting medications in the feed of the fish. Other treatments involve dipping fish into a chemical solution or bathing the fish in a solution. Valuable broodfish can be injected with medications. The U.S. Food and Drug Administration (FDA) regulates the medications that can be used on fish that will be used for human food in the United States. Always follow the regulations.

12-32. Double-crested cormorants are major pests on some fish farms.

PREDATORS

A *predator* is an animal that hunts and kills other animals. Large birds can prey on aquacrops in ponds or other growing facilities. For example, the cormorant (a water bird) can cause large losses in a fish production facility by eating a pound or so of fish a day. Other predators include alligators, turtles, frogs, other fish, snakes, and insects.

With ponds, keeping the grass mowed around the edge of the water will discourage some predators. Air cannons are sometimes used to frighten law-protected birds away. Small ponds and tanks can be covered with netting to keep predators away. Constructing ponds with a minimum area of shallow water where birds can stand will help keep some of them away.

HARVESTING

Harvesting is capturing the aquacrop so it can be hauled and sold to a processing plant or other facility. The methods used must result in a product free of injury.

12-33. A small pond is easily harvested with a seine.

How an aquacrop is harvested depends on the crop and the water facility. Large seines may be put around ponds and the fish confined to a small area in the pond. This is accomplished by drawing the seine through the water. After seining, the fish are dipped and loaded into a haul truck. Frequently, the basket used to dip the fish has a scale for weighing the fish as they are being loaded.

Fish in tanks or raceways may be harvested similarly to those in ponds. A net or grading device may be put across the facility and moved toward one end where the fish are confined. The confined fish are dipped from the water.

12-34. Powered seines are used in harvesting large ponds.

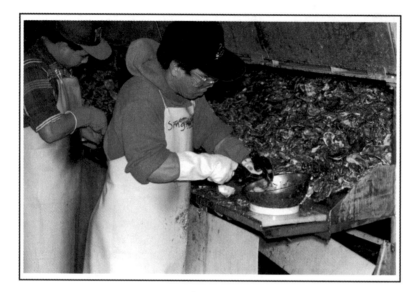

12-35. After harvest, crops are processed. This shows cultured oysters ring shucked, which is opening the shell and removing the oyster in one piece.

Sometimes, the water level is lowered so the aquacrop is easier to confine. This creates the need for additional water to again fill the facility. Water is expensive!

Harvested aquacrops must be kept alive until they reach the processing plant. This requires hauling in a tank of water with oxygenation equipment.

REVIEWING

MAIN IDEAS

Aquaculture is the production of aquatic organisms. Plants, animals, and other species, such as algae (kelp), are produced. The major emphasis in the United States is on aquatic animal production.

The most widely produced aquacrops are fish, crustaceans (including shrimp, prawns, and crawfish), and mollusks (including clams, oysters, and mussels). Freshwater fish farming has grown most rapidly in recent years. The species grown are catfish, trout, tilapia, salmon, hybrid striped bass, and a few others.

Water is important in aquaculture. Freshwater, saltwater, and brackish water are used, as appropriate to the species. Water is obtained from runoff, wells, and waste from industry. Water quality is the suitability of water for a particular aquacrop. A critical factor in any aquaculture is dissolved oxygen (DO). Most aquacrops prefer a minimum DO level of 5.0 ppm. The common water facilities (impoundments) include ponds, raceways, tanks, vats, and aquaria.

Production cycles of aquacrops include broodfish and spawning, hatcheries and fry rearing, fingerling production, and food fish production. Nutrition and feeding are important at all stages. Controlling diseases and predators is essential. Harvesting is a part of the final production cycle with food fish.

QUESTIONS

Answer the following questions using correct spelling and complete sentences.

1. What species of aquacrops are commonly produced in the United States? List any five and give one item of descriptive information about its requirements.

2. What are the general requirements for aquaculture?

3. What is production intensity? Why is it important in the production of aquacrops?

4. Distinguish between the kinds of water based on salinity and temperature.

5. Draw a fish and label its external anatomy.

6. What kinds of water facilities are used? Describe each.

7. Name and briefly describe the four phases in the production cycle of aquacrops.

8. What are the major sources of water?

9. What is water quality?

10. What is dissolved oxygen? Why is it important?

11. What are the major areas of nutrition and feeding of aquacrops?

12. What are the major areas of health management with fish?

13. How do predators cause losses?

14. What is harvesting? How is it done?

EVALUATING

Match the term with the correct definition. Write the letter by the term in the blank provided.

a. predator   e. production intensity   i. fish
b. aquaculture   f. freshwater   j. crustacean
c. aquacrop   g. brackish water
d. pond   h. saltwater

_____ 1. Arthropoda phylum.

_____ 2. Contains little or no salt.

_____ 3. A mixture of freshwater and saltwater.

_____ 4. Water with more than 16.5 ppt salt.

_____ 5. Production of aquatic plants, animals, and other species.

_____ 6. The number or concentration of aquatic species in a volume of water.

_____ 7. Commercially produced aquatic plants, animals, and other species.

_____ 8. Chordata phylum.

_____ 9. A water facility made with an earthen dam or levee.

_____ 10. Animals that hunt and kill other animals.

EXPLORING

1. Set up an aquarium or fish tank in the classroom. Select a species that will survive in the environment. Goldfish are very hardy. Develop a plan for maintaining the water facility and caring for the fish. (Reference: *Aquaculture: An Introduction* available from Interstate Publishers, Inc.)

2. Tour an aquafarm or hatchery in the local area. Determine the species produced and the management needed for successful production. Also, learn how the aquacrop is marketed.

3. Invite an aquaculturist or fisheries specialist to serve as a resource person in class and discuss the locally adapted species and how to culture them.

13

HORSES

"Let's go horse back riding!" Horse enthusiasts across the nation regularly ride, care for, and train their horses. Some people do their own training and daily care. Others have a stable perform these services.

Today, most horses are owned because of their recreational and personal value. Horses have been useful in many ways. They are used for food and pet feed. They have been used for military purposes. Horse racing is the largest spectator sport in the world. Some horses are used to work cattle on western ranches.

In the 1800s and early 1900s, horses were used for power and as a major mode of transportation. The United States had 20 million horses in 1900. This dropped to 9 million in 1977 and to less than 6 million in the late 1980s but has increased to 6.6 million today.

13-1. Horses are popular in many kinds of recreation. This shows barrel racing at a rodeo.

OBJECTIVES

This chapter has basic information on the care and management of light horses. It has the following objectives:

1   Describe horses as organisms

2   Explain breeding practices with horses

3   List the nutritional requirements of horses

4   Explain important health management practices with horses

5   Describe facility and equipment requirements of horses

6   Describe types of equitation

TERMS

draft horse	jog
driving horse	light horse
equitation	lope
farrier	mare
filly	plug
floating	polo mount
foal	pony
foaling	racehorse
frog	riding horse
gait	stallion
gelding	stock horse
hand	stud horse
horsemanship	walk
hunting and jumping horse	

HORSES AS ORGANISMS

Horses are similar to other animals in many ways. Yet, they have differences that make them unique. The scientific name of horses is *Equus caballus*. Horses are sometimes called equine, which is based on the scientific name. Riding and managing horses is **equitation**.

AGE AND SEX CLASSES OF HORSES

Several terms are used to describe horses. A mature female horse over four years of age is a **mare**, while a mature male horse over four years of age is a **stallion**. In Thoroughbreds, both classes begin at five years of age.

13-2. Horses provide fun and companionship. (Courtesy, DeShannon Davis, Mississippi)

A mare that has never been bred is a maiden horse. A sire is the male parent. A female parent is a dam.

The gestation period in horses is longer than other livestock at 336 days. Giving birth (parturition) is **foaling**. A young horse of either sex that has not been weaned is a **foal**. A **gelding** is a male horse castrated before reaching sexual maturity. A **filly** is a female horse under three years of age (except Thoroughbreds, which are under four years of age). A **stud horse** is a male kept specifically for breeding purposes. A stag is a male castrated after reaching sexual maturity.

Horses are known as intelligent animals. They can be taught and directed in many ways. Some horses fail to meet the notion of a beautiful, intelligent animal. A horse that has an ugly head or behaves oddly is known as a jughead. A horse with a large, coarse head is called a hammerhead. A **plug** is a horse with poor conformation and common breeding. Older horses that show signs of aging may also be known as plugs.

Conformation tells much about the desirability of a horse. Healthy, disease-free horses should be selected. Avoid horses with defects that detract from your intended use. Figure 13-3 shows the desired conformation of a horse.

I3-3. Conformation of a desirable horse.

SIZE CLASSES OF HORSES

Horses are classified by size as either light horses, ponies, or draft horses. Draft horses are suited to drawing loads and performing heavy work. This chapter includes light horses. Draft horses are covered in Chapter 16.

Horse classification is based on height. Height is the distance from the highest point of the withers to the ground. Horses are measured by hands. One *hand* is equal to 4 inches (10.2 cm). If a horse is 15.2 hands, then it is 61 inches (155 cm) tall.

A *light horse* stands 14.2 to 17 hands high and weighs 900 to 1,400 pounds (408–635 kg). They are used mostly for riding, driving, and racing. A *pony* stands under 14.2 hands high and weighs 500 to 900 pounds

I3-4. Measuring the height of a horse.

(227–408 kg). A ***draft horse*** stands 14.2 to 17.2 hands high and weighs over 1,400 pounds (635 kg). Donkeys and mules are again becoming popular. Mules are being used for trail rides and working with cattle.

SELECTING A HORSE

Selection of a horse depends upon many factors. Several major factors are the intended use, price range, disposition, size, gait, breed, and color.

Disposition is important for everyone, especially for an amateur or child. They need a horse or pony that is quiet, gentle, and well broken.

The size of the horse should fit the size and weight of the rider. A person that is tall will overpower a small horse. A small person will look tiny on a tall horse. The breed and color of the horse are of personal preference.

Colors range from white to black. Bay, brown, and chestnut (sorrel) are popular. Dun, gray, palomino, paint, and roan colors are selected by some people.

When selecting a horse, bring an experienced horseperson along for his or her opinion. When selecting for breeding purposes, check the horse's progeny and pedigree.

13-5. Four breeds of horses are (clockwise from top left) Quarter Horse, Paint, Appaloosa, and American Saddlebred. (Paint courtesy, BarLink Paint Horses, Oregon)

Connection

HORSES FOR MEAT

Horse meat is known as chevaline. It is not available in most U.S. supermarkets, but chevaline is popular in France, Italy, and other European countries. Many horses are used for meat products.

The United States has a Horse Meat Act. This law covers inspecting and labeling horse meat for human consumption. Inspection stamps use green ink rather than purple as on carcasses of other animals. Most of the horses slaughtered in the United States for human consumption are exported to other nations. Some years the export of horse meat has been among the largest of the exported meat products.

Meat markets may offer attractively packaged horse meat cuts. It is wholesome meat high in protein.

TYPES OF LIGHT HORSES

Light horses are classified by type. Type refers to the best use of the horse based on its attributes. Specific breeds of light horses are available for each type.

RIDING HORSES

A *riding horse*, as the name implies, is a horse ridden for pleasure or work. These horses have a definite utility value. Types of riding horses include gaited horses, stock horses, polo mounts, hunters and jumpers, and ponies. The most popular riding horses are the Quarter Horse, Arabian, Appaloosa, Morgan, Thoroughbred, American Saddle Horse, and Tennessee Walking Horse.

GAITED HORSES

Gait is the way a horse moves—walks or runs. It is associated with the rhythmic movement of the feet and legs. The run or gallop is a fast gait. The two hind feet leave the ground at different times followed by the two front feet. The walk is a gait in which each foot leaves and touches the ground at different intervals.

Three-gaited horses are known for their gaits of walking, trotting, and cantering. They are used chiefly for pleasure riding. After the development of New England towns and the opening of roads, these horses greatly increased in popularity. Gaited horses lack the style and action of a stock horse, yet they have their own unique style.

Other gaited horses are five-gaited saddle horses. The five-gaited horse must also perform a slow gait and the rack or single-foot. The slow gait is either the running-walk, fox trot, or slow pace. These horses were developed on the southern plantations by people who spent long hours on horseback. They chose animals that had an easy, springy step. Gaited horses are ideal for the amateur rider because of their easy disposition and easy gaits.

STOCK HORSES

*A **stock horse*** is a type of horse that is used in managing cattle. It is usually of mixed breeding and a descendant of the Mustang. A stock horse is the most popular type of horse in the United States. It is short coupled, well muscled, and deep bodied. The stock horse needs to be hardy, agile, sure-footed, fast, deep, powerfully muscled, durable, and must have good feet and legs. Stock horses must also possess "cow sense."

13-6. A stock horse may be used to herd cattle.

POLO MOUNTS

The **polo mount** is a type of horse used in playing the game of polo. Polo is played by four people on horseback who try to drive a wooden ball between a goalpost. A polo mount is of smaller size than hunters and Thoroughbred types. It must be quick and clever in turning. It must be able to dodge, swerve, and turn while running. This sport is very expensive because up to six years is required to train a polo mount and a person may use four to six mounts during a single game.

HUNTERS AND JUMPERS

A **hunting and jumping horse** is a type that is used in fox hunting. They are large, clean-cut horses that perform well in cross-country riding and jumping. Again, Thoroughbred blood is predominantly found in these breeds. Hunters and jumpers must be of good size and height to jump tall fences and wide ditches. They also require stamina and good conformation to keep up with the pack as the hounds track the fox.

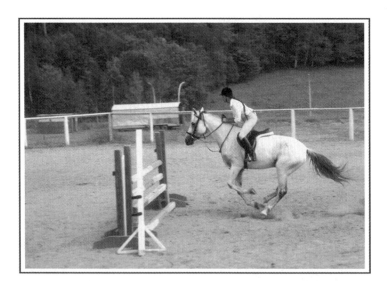

13-7. Horses may be trained to jump and perform in other competitive events.

PONIES

A pony is a type of horse that is typically raised for children. Ponies are not only unique in their small size, but must also possess many other characteristics. They must be gentle, sound in feet and legs, symmetrical, have good eyes, and possess endurance, intelligence, patience, faithfulness, and

13-8. A Shetland Pony is used for a fun activity.

hardiness. The most popular breeds of ponies are the Welsh, Shetland, and the Pony of the Americas.

RACEHORSES

A *racehorse* is a type of horse used for racing—either a running racehorse or a harness racehorse. Running racehorses are almost exclusively Thoroughbreds or Quarter Horses. They are used for other purposes, such as for polo mounts, hunters, and calvary horses. From the earliest recorded history, racing has taken place between horses.

Running racehorses must be extremely refined, have oblique shoulders, well-made withers, heavily muscled rear quarters, and straight hind legs. Harness racehorses need to be fast and light. The two main breeds of harness racehorses are the Morgan and the Standardbred.

DRIVING HORSES

A *driving horse* is a type used to pull two-wheel and four-wheel light vehicles. They are important in the ring. Showing is competitive and entertaining to some people.

BREEDING PRACTICES

Conception rates among mares are the lowest of all farm animals, with average conception rates of 50 percent. Normal breeding, artificial insemination, and embryo transfer are used. General procedures for normal breeding are discussed here.

NORMAL BREEDING OF MARES

Mares will begin coming into heat at 12 to 15 months of age. The general recommendation is to breed a mare at three years of age, so she will foal at four years of age. This way, she will not be training while heavy in foal and she will be more mature, taller, and fuller grown. Mares will continue to foal up to 14 to 15 years of age.

Heat periods are around 21-day intervals and last four to six days. Signs of estrus include relaxation of the external genitals, increased frequency in urination, teasing of other mares, desire for company, and a slight mucous discharge from the vagina.

The average gestation period of a mare is 336 days or slightly over 11 months. Pregnant mares should be kept separate from others because they are usually more sedate. The pregnant mare needs exercise and is best kept in an open pasture.

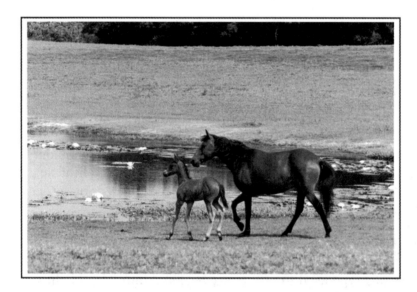

13-9. A mare with foal exercising in a pasture. (Courtesy, Paso Fino Horse Association, Inc.)

Parturition

The most obvious sign of nearing parturition is a distended udder. This may happen from two to six weeks before foaling. Foaling is the act of a horse giving birth. There will be a falling away of the muscular parts of the top of the buttocks, near the tailhead, and in the abdominal area about 7 to 10 days beforehand. The vulva will become loose and full. As foaling comes even closer, milk will fill the ends of the teats, she will become very restless, and may break into a sweat and urinate frequently. These signs are not 100 percent foolproof. Always watch the mare and note changes in her behavior.

When the weather is warm, foaling may take place in a clean, open pasture away from other animals. There is less chance of injury and infection. During poor weather, the mare should be kept in a roomy, well-lighted, ventilated, quiet stall.

Just before parturition, the mare will exhibit extreme nervousness and uneasiness, biting of her sides and flanks, switching of the tail, sweating, and more frequent urination. When the outer fetal membrane breaks and a large amount of fluid flows out, the water bag has broken. Foaling should take no more than 15 to 30 minutes.

A foal should come front feet first with heels pointing down. The nose should follow as it rests on the forelegs. Then the shoulders and the rest of the body should come. If the foal is presented in any other manner, call a veterinarian.

Care After Parturition

Once the foal is out, check to see that it is breathing and that no membranes are covering the mouth or nostrils. Dip the navel cord in iodine to prevent infection. The mother should be lightly fed and given small amounts of lukewarm water at frequent intervals. Frequently clean the stall and observe both mother and offspring for signs of illness.

Feeding colostrum to the foal immediately after birth is very important. Colostrum contains antibodies that protect the foal against certain infections. The healthy foal should be ready to nurse within 30 minutes to two hours, but they may need assistance and coaxing.

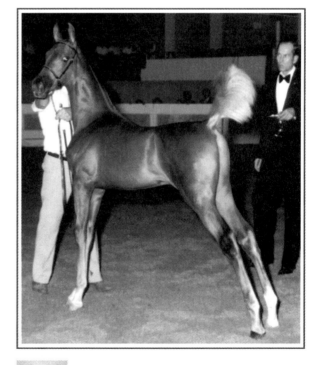

13-10. An American Saddlebred weanling has the traits of being a champion adult. (Courtesy, American Saddlebred Horse Association, Inc.)

CARE OF THE STALLION

Any stallion bought or used for breeding purposes must be a guaranteed breeder. The number of healthy foals that a stallion has sired is more important than the number of times that the stallion has serviced mares. Some farms will periodically check the semen of stallions to measure the fertility of the animal. Semen must be of good volume, have a high sperm count, the sperm must move about, and they must be morphologically correct.

Because stallions are more nervous, they need to be kept in separate quarters. The best arrangement for stallions is a roomy box stall with a two- to three-acre pasture. The stallion must receive daily exercise to maintain thriftiness.

BREEDING THE MARE

The mare should be in good condition at breeding time. She should not be too thin nor too fat. She can be bred by hand breeding, corral breeding, or pasture breeding. The most accepted method is by hand. This will decrease the chance of injury to the mare or stallion. With corral breeding, they are turned loose into a corral. It is best to stand in a spot where you can see, but not be seen. Do not stand in the corral. Once the mare is bred, they are returned to their separate quarters.

Pasture breeding is hard to control and monitor. Valuable animals should not be pasture bred as there is an increased incidence of injury. With pasture breeding, the stallion is turned out with a band of mares. As the breeding season progresses, the stallion may become sterile.

13-11. A mare in good condition. (Courtesy, Paso Fino Horse Association, Inc.)

After breeding, both the stallion and mare should be checked for injury. The mare should be pregnancy checked shortly thereafter.

FEEDING

Horses require nutrients for maintenance, growth, reproduction, lactation, and work. They require all six of the basic nutrients: carbohydrates, fats, water, protein, vitamins, and minerals, just like any other animal. They require energy, or carbohydrates and fats, in the largest amounts. Horses receive these nutrients from concentrates, forages, free-choice vitamins and minerals, and water.

CONCENTRATES

Concentrates consist of various grains and supplements. They mainly supply energy and protein.

Grains

Grains are concentrates, such as oats, corn, or barley, that provide energy. Grains provide protein. Protein is necessary for the formation of bones, ligaments, hair, hooves, skin, organs, and muscles. Horses are pseudoruminant so they cannot significantly synthesize their own protein. Young horses espe-

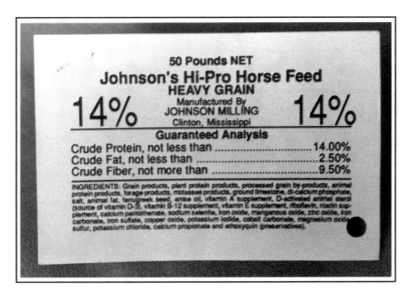

13-12. Label from a commercial horse feed.

cially need high-quality protein because their cecum is not fully developed. Maintenance requirements of protein are about 1 pound of protein for every 1,000 pounds (453 kg) of weight.

Oats are the most commonly used grain for horses. Oats are low in energy, but are easier on a horse's sensitive stomach. Dusty oats can cause colic, which is the number one loss of horses.

Corn is usually mixed with the oats because it is higher in energy. This is especially done in the winter. Corn should be cracked, rolled, coarsely ground, or shelled. Care should also be taken to avoid dusty or moldy corn. Moldy corn is very dangerous for any animal. If barley is fed, it must be ground or rolled because it is hard.

Supplements

The most common supplements are soybean meal and linseed meal or oil. Soybean meal is very high in protein content. Linseed meal or oil puts a shine in a horse's coat.

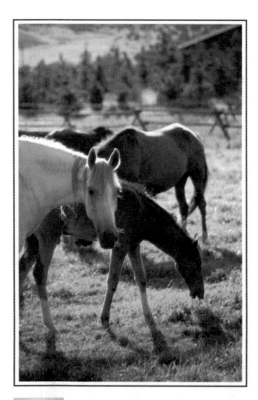

13-13. Horses benefit from good pasture. (Courtesy, Paso Fino Horse Association, Inc.)

FORAGES

Forages for horses include pasture and hay. Forages should be fed at 1 to 2 pounds (0.91–1.82 kg) for each 100 pounds (45.36 kg) of body weight each day. One horse should be allowed three acres on which to graze. Hay should be of good quality and free of dust or mold.

MINERALS AND VITAMINS

Minerals are necessary for the growth and development of bones, teeth, and other tissues. The most essential and most likely to be deficient minerals are calcium and phosphorus. Steamed bone meal, dicalcium phosphate, salt, or trace minerals are the best sources of minerals.

Vitamins are needed for proper growth and development, health, and reproduction. Vitamin deficiencies are not a problem if the horse is fed a well-balanced diet. Supplemental vitamin A should be fed if the horse is not in a pasture to consume green grass. Vitamin D may be needed if the horse is kept inside.

WATER

Water is the most important nutrient that horses and all animals need. An average horse will consume 10 to 12 gallons (37.9–45.5 L) of water each day. They may require more while working or in the summer. They should not drink excessive amounts of water during the summer because they will founder.

HEALTH MANAGEMENT

Daily health management is imperative. Prevention is the key. Work closely with a veterinarian to devise a vaccination schedule. This will result in fewer health problems, lower veterinarian fees, and decreased animal losses. Cleanliness is necessary to prevent disease and for the well-being of the animal.

COMMON AILMENTS

Several common ailments that may afflict horses are listed here. Refer to Chapter 4 for more information on animal health.

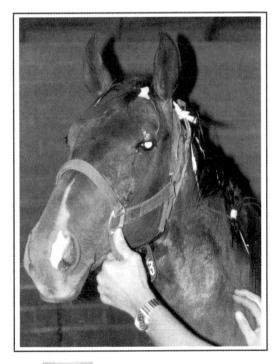

- Colic—Colic is the leading cause of death in horses. It may be caused by internal parasites, improper feeding, and excessive water intake. Signs of colic include a distended abdomen, kicking or rolling, heavy perspiration, constipation, and refusal to eat or drink. Walk the horse and call the veterinarian immediately.

13-14. Diseased eye being treated.

■ Encephalomyelitis (Sleeping Sickness)—Encephalomyelitis is caused by a virus and is carried by mosquitos. Vaccinating and applying insecticides is the best prevention of this disease. Signs of encephalomyelitis are a sleepy attitude and localized paralysis of the lips and bladder.

■ Equine Infectious Anemia (Swamp Fever)—Equine infectious anemia (EIA) is another viral disease transmitted by biting insects. It can also be spread by infected needles. There is currently no vaccine for EIA. Signs of EIA include a high, intermittent fever, stiffness, weakness, loss of weight, anemia, and swelling. The Coggins test is used for diagnosis.

■ Equine Influenza—Equine influenza is highly contagious. Symptoms include fever, loss of appetite, depression, rapid breathing, cough, weakness, and eye and nasal discharge. If influenza is suspected, consult the local veterinarian; there is a vaccine available.

■ Founder—Founder affects the tissue connecting the hoof wall to the foot. Severe founder cannot be cured. Founder is caused by grain or forage overeating, consuming too much cold water, overwork, rapid change in diet, or inflammation of the uterus following foaling. Signs of founder include pain in the feet, fever, and a reluctance to move.

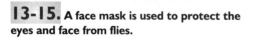
13-15. A face mask is used to protect the eyes and face from flies.

■ Tetanus (Lockjaw)—When an open wound becomes infected with *Colostridium tetani*, a bacterium, tetanus will result. These bacteria are found in manure, making it necessary to keep stalls and barns clean. Symptoms include stiffness about the head and slow and awkward chewing and swallowing. An annual vaccination is recommended.

■ External Parasites—External parasites common to horses include flies, mosquitos, mites, lice, ticks, and ringworm. They are not only annoying to the animal and owner, but also carry diseases. Application of insecticides and regular cleaning of stalls will greatly decrease the external parasite problem.

■ Internal Parasites—Internal parasites include roundworms, bots (flies), pinworms, and strongyles (bloodsucking worms). To prevent internal parasites, do not allow the horse to eat off the ground or drink from puddles. Rotate pastures frequently. Regular worming may be needed. If internal parasites are suspected, get a veterinarian to examine the feces of the animal. The extent and type of infestation determines the plan of action.

CARE OF THE TEETH

Determining Age by Teeth

Horses live to be 20 to 25 years old. Their best years are from age three to twelve. Age can be decided by examination of their teeth. Time of appearance of temporary and permanent teeth, the shape, and the degree of wear are important in aging horses.

Temporary teeth are smaller and whiter than permanent teeth. A mature male has 40 teeth, while a female has 36. As horses age, the cups wear on the inside and center of the tooth. After 12 years, the teeth change from an oval to a triangular appearance and they slant forward more.

13-16. A leg injury has been treated by an equine veterinarian.

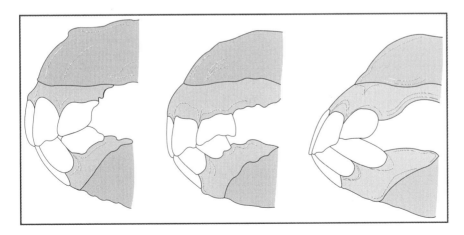

13-17. Teeth slant forward as horses age. This side view shows the slant in teeth of horses that are 5, 7, and 20 years old.

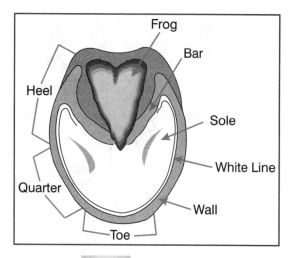

13-18. Parts of a hoof.

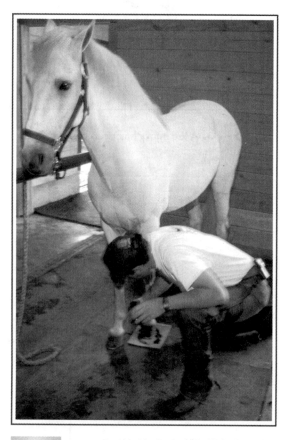

13-19. A farrier dressing the foot of a horse. (Courtesy, Paso Fino Horse Association, Inc.)

Floating the Teeth

Because the upper jaw of a horse is wider than its lower jaw, teeth will wear unevenly. This causes a sharp edge to form. Teeth should be floated each year to reduce the discomfort of the sharp edges. *Floating* is using a file to smooth the sharp edges of the tooth. Only a trained person should attempt floating.

HOOF CARE

The ability of a horse to move will determine its worth. Proper care and maintenance of the hoof is very important. Through the domestication of horses and the change in their living environment, the care that a horse's hoof requires has significantly increased. The hoof should be kept clean, prevented from drying out, maintained at the proper length and shape, and protected from the hard surfaces with which it comes in contact.

A hoof pick should be used daily to clean out the hoof. Carefully clean around the frog. The *frog* is the V-shaped, elastic-like pad in the middle of the sole of the hoof. Keeping the hoof wet may be done by simply keeping the area around the water tank wet, applying dressing to the hooves, or attaching burlap to the hooves.

A *farrier* (a person who puts shoes on horses) should trim the hooves every four to six weeks. The sole of the hoof and jagged edges of the frog should be trimmed with a hoof knife. The sole of the hoof should not be trimmed unless necessary. Protection from hard surfaces is accomplished with shoes. Shoes should either be replaced or reset after the hoof is trimmed. Horses that are seldom ridden or that spend little time on hard surfaces may not need shoes.

FACILITY AND EQUIPMENT REQUIREMENTS

Horse facilities do not need to be expensive or fancy. Again, no more money should be spent on facilities than can be afforded.

BARN

The facilities need to protect the horse from the wind, sun, and other poor weather. The barn should have enough space for equipment and feed. Within the barn, the horse is given a stall. Standard recommendations for stall size are at least 10 feet by 12 feet (3 × 3.8 m) with a height of 8 feet (2.62 m). Tie stalls should be 5 feet by 12 feet (1.6 × 3.8 m). The bottom 5 feet (1.6 m) should be solid boards that can withstand kicking. The horse should also have access to a paddock or pasture to get plenty of exercise.

Keep the stalls clean and dry. If clay is used as flooring, remove and replace the top foot yearly. This decreases the incidence of diseases and parasites.

13-20. Good facilities are needed with horses. These photographs show a horse in a secure stall and a tack room where equipment has been properly stored.

FEEDING EQUIPMENT

Hay or other forage should be fed in a hayrack or manger that is positioned above the ground. Grain should be fed in a box, pail, or tub. Watering is best with buckets or automatic waterers. Do not allow horses to consume large amounts of water at one time.

GROOMING

Horses should be groomed regularly. The grooming equipment needed includes a hoof pick, body brush, and a mane and tail comb. A curry comb and a sweat scraper work well when bathing the horse.

Each horse should be provided with its own halter. Make sure that the halter fits the animal and that the throatlatch is not too tight. When a horse is let out to pasture, remove the halter. Otherwise, the horse may get caught and strangle itself.

SADDLE AND BRIDLE

There are two types of saddles: western and English. The type chosen depends upon the type of riding that will be done. The saddle should be com-

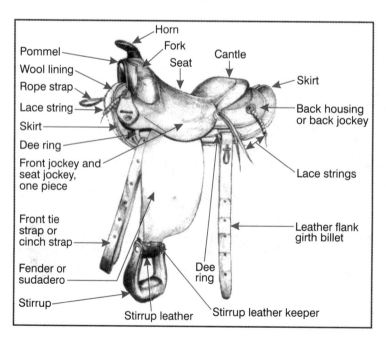

13-21. Parts of a Western saddle.

fortable for the rider and the horse. A saddle pad is placed under the saddle to comfort the horse.

A bridle is also needed when riding. A bridle is the part of the harness placed on the horse's head. It is used to direct and control a horse. There are dozens of styles and dozens of bits to choose from. A bit needs to be wide enough not to pinch the cheekbones.

EQUITATION

Equitation is riding and managing horses. It is often called **horsemanship**. Development of good horsemanship and patience is essential for working with horses. Each animal should be treated as an individual. Horse and rider should work together as a team.

13-22. An English saddle.

Mounting and Dismounting

A horse should be approached, mounted, and dismounted from the left side. The best position to stand when mounting is to place the left foot in the stirrup facing the saddle squarely. Hop and swing up and into the saddle using momentum from the right leg.

When dismounting, hold the saddle horn with the right hand and balance with the left hand on the horse's neck. Making sure the horse is steady, loosen the left foot in the stirrup and shift your body weight to the left foot. Free the right leg and gently let down.

Seat Position

Sit tall in the saddle. Your back should be erect not slouched. Your position in the saddle should allow for comfortable control of the horse and use of aids. Your shoulders should be back and your arms held close to the body.

From the left, gather reins in left hand, and place left hand on or just in front of withers.

Turn stirrup iron one-quarter turn, steady stirrup with right hand and shove left foot into it.

Spring upward and lean on left arm, shift right hand from cantle to pommel of saddle; then swing extended right leg over horse's back and croup.

Ease down and sit easily, be alert and keep head up, and allow legs to hang comfortably with heels well down and toes turned out slightly.

13-23. How to mount a horse.

HORSE CONTROL

Four basic aids are used to control a horse. They are hands, legs, weight, and voice.

Hands

Your hands control the horse's forehead with use of the reins. Hands should be soft, light, and steady, yet firm. Your arms, elbows, wrists, hands, and fingers should be relaxed. Holding the reins with a little slack will keep control, yet relieve pressure on the bit.

13-24. How to hold the reins.

Legs

Your legs are used to control the hindquarters and the forward movement of the horse. The horse should learn to move when your legs are squeezed.

Voice

Successful use of the voice means being consistent. Words such as "back," "easy," and "whoa" are often used. The words used, as well as the tone of the voice, are important. Remain calm and do not yell at the horse.

Weight

Shifting your body weight signals the horse to shift its weight. This causes the horse to either be free to move or to be held in place. Signaling a horse by use of weight can be done by shifting your seat, moving weight from one stirrup to another, or being in rhythm with the horse by leaning your upper body forward.

MOVEMENT

Movement is initiated by using the four aids. Overuse of these is a common fault of many beginning riders. There are various types of movement. The horse will become ready to move when you squeeze your legs. Movement is governed by the actions of the reins.

Walk

A *walk* is a four-beat gait. During a walk, both you and the horse are relaxed. Slightly squeeze your legs and move the reins to begin walking.

Jog

A *jog* is a two-beat gait. You should be sitting deep in the saddle with just enough weight on your ankles to absorb some of the motion. Your upper body should be leaning forward to maintain balance.

Lope

A *lope* is a three-beat gait. You should sit very deep in the saddle and again lean forward to maintain balance.

Backing

To back the horse, sit erectly in the saddle with your weight forward. Collect the horse and squeeze with your thighs. Tell the horse to "back." Alternately put pressure on each rein.

Stopping

For proper stopping, a horse should be cued ahead of time by either saying "whoa," squeezing with your thighs, tugging on the reins, or gently touching the horse's neck. To bring the horse to a stop, squeeze with your thighs and firmly pull on the reins. Sit erect in the saddle, grip with your thighs, and your body weight should be forward.

13-25. Riding in small groups is a fun activity. (Courtesy, Paso Fino Horse Association, Inc.)

Some inexperienced horse riders are frightened by not being able to quickly stop a horse. Until you gain good experience in handling a horse, always ride one that readily stops.

REVIEWING

MAIN IDEAS

Horses are chiefly used today for their recreational and personal value. They are classified as light horses, ponies, or draft horses. The types of light horses are riding horses, gaited horses, stock horses, polo mounts, hunters and jumpers, racehorses, and driving horses.

Selection of a horse should be based upon an affordable price range, the animal's disposition, size, gait, breed, color, and the work to be performed. Purchase a horse that suits its rider.

Breeding requires careful management. Conception of mares is the lowest of any farm animal. Careful and precise management of horses is a must.

Horses require concentrates in their rations to provide energy and protein, minerals, and vitamins. Good water is needed. Oats, corn, and barley are the most common sources of carbohydrates and fats. Supplements that are widely used include soybean meal and linseed meal. Mineral and vitamin supplements are mixed with the total feed ration.

Horse health should be checked each day. A horse's teeth may need to be floated. Their hooves should be cleaned daily and if shoed, they should be trimmed every four to six weeks.

Facilities for horses do not have to be expensive and extravagant. A barn with a good-sized stall is needed. Basic equipment includes a hayrack or manger that is raised off the ground, hoof pick, body brush, and a mane and tail comb.

Equitation (horsemanship) is skill in riding and managing horses. A beginner must learn how to properly mount, dismount, sit in the saddle, use the aids (hands, legs, weight, and voice), coordinate movement, and stop a horse.

QUESTIONS

Answer the following questions using correct spelling and complete sentences.

1. What is the purpose of horses?
2. How are horses classified?
3. How are light horses broken down in types?
4. What factors should be considered when selecting a horse?
5. Why is good management of utmost importance in horse reproduction?

6. Name six things that should be considered in the breeding management of horses.

7. What does an average horse's diet consist of?

8. When managing the health of a horse, what should the owner look for?

9. How should hooves be cared for?

10. What facilities and equipment are needed for horses?

11. What is horsemanship? Why is proper horsemanship important?

12. How are the four aids used in horsemanship?

EVALUATING

Match the term with the correct definition. Write the letter by the term in the blank provided.

a. mare   d. lope   g. foal
b. gelding   e. farrier   h. stud horse
c. plug   f. gait

_____ 1. A young horse that has not been weaned.

_____ 2. A male horse kept for breeding.

_____ 3. A person who puts shoes on horses' feet.

_____ 4. Slow, three-beat gait in which the head is carried low.

_____ 5. Horse with poor conformation and common breeding.

_____ 6. Sexually mature female horse.

_____ 7. The way a horse moves, walks, or runs.

_____ 8. Male horse castrated before reaching sexual maturity.

EXPLORING

1. Visit a horse farm. Explore the breeding, feeding, health, and facility management of the farm. Prepare a written or oral report for the class on what you learn.

2. Study common management practices used with horses and explore how this is changing. Interview authorities on horses. Use several references with your study, including *Equine Science and Management* and *Horses and Horsemanship* (available from Interstate Publishers, Inc.).

3. Attend a horse show. Observe the breeds of horses, how they are groomed and ridden, and other areas related to horse management.

14

DRAFT ANIMALS

If you have a heavy load that needs moving, a draft animal might do it for you! No gasoline or tires are needed. Hay, grain, and care will handle routine maintenance. Not many people use draft animals to do work anymore. They are used as hobbies, in animal-pull events, and as companions.

Draft animals had a major role in the settlement of America and farming of the land. They are still important in some countries. In early America, draft animals supplied the power to farm and clear land. Draft animals were replaced with engine-powered machinery and equipment.

Draft animals are still around today. They have only limited use in farming in the United States. They are no longer major power sources in the military. When military tactics changed and gun powder was developed, the horse was no longer vital in battle. Today, draft animals are largely kept for fun.

14-1. A camel is being used to plow land in India.

393

OBJECTIVES

This chapter provides basic information on draft animals. It has the following objectives:

1   Describe draft animals and list major kinds and breeds

2   List the possibilities in draft animal production

3   Explain important management practices with draft animals

4   List the nutritional requirements of draft animals

5   Describe the facility and equipment needs with draft animals

TERMS

automated feeding
breeding period
curry comb
draft
draft animal
feather
gregarious behavior
hand-feeding
harness
mule

ox
power
regurgitate
roan
ruminate
self-feeding
social ranking
tri-purpose animal
yoke

KINDS OF DRAFT ANIMALS

Several different animals are used for draft. A ***draft animal*** is an animal that has been trained and is used for pulling heavy loads. Most often, draft animals are horses or oxen. However, donkeys, mules, camels, and buffalo have been used as draft animals and still are used today in some countries.

When a draft animal is selected, consider the climate, the availability of the animal, the cost of the animal, the type of work to be done, and the social and religious traditions, which might limit the ownership or use of some animals.

14-2. A few farms in the United States continue to use draft animals, such as this team of horses on an Amish farm.

OXEN

The most popular draft animal, the *ox* is any animal of the bovine (cattle) family. Oxen are primarily castrated males or bulls, but sometimes females are used. Oxen are preferred over other draft animals because they are ruminants and use their feed better, but they are much slower than the other animals. Chapter 7 gives you more information on beef cattle and Chapter 8 gives you more information on dairy cattle. Oxen are classified as a ***tripurpose animal*** because they served three purposes: work, milk, and meat.

DRAFT HORSES

Draft horses are common draft animals in the United States. Some breeds of horses are defined as draft horses. They are larger and heavier than

14-3. A team of Shorthorn steer oxen is being used to pull a plow on a Michigan farm.

the horses used for riding and pleasure. All horses and horse-like animals are from the equine family. Draft horses were developed to be 1,500 to 2,500 pounds (680–1,134 kg); have a low center of gravity; and be wide, deep, compact, strong, and large boned. Other members of the equine family, such as mules, are used because they are cheaper to buy than horses.

Five breeds of draft horses are found in North America: Belgian, Percheron, Shire, Clydesdale, and Suffolk.

Belgians, which are a Flemish breed, are brown to light blonde and sometimes even **roan**, which is a mixture of white and colored hairs, and became popular because of their size, endurance, tremendous power, excellent mus-

14-4. Cyldesdales are among the best-known draft horses.

14-5. Belgian horses plowing on an Ohio farm.

cling, and style. Belgians are the number one breed of draft horses in the United States today, replacing the Percheron that was previously number one.

Percherons, which are from France, are more refined and balanced than some of the other draft horse breeds. Fifty percent of Percherons are black and the other 50 percent are black and gray.

Clydesdales are the third most popular breed in the United States. They are brown or black with white faces; have long white hair on their legs, known as *feathers*, which sometimes extend all the way up to their bellies;

14-6. Donkeys have often been used as pack animals.

and black manes. Clydesdales originated in Scotland and are well known for pulling heavy wagons.

Shires are black in color and, like Clydesdales, have feathers on their feet, but they have greater bulk than the Clydesdale. Shires were developed in England.

Suffolks were also developed in England. Unlike the other four breeds, their development was solely for agricultural reasons. They are chestnut color (sometimes referred to as sorrel) and have very little white on them. They have very little or no feathers. Suffolks are easy keepers, which means they stay in good condition with less feed than other horses require for the same condition. Suffolks have a very fast walk, great stamina, longevity, and a willingness to work.

Mules

A *mule* is the hybrid offspring of a male ass (donkey, jack, or jackass) and a mare (female horse). Mules physically resemble both donkeys and horses. They have the large ears of donkeys, are strong like a horse, and tend to pace themselves for long-term power like a donkey. Horses tend to do work at a fast pace and do not hold up for the same long hours as mules. Mules are usually larger in size than donkeys, but smaller than draft horses. As hybrids, mules do not reproduce, though they are born as males and females. Male mules are gelded (castrated) as young animals.

14-7. A mule is being used to drag a log. (Courtesy, Lucky Three Ranch, Colorado)

Mules have advantages over other draft animals. Mules endure heat better than horses. They are less sensitive to their feed than horses and have fewer feed-related diseases. Mules eat lower-grade feed, which saves money.

Mules are physically sound, have fewer defects than horses, are surefooted and careful, and live a long time. When treated with force, mules tend to have stubborn dispositions somewhat like a donkey. Some people like mules and get along well with them. Other people prefer horses to mules. In some cases, mules

14-8. A team of mules is pulling this covered wagon. (Courtesy, American Donkey and Mule Society, Texas)

14-9. Mules may be trained for showing and competitive events. (Courtesy, Lucky Three Ranch, Colorado)

will turn aggressive toward other animals, such as cattle, and chase them through a pasture.

Many farmers in the United States relied on mules to provide power for long hours of the day doing tedious crop cultivation. In the late 1800s and early 1900s, the large cotton plantations of the South often used hundreds of mules for power.

Today, some people view mules much as horses. They train them for show and dress them in fancy ways formerly reserved only for horses. The mules are pets!

DRAFT ANIMAL PRODUCTION

After a period of decline, draft animals are increasing in numbers. Special events feature draft animals performing power events as well as skill events.

Connection

HORSEPOWER SCIENCE

Horsepower is used in many ways today. Engines have horsepower ratings, as do electric motors. But what does the rating of engines and motors have to do with horses?

Horsepower was developed in Britain as the unit of power. It refers to work done at the rate of lifting 550 pounds a distance of 1 foot in a time of 1 second. A strong, fast horse would be needed to meet this definition of horsepower. Animal pull competitions use a different measure of power, as stated in the chapter.

Horsepower has been converted to other measurements. Electricity, for example, may use horsepower. One horsepower is equal to 746 watts of electricity.

1 foot in
1 second

550
lbs

Many enthusiasts show their draft horses in halter classes. The animal is judged on conformation. Hitch classes are used where one to six horses may be used in a class. There are also clubs of draft horse owners who put on old-fashioned farming demonstrations for the public. Loggers in the north are starting to use the draft horse again. Horses cause much less damage to the environment than tractors, especially crawler-type tractors. Teams of horses also compete in horse pulls. The team is hooked either on a boat or on a machine and they pull the weight. The team that pulls the most weight is the winner.

DRAFT ANIMAL REPRODUCTION

Reproduction is a key component in the draft animal industry. Strong, well-bred animals are needed.

Estrus is the time when the female animal is most receptive to be bred by the male animal or artificially. The time of the year when the female has her estrus periods and estrous cycle is the ***breeding period***.

Oxen

Oxen are raised similarly to beef cattle. The females reach puberty at 8 to 15 months of age. The duration of their cycle is 12 to 20 hours every 21 days year round. Breeding must occur during this time.

Even though cows cycle year round, most producers prefer to have the females bred about the same time so they calve about the same time of the year and all of the young are close to the same age. When it comes time to make management decisions, such as vaccinating or castrating of the male calves, it is easier when the whole group can be done at once. Oxen in estrus stand to be mounted by herdmates, lack an appetite, act nervous, restless, and are very vocal. After the cow has been bred, it takes about 9 months or 270 to 295 days for her to calve.

Draft Horses

Draft horse females reach puberty at 12 to 15 months of age. Unlike oxen, females have a cycle that lasts 4 to 6 days and comes every 21 days during the estrous cycle, which begins in the spring and stops in midsummer. Females must be bred during this time of the year.

When females are experiencing estrus, they are very vocal, get nervous, restless and irritable, squat and urinate frequently, and seek out the stallion. Mares should be bred on the second day of estrus and every other day for two to four days. After the mare has been bred, either artificially or by the stallion it takes 11 to 11½ months, or 330 to 340 days, for her to foal.

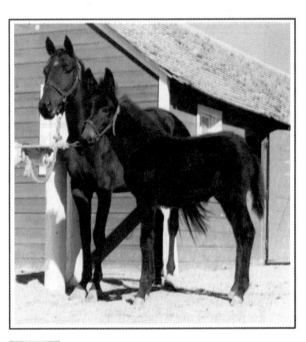

14-10. Horse mare with mule colt. (Courtesy, Lucky Three Ranch, Colorado)

YOUNG ANIMAL CARE

Both female horses and oxen clean their young with their tongues after giving birth. This gets blood circulating in the newborn and encourages the

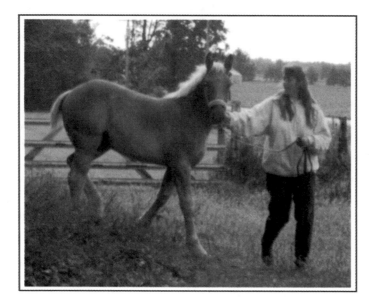

14-11. A five-month-old Belgian filly is being broke to lead.

newborn to stand and nurse. Instinct results in mothers being aggressive in protecting their young. Even a calm pet may become ornery with a baby.

DRAFT ANIMAL
MANAGEMENT PRACTICES

Understanding animal behavior is essential for good management of draft animals. First of all, *draft* is the pulling of an animal to move any object. For an animal to move an object, it must exert a force equal to or greater than the weight of the object.

DRAFT CAPACITY

The draft capacity of an animal increases with the weight of the animal. The rule is that an animal can exert a constant pull at its normal pace of one-tenth its body weight. If an animal is required to exert a concentrated effort, the time the animal can work will greatly diminish. Horses, mules, and oxen are preferred because they can pull loads at a normal speed over long time periods and exert extra energy when needed.

Power is the combination of pulling capacity and speed. A draft horse will pull a 150-pound (68 kg) load at a steady rate of $2^1/_2$ miles per hour, which is defined as one horse power. One horse power may also be defined as pulling

14-12. These Belgians are being used to rake hay. They are well conditioned for the work.

150 pounds (68 kg) out of a 220-feet (67 m) deep hole in one minute. An ox pulling the same weight will travel at 1¹/₂ miles per hour, so at the end of the day the horse has traveled further. Also, shorter animals have a lower center of gravity, so they must be hitched lower so less work is required for them to get the task accomplished.

All animals have certain behavior patterns, which must be recognized for proper health and overall well-being of the animal. Some animals pull better and reach maximum capacity when worked alone, but not when worked with other animals. Animals work well only when they have been properly trained and properly outfitted with the correct harness (for more information on the harness see the equipment section of this chapter).

An ox can deliver 25 to 50 percent more horsepower when harnessed with a collar instead of a yoke. This is because of the ox's lower center of gravity and the wider pushing area of the collar.

DRAFT ANIMAL SELECTION

The type of draft animal selected depends on the job and the time in which to do the job. For example, horses work much faster than oxen, but oxen use their feed more efficiently because they are ruminant animals. If horses are considered, there are other animals in the equine family that may be cheaper to purchase than a horse, such as a mule. All of these factors need to be taken into account by the owner when a draft animal is selected.

When selecting animals, it is also important to note that horses and oxen in herds develop social rankings. *Social ranking* is the order the animal falls within the herd. Social rank is determined by the age, size, strength, genetics, if they have horns or are polled (which means they are bred to be born without horns), and previous experience of the animal.

There are many sources available to assist producers of draft animal owners and producers. Veterinarians, the Cooperative Extension Service, state universities, state experiment stations, feed suppliers, other animal owners, animal associations, and libraries are all important sources for people to gather information pertinent to their animal and situation. Since animals and situations are different, it is impossible to answer every question that may occur.

NUTRITIONAL REQUIREMENTS

Animals vary in their feed needs, just as they vary from breed to breed and horse to oxen. A good feeding program is vital in maintaining an animal's health and strength. In determining the feed ration for a draft animal, first the species, development stage of the animal, weight and size of the animal, if they are or are not pregnant, and how often and hard they are worked must be decided. Many people use a commercial feed prepared especially for draft horses.

Water is important for all animals and should be available to them at all times, except during and right after working. Draft horses and oxen both require approximately 8 gallons (30.3 l) of water per head per day. Some may drink more, depending on activity.

All draft animals need protein. It may be fed in the form of alfalfa hay to both horses and oxen and/or cottonseed or cottonseed meal to oxen. The feed ration needs to be complete. The cost of the ration must be considered.

High-fiber and low-fiber feeds are also important to draft animals. In the high-fiber category, horses and oxen both benefit from oat hay. Oxen also benefit from corn silage and haylage. In the low-fiber category, both oxen and horses eat corn, grain, barley, and oats. By-product feeds are also used, such as pulp from citrus or beets, barley malt, and others. The availability and types of by-products depend upon the area in which the producer lives.

*F*EEDING OXEN

Oxen are ruminant animals, which means they have four compartments in their stomachs. They swallow their food as soon as they bite and salivate on it. Later, often while lying under a shade tree, they ***regurgitate*** the food. This is returning eaten food to the mouth for chewing. The food in the mouth is a cud. Oxen graze from four to nine hours per day and ***ruminate*** (chew their cud) for four to nine hours per day.

Extreme heat or cold lessens the amount of food consumed by the animal. Oxen are similar to beef cattle in many nutritional needs. An exception is that they need feed with higher energy when they are being worked.

FEEDING DRAFT HORSES

A horse is a single-stomached (monogastric) animal. It eats and chews its food immediately and then the food is used to give the animal energy. Ruminants are sometimes preferred because it has been said they use their food more efficiently.

Vitamins and minerals are important to assure the draft animal remains healthy and strong. The three minerals often added as supplements to the feed ration are salt (sodium chloride), calcium, and phosphorus.

Draft horses need the following seven major elements: calcium, phosphorus, magnesium, sodium, chloride, potassium, and sulfur. There are also some trace minerals that are required: chromium, nickel, silicon, vanadium, tin, copper, cobalt, and zinc. The vitamins needed by horses are A, D, E, and C. In most cases, good-quality hay will eliminate the need for vitamin supplements.

Horses should be allowed to cool down after working before they are fed. They may be given some water after they have cooled, but horses will get lame or foundered if they eat or are watered when they are too hot. Give a horse an hour or so to cool down before feeding or watering it. Also, give a horse a ½ hour to 45 minutes to digest food before harnessing the horse and making it do hard work. When feeding horses, make sure the hay is dust and mold free to prevent illness. The harder the horse works, the more feed the horse should be fed.

FEEDING SCHEDULES

In addition to feeding the correct ration, it is important to develop a feed schedule so animals are fed at the same times each day. The animals may be fed by ***hand-feeding***, which is giving animals the same amount of feed either one or two times daily. This would entail their grain and hay. For example, hay may be given two or three times daily and grain only once or twice.

Self-feeding is having feed in front of animals at all times. In this system, the producer needs to ensure some way of having feed available at all times. Many times, hay is available in the self-feeding system and grain is given through the hand-feeding system. This tends to work well since draft horses will overeat and make themselves sick on sweet feeds, such as grain.

The third option of feeding is **automated feeding**. Animals are fed mechanically so no one needs to haul feed at feeding time. Usually, an auger is used to bring the feed to the animals. This is a common way of feeding silage to cattle.

Draft animals ready for weaning from their mother should have their mother's milk supplemented with hay and grain. When the weaning takes place, the young animal will do better if it is used to eating food along with its mother's milk. It will cause less stress to the young animals if two or more of them are weaned at a time. The companionship is important to these animals and they will do better if they are not alone.

HEALTH PRACTICES

Animals respond differently to their environment. Some things may cause stress in one animal while not in another. It is vital that owners understand their animals' attitudes and behaviors. If an animal is behaving strangely, it may be ill, under stress, ready to be bred, or have other problems that may require attention.

There is no one who knows an animal better than the person who cares and works with it everyday. This person must always pay attention to the actions and attitudes of the animal.

CONDITIONING

An important health practice is to condition a draft animal before it is used for heavy pulling. If the animal does not work hard over the winter, start the spring with a smaller work load and shorter days until the animal has gotten back into condition.

A draft animal must physically train much as a human athlete. Because they do not have the power to reason and decide to stay in shape on their own, it is the responsibility of the owner to ensure that an animal is in shape before hard work. Animals may pull a muscle or physically hurt themselves if they are not conditioned. The least that will happen will be a break in their spirit because they will experience defeat.

CONSIDER HEALTH

Health begins with animal selection. To have a good, strong, sound draft animal, buy that type of animal. The stock purchased will make a big difference in the amount of work actually done. It is an accepted practice to take a veterinarian to inspect an animal before it is bought. The buyer is usually the one who has to pay for the veterinarian's visit, but the buyer then knows the soundness of the animal.

General Health

A draft horse has a normal heart rate of 28 to 50 beats per minute and an average temperature of 100.5°F (38°C) with ranges of 99.0 to 100.8°F (37.2–38.2°C). An ox has a normal heart rate of 40 to 70 beats per minute and an average temperature of 101.5°F with ranges of 100.4 to 102.8°F (38–39.3°C) temperature. When the owner has a feeling the animal is not feeling well, a higher temperature is a good indication that something is wrong. There are many possibilities of what can go wrong with an animal; and a veterinarian is a good person to call for assistance.

Nutritional diseases may affect draft animals. If animals are not fed the proper ration or supplements, they may have deficiencies. Too much feed can also cause problems. Metabolic disease may result from erratic feeding times and feeds themselves.

Parasites

Internal parasites may cause major problems in animals. They live at the expense of the animal they are living inside. External parasites get their nourishment from the outer surface of the animal's body.

Deworming and vaccinations of draft animals are important to stop or prevent parasites. Vaccinations are the injection of an agent into the animal for the purpose of developing resistance to what the animal is treated for. Animals pick up worms through their feed and pasture. As long as the animals are treated for parasites periodically, they will be okay.

Small precautions, such as brushing and grooming draft animals, will cut down on any manure on their coats. Manure attracts flies and other insects that may carry disease. A curry comb and a brush are both needed to groom the draft animal. The *curry comb* is an oval-shaped plastic or metal device used to loosen sweat, manure, and other foreign materials from the animal's coat. The brush is used to remove the materials from the animal's coat. If an animal is not bothered by flies and insects, it will be happier and healthier.

Foot Care

Another important area of health care is the feet of the draft animal. Since the animals are used to do heavy work, their feet must be maintained in good condition. If an animal has sore feet it will work much less and put forth minimal effort. Feet must be trimmed about every six months, depending upon the growth of the animal's hoof.

Horses have metal shoes formed to fit their feet. The shoes, which are nailed on, give the animal better traction for pulling loads. Proper shoes assist the horse in digging its feet into the ground to pull. Proper shoes and trimming lessen the possibility of the animal falling or pulling a muscle or developing a foot disease.

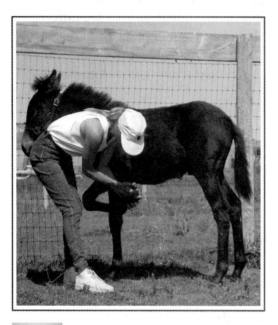

14-13. The foot of a mule is being examined. (Courtesy, Lucky Three Ranch, Colorado)

FACILITY AND EQUIPMENT NEEDS

To develop proper housing for a draft animal, the personality and characteristics of the animal must be understood.

Gregarious behavior is the instinct of animals to flock or herd. When an animal that normally is in a flock or herd cannot be with other animals, this causes stress in the animal and makes it difficult to deal with. Although draft animals prefer to be in herds, they also need to have enough space to live comfortably, because overcrowding of animals is also a source of stress.

When constructing windbreaks or barns, make sure the builder takes into account that the area needs to be easily cleaned and well drained. By doing so, animals can stay dry and out of the mud.

Draft horses need shelter or a windbreak, especially in the winter. Mares who are going to foal need to be kept separate from barren mares and need to be in an area where they can get plenty of exercise to stay in peak physical condition. If they are bred to foal in the winter, they need a warm, dry place to give birth. Some horses do foal outside, but the survival rate of the newborn is better and there are less complications if it is born inside.

Draft horses prefer to be in the company of other horses. If a person plans to work a horse—one that is usually worked as part of a team—alone, he or she may have trouble. Many times, when horses are worked as a team, it is difficult to separate the team without the horses getting unruly or upset. Horses tend to be naturally curious. Owners may put toys in the horse's stall to keep it entertained or leave it out in the pasture longer so it does not get bored in a stall. Remember, oxen are ruminants; they spend many hours per day laying around and chewing their cuds. Horses have to search for pasture longer since they only have one stomach.

Horses may be kept out on pasture where they have access to a windbreak, or kept in box stalls or tie stalls. Tie stalls are okay for short periods. After long periods in a tie stall, a horse will get bored and start causing trou-

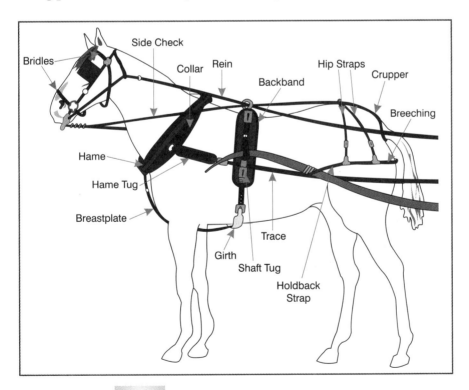

14-14. Parts of the harness of a draft horse.

ble, such as chewing on the boards it can reach, pawing or whinnying, and being vocal. These are all signs of boredom in the horse. If horses are pastured in lots next to each other, they may run the fence together and be very vocal with one another.

STRONG FENCES AND STALLS

Oxen, on the other hand, because they are so big, tend to break through fences to get with other cattle if they are separated and penned alone. Oxen prefer shady areas in the warm weather and gather under shelter or windbreaks in the cold weather.

A cow ready to calve in extremely cold weather should be kept inside. The newborn has a better survival rate if it is born inside, out of the elements. A cow and calf require about 300 square feet (27.9 m^2) of space and a steer needs 150 to 200 square feet (13.9 to 18.6 m^2) of space. The fences should be 50 to 60 inches (1.3 to 1.52 m) high; however, this depends on the size of the cattle.

Oxen are followers and they will tend to follow other cattle in a herd. When animals must be moved, get the leader moving and the others will follow. When moving either horses or oxen, it is necessary to move slowly and quietly so the animals do not get more nervous and have extra stress. If they are to be loaded, make sure they only see one way to go.

Draft animals have a keen sense of smell and notice loud, unfamiliar noises and strange objects they may see. It is best to avoid exposing them to

14-15. A Percheron team is harnessed and ready for work.

many strange, unfamiliar things when moving them. Many animals get scared and even unruly in strange surroundings with other animals and people they do not know. Working calmly and slowly will quiet the animal and eventually pay off because they will do what is expected sooner than if they are scared and try to run through things and over people.

EQUIPMENT

The harness and bridle are the most important equipment in using draft animals. Some people consider the bridle a part of the harness. The *harness* is the equipment put on a horse or other draft animal that enables it pull with its shoulders. The harness is the key to the amount and efficiency of the power of the animal. There are different types of harnesses, from fancy to plain and from leather to nylon.

A harness must fit properly. A good-fitting harness will allow the animal to maximize its power. In actuality, an animal is pushing instead of pulling. The horse pushes on the collar and this pulls on the hames and the tugs, which causes the load to move.

Each animal usually has its own collar, harness, and bridle, which are adjusted to fit perfectly. The collar goes over the horse's head and rests around the neck and on the shoulders. The collar should fit snugly, but not bind on the horse. If the horse is worked, and a sore results, the collar is not properly fitted.

The harness is put over the top of the horse and the hames are attached around the collar. It is vital the hames are a good fit with the collar. If the fit is improper, a sore will once again result. All of the power comes from the collar, hames, and to the tugs, which run the length of the horse and are attached to the load at the back end of the horse. The bridle goes over the head of the horse and is hooked to the lines, which are used to steer and stop the horse.

After the horse is harnessed, it is hooked onto a pair of whipple trees. The tugs hook onto the whipple tree, which is then hooked onto the load. It is imperative to hook the animals the proper distance from the load; otherwise, they may get their heals scrubbed or be so far out and away from the load that they lose the advantage of being hooked low and close.

If one has never harnessed an animal before, it is best to find someone who knows how and ask for assistance. Many times, the people the animal is bought from can give good advice as to the collar size and the way it was hitched before.

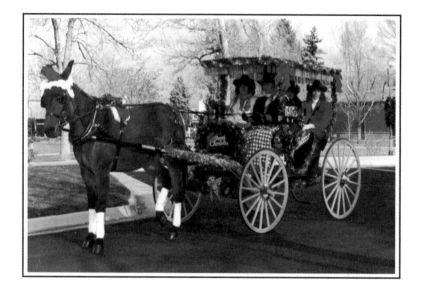

14-16. A holiday season parade entry of a mule. (Courtesy, Lucky Three Ranch, Colorado)

A harness can be a very expensive piece of equipment and must be cared for. A properly cared for harness will last virtually forever. All that is required is to keep the harness clean and well oiled with a good harness oil. Some communities have harness shops that can dip harnesses in oil, eliminating the owner's mess and problems of what to do with the old oil.

Just as horses push on their collars, oxen push through their yoke to move their load. The *yoke* is a wooden bar that hooks two animals together with a bar between them. The oxen are hooked around their necks or horns with a bar between their heads. They hold their heads low and since that is where their strength is, they can move more hooked this way. Oxen do not have bridles like horses do. Some oxen are steered with lines hooked to a halter and some have nothing on their faces.

REVIEWING

MAIN IDEAS

Draft animals have been trained to pull heavy loads. Once used for power in agriculture, they have been replaced in North America by machinery. Today, people enjoy having draft animals as pets and for competitive events.

Draft horses, mules, and oxen are the most common draft animals. The horses and mules are often trained for showing and groomed in fancy ways.

To effectively work with and raise draft animals, many considerations should be taken into account. The appropriate breed needed for the work to be accomplished, good management practices, a nutritionally complete diet, proper health program, and the correct facilities and equipment are all important to healthy draft animals.

A healthy draft animal will work hard to accomplish the tasks needed to be done. Diseases, parasites, and nutritional problems affect most draft animals in one form or another.

Appropriate facilities and equipment are needed. Since draft animals are very strong, all pastures, pens, and other structures around them should be built to withstand their power.

QUESTIONS

Answer the following questions using correct spelling and complete sentences.

1. What is a draft animal? Name three and compare their performance.

2. What are the major differences between reproducing oxen and draft horses?

3. What are the power and drafting potential of animals?

4. What are the nutritional needs of draft animals?

5. What are the major health practices with draft animals?

6. What facilities and equipment are needed with draft animals?

7. In constructing facilities, why is the power of the animal important?

8. How is an animal or team harnessed to its load?

EVALUATING

Match the term with the correct definition. Write the letter by the term in the blank provided.

a. mule	d. harness	g. yoke
b. power	e. social ranking	h. gregarious behavior
c. draft	f. regurgitate	

_____ 1. The pulling of an animal.

_____ 2. Hybrid of ass and horse.

_____ 3. The combination of pulling capacity and speed.

_____ 4. The instinct of animals to flock or group together.

_____ 5. A wooden bar that holds two harnessed animals together.

_____ 6. Returning undigested food to the mouth for chewing.

_____ 7. The order an animal has in a herd.

_____ 8. Attachments put on an animal so it can pull.

EXPLORING

1. Attend a draft horse, mule, or oxen pulling event. Observe how the animals are handled and the kind of work they do. Give a report in class on your observations.

2. Assist in harnessing a team of animals. Study the different parts of the harness and how they are connected. Be very careful; draft animals can be dangerous.

15

DOGS

Nearly everyone has a favorite dog. Cuddly puppies, trained companions, and smart hunting dogs create excitement. Having a puppy or dog is more than adoring it. "Having" involves care, training, and attention on a daily basis.

Companion dogs are important to people. Dogs provide fun as well as make people's lives better. People relax when they have an enjoyable dog. Medical experts say that a person's heart rate decreases when petting a dog. The touching and sharing of affection have a positive effect on the person and the dog. Some dogs help people in daily living, such as leader dogs for the blind.

In addition to dogs at our homes, some people have businesses involving dogs. Pet stores, kennels, animal shelters, feed manufacturers, and other places offer career opportunities for people who know about animals. Some people are entrepreneurs and start their own businesses.

15-1. Thanks to her owner, this Italian greyhound has its own special outfit for the holiday season.

OBJECTIVES

This chapter presents basic information on dogs. It has the following objectives:

1 Describe the biology of a dog

2 List and describe the classes of dogs

3 Identify factors to consider in selecting a dog

4 Explain management practices with dogs

5 Describe reproductive practices with dogs

6 Describe nutrition and feeding of dogs

7 Identify important health practices with dogs

8 Identify grooming practices with dogs

9 Describe facility and equipment needs of dogs

10 Assess sanitation and cleaning practices used with dogs

TERMS

bitch

grooming

herding dog

hound

mixed-breed dog

nail bed

non-sporting dog

orphaned puppy

sporting dog

stud dog

terrier class

toy breed

training

weaning

whelp

whelping box

working dog

working dog

THE BIOLOGY OF DOGS

Dogs are among the animals that have been pets the longest. Their use as companion animals began more than 10,000 years ago. In the centuries since, many breeds and varieties have been developed.

Dogs are carnivores, meaning that they are flesh eaters. Dogs are also mammals and monogastric. The scientific name for the domestic dog is *Canis familiaria*. They are often called canines (or K-9) because of their family (*Canidae*) and genus (*Canis*). Understanding their nutritional needs and digestive systems helps provide better care.

Dogs are endothermic animals. They have well-defined, sturdy skeletons. Many are adapted to specific uses, such as hunting and retrieving.

The body structure of a dog varies with the size and breed of the dog. The weight of dogs varies. A mature Chihuahua may weigh less than 6 pounds (2.7 kg). A mature Saint Bernard may weigh as much as 200 pounds (90 kg).

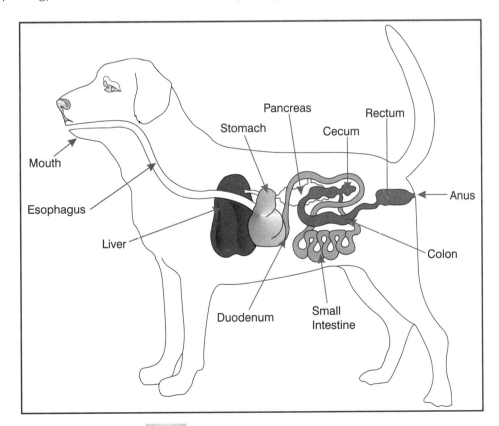

15-2. Gastrointestinal system of a dog.

15-3. Skeleton of a dog.

Most dogs have similar skeletons of about 320 bones. The major difference is in the size of the bones.

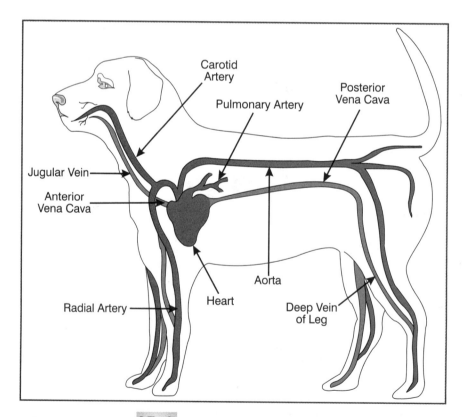

15-4. Circulatory system of a dog.

Dogs may live many years. The smaller breeds may live 15 to 18 years. The larger breeds, such as the Saint Bernard, have a shorter life span, often only 8 to 10 years. Babies are born in groups of 1 to 10 puppies, known as litters. The female, known as a ***bitch*** or dam, has a gestation period of about 63 days, depending on the breed. The male is the sire or ***stud dog***. (More information on dog care is given later in the chapter.)

CLASSES AND BREEDS

About 300 breeds of dogs are found throughout the world. In the United States, the American Kennel Club (AKC) classifies and maintains records on dog breeds. Seven classes have been identified, based on the use and characteristics of the breed. This information is helpful in selecting a breed.

■ Herding Dogs—A ***herding dog*** is in a class that is popular with many people who have sheep and cattle. The dogs are easily trained to assist in herding animals into pastures and barns. The common herding species include the Australian cattle dog, Shetland sheepdog, collie, and German shepherd dog. Select these dogs as pets only if space is available for them to run and enjoy. A herding dog does not usually like to live in confinement.

15-5. This Australian cattle dog is working in a barn.

■ Sporting Dogs—A ***sporting dog*** is in a class that is used for sporting purposes. Breeds include pointers, setters, spaniels, and retrievers. The pointers and setters are hunters that run and then stop when they find their prey.

15-6. The cocker spaniel (left) and golden retriever (right) are examples of sporting dogs.

Spaniels run and scare their game out of the cover. Retrievers swim and retrieve game in both the water and on the land.

■ Tracking Dogs (Hounds)—A *hound* is a class of dogs used for tracking. They have good abilities to follow the scent left by animals or people. Hunters may use hounds to locate game. Law officials use them to find escaped prisoners or lost people. The most common breeds are the beagle, basset, dachshund, and greyhound. Some breeds run quickly to track and others work slowly and thoroughly to find their catch.

15-7. The beagle (left) and basset hound (right) are examples of tracking dogs.

- Working Dogs—A **working dog** *is* a class used by people to get something done. Examples of the work done by these breeds include pulling sleds, protecting property and other animals, and sniffing for bombs and drugs. Some of these dogs are used to guard property and for human protection. The breeds of working dogs include the Alaskan malamute, Doberman pinscher, Rottweiler, Saint Bernard, and Samoyed. These dogs are sometimes classified as service animals.

- Terrier Dogs—The **terrier class** is a group of 25 different terrier breeds that are recognized by the AKC. The name, terrier, is from the Latin word, *terra*, meaning "earth." The breeds are noted for following the animals they are

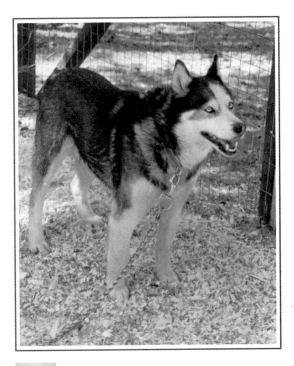

15-8. This female Alaskan malamute (also known as a husky) has been trained as a sled leader dog in a Tok, Alaska kennel.

15-9. A Bedlington terrier (left) and a Scottish terrier (right) are 2 of the 25 terrier breeds. Scotties can be white or black.

15-10. A proud owner enjoys her Yorkshire terrier—a very small dog.

chasing down into the earth. They dig in the earth to capture their prey. Miniature Schnauzers, Airedales, bull terriers, and Scottish terriers are representatives of this group.

- Toy Breeds—A *toy breed* is a class of small dogs that weigh between 4 and 16 pounds. Dogs in this class are known for their companionship and their long lives. Examples of toy dogs are Yorkshire terriers, toy poodles, English toy spaniels, and Chihuahuas.

- Non-Sporting Breeds—The *non-sporting breed* class of dogs was developed for specific purposes. They are now primarily pets. Examples of these are Dalmatians, bulldogs, and poodles (miniature and standard).

15-11. A Dalmatian (left) and an English bulldog (right) are examples of non-sporting breeds.

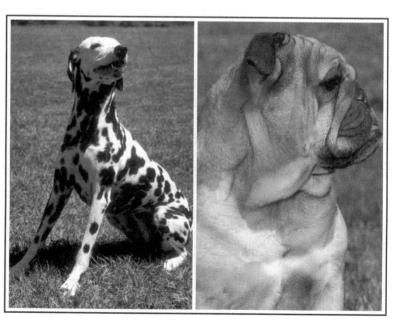

SELECTING A DOG

Selecting a dog is an important decision. Good information is needed. Careful consideration should be given to the traits of various breeds and an individual's personal wishes.

If a puppy is obtained, it should be between 8 and 12 weeks of age. During this time the puppy is old enough to be removed from its mother and is still forming its personality. If the owner wants to have a dog that is an average size for the breed, then an average-size puppy should be selected.

Some considerations in selecting a dog are discussed here.

- Mixed or Purebred—Many people have mixed-breed dogs. A ***mixed-breed dog*** is one of unknown ancestry. In some cases, owners know the breeds of the parents, while in other cases they do not. Some people want purebred dogs. They want a dog whose ancestry is known in terms of breed. In some cases, the dog may be bred to another of the same breed and a litter of puppies produced that is entitled to registration. A problem with a mixed-breed dog is knowing how large the mature dog will be. A young puppy can change significantly as it grows and matures. The mixed-breed dog cannot be shown in breed shows nor can pups to be sold as purebreds. Of course, with purebred dogs, keeping records and making application for breed registration requires effort.

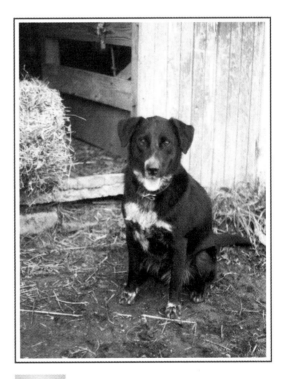

15-12. Mixed breeds often have desirable traits. This puppy was selected for farm life because of the breeds of its parents.

- Space Requirements—Dogs need a good environment. Some are kept inside; others roam outside. The size and nature of the dog may determine if it is kept inside. A large dog in a home is much more difficult to manage. A hunting dog is best kept outside. In cities and restricted residential areas, most dogs must be confined. Non-dog owners do not like the dogs of other people on their property or threatening them. Decide if the dog will live inside or outside before selecting and bringing it home. Most breeds can live healthy lives outside if they have shelter. A dog should not be kept inside for a while and then put outside. Temperature changes make it difficult for the dog to adjust. Also, a dog that is allowed outside part of the time will bring dirt into a home. Inside dogs require much more training than outside dogs. This is because dogs have to learn what they can and cannot do. Owners must be involved in the training. Outside dogs need proper housing and may need to be in a pen appropriate to their needs.

■ Hair Length—Hair length is important in selection. Disease, parasites, and dirt are associated with longer hair. Dogs with short hair are not as likely to develop diseases of the skin and bring dirt inside. Long-haired breeds of dogs are more likely to collect plant seed, feces, mud, and other dirt. The owner must note the extra time required to properly care for and groom a long-haired dog. A short-haired dog does not need to be brushed as often and it does not take as long when the dog is brushed.

■ Gender—The sex of a dog should be considered in the selection process. Most dog breeds reach sexual maturity between 9 and 10 months of age. There are positives and negatives with both males and females. If the dog is a female, steps are needed to keep it from getting pregnant. A male dog may pick up the scent of a female that is outside and breed her. To avoid breeding, the female can be spayed. Male dogs will try to break out of pens or off of leashes if a female in heat is nearby. In some cases, the males may run away for several days looking for females. Male dogs also fight, so the dog may come home injured. Male dogs may be castrated to avoid breeding. As in the case of the female, if the owner intends to enter a male dog in breed shows, it may not be shown if it is castrated. Neutered or castrated dogs may be shown in obedience classes.

REPRODUCTION

Dogs should breed and raise puppies only if there is a demand for the puppies. Many dogs are put to sleep (humanely killed or euthanized) each year at shelters because they are not wanted.

BREEDING

The female dog comes into her first heat at 6 to 12 months of age in small breeds and 8 to 18 months in larger breeds—they mature slower. The owner can tell when the bitch is in heat because blood is discharged from the vulva. After the discharge, the bitch should be exposed to a male in 10 or 11 days. If puppies are not wanted, the female should be confined, because male dogs from the neighborhood will breed her when she is outside.

When selecting a male to breed the female to, it is important to have parents whose weak points compliment each other. Although the bloodlines of registered dogs are on their papers, about 50 percent of the genetic make-up of a puppy comes from one parent.

Whelping

The bitch will give birth, or *whelp*, 63 to 67 days after she is bred by the male. Most bitches will take care of their puppies with little or no assistance. A **whelping box** should be used. It is a box specially designed for the birthing process. It can be constructed about two or three weeks before whelping so the female can get used to it.

The box should be large enough for the female to lie down and stretch out comfortably. Bedding from the female's bed should be put into the whelping box, along with newspapers, shredded paper, or carpet, for her to use at whelping time. The box should be placed in a warm, quiet, secluded place where there is very little, if any, traffic.

15-13. A five-day-old bulldog puppy with eyelids sealed.

Caring for Puppies

The puppies should be handled as little as possible for the first 14 to 21 days. For the first few days, they need a temperature of about 85°F (29°C).

Newborn puppies can neither see nor hear very well for 10 to 15 days following birth. Their ears and eyelids are sealed. Puppies begin walking at about two weeks of age when their ears and eyes open.

Newborn puppies cannot urinate or defecate. They do so when their mother licks under the tail. Puppies begin to wag their tails and bark at about three weeks of age. When they are so young, they need warmth, their mother's milk, and plenty of sleep.

NUTRITION AND FEEDING

15-14. Many kinds of dog foods are available. Read the label in selecting the type best for your dog's age, sex, and condition.

Dogs and puppies need food that meets their nutritional needs. Puppies need foods higher in protein than adult dogs. Food intake is also regulated by the activity of the dog. Working dogs need more energy than those that get little exercise.

FEEDING AN ORPHANED PUPPY

An ***orphaned puppy*** is a puppy whose mother has died or is not able to nurse. They will quickly starve unless fed. Puppies can be fed a milk formula. Bitch milk is richer than that of a cow or goat. Therefore,

Connection

DOGS SAY "NO" TO CHOCOLATE

Dogs like chocolate. They welcome a small piece shared by an admiring human chocolate-lover. But, chocolate is dangerous to dogs!

Never allow a dog to eat chocolate. The animal will likely suffer chocolate poisoning. A small dog can die within a few hours from eating chocolate. Larger dogs have vomiting, diarrhea, and increased heart rate. This can be followed by tremors, seizure, and a coma.

Prevention: Don't feed chocolates. Treatment: Induce vomiting or administer activated charcoal to prevent absorption of the toxin by the body.

cows' milk does not provide sufficient nutrients. A suggested formula: 1 cup homogenized milk, 3 egg yolks, 1 tablespoon corn oil, and 1 dropper liquid vitamins (pediatric). The materials should be thoroughly mixed. Store the formula in the refrigerator. Warm the formula until a drop on your wrist cannot be felt. It should never be overheated and given to a puppy.

Orphaned puppies can be fed with an eye dropper, a bottle and nipple, or with a feeding tube. An eye dropper is used with the smallest and youngest puppies. With a bottle, open the puppy's mouth, insert the nipple, and hold the bottle at a 45-degree angle. This will keep air out of the stomach. With a feeding tube, attach a small plastic tube about the size of a small straw to a hypodermic syringe. The formula is released from the syringe.

A puppy's stomach is quite small and may hold only a small amount. It is

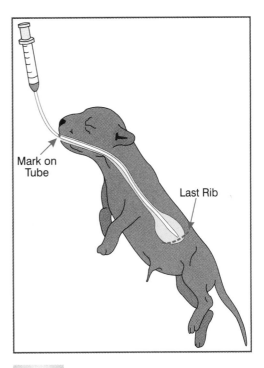

15-15. How a feeding tube is used with a puppy. (Never insert the tube too far nor feed too much.)

usual practice to feed puppies four times a day. The rate of formula is approximately 60 ml per pound of weight. As an example, an 8-ounce puppy would need 30 ml per day. Divided into four feedings, the amount per feeding is 7 to 8 ml. At a couple of weeks of age, the number of feedings can be reduced to three per day. Overfeeding can cause formula to get into the lungs resulting in pneumonia.

Weaning

Weaning is withdrawing the need or opportunity for a puppy to nurse. Puppies can be weaned at the age of three to six weeks, depending on the species. A puppy should be eating solid food. Its digestive system should be developed sufficiently to handle the food. Special puppy feeds should be provided.

A bitch will nurse puppies for several weeks, and sometimes longer than necessary. Most puppies will begin to lap food from a bowl at about three weeks of age. The puppies should be introduced to foods suitable to them. A

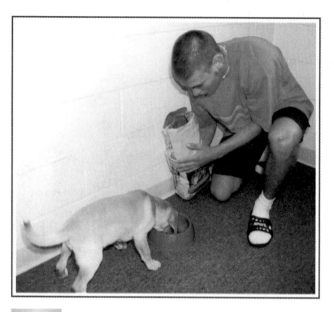

15-16. Use a food container suitable for the size of the dog and designed so it is hard to turn over.

commercial food for young puppies is likely appropriate. Puppies should be eating well at the time of weaning.

In some cases, the puppies may need to be kept in a separate place from their mother. This will encourage weaning. Puppies are often sold at weaning age.

Feeding a New Dog

Dogs and puppies are fed differently. For a new owner, the key to feeding a puppy is to get food from the person you are getting the puppy from. The same feeding schedule and food should be used the first days your puppy is in its new home, because it has enough other new things to get used to. Starting the second day, gradually switch your puppy over to the food you will be feeding it. After about four days, the puppy should be completely switched over.

Feed the puppy three times a day. The amount of food should not exceed what a puppy can consume in 15 minutes. A high-quality, balanced puppy ration is the best food for puppies and there are many commercial types on the market. If the puppy is eating dry food, moisten it with water so it is easier for the animal to eat. It is important to get the puppy on dry food now. This makes them content eating it the rest of their lives.

Feeding the Growing and Mature Dog

Dry food for dogs and puppies is much more economical than other foods. The dry food contains less water, so the animal is getting more nutrition in dry food. The feeding of dogs and puppies is a major factor in keeping the animal healthy from puppyhood into adulthood. A two-pound dog will eat 1.3 ounces of dry dog food each day. A 10-pound dog will eat 4.3 ounces of dry dog food each day. A dog weighing 75 pounds will eat 20.4 ounces of dry dog food each day.

Commercial dog food has been tested and is a complete ration. The bag of food will indicate how much food the dog should be fed; it should be what the dog will consume in 15 to 20 minutes. The dog should be fed once per day. The exception is a very large dog or a hard-working dog; then, twice a day is acceptable. The dog should be fed on the same schedule each day as well as the same feed. Table scraps are sometimes added to dog food to encourage the animal to eat. It is best not to use them to add anything to the dog's diet except fat and energy.

HEALTH CARE

Health care begins with sanitation and general condition of the environment in which a dog is kept. Dirty, cluttered pens and houses harbor disease.

Good health care includes a vaccination program. Vaccinations help a puppy develop and maintain immunity to disease. Vaccinations are given for the following diseases: distemper, infectious hepatitis, leptospirosis, coronavirus, rabies, and canine parainfluenza (kennel cough). A possible vaccination schedule is shown in Table 15-1. Consult a veterinarian for the needs of your puppy or dog.

Dogs and puppies should have routine visits to the veterinarian to make sure they are healthy and to receive all of their vaccinations against diseases and wormings against parasites. Many times, a breeder will take the bitch after she is bred for a re-vaccination. This allows the puppies to be born with antibodies to those diseases the mother was vaccinated for.

The external parasites of dogs include fleas, lice, ticks, mange, and ringworm. There are also some diseases dogs can transmit to humans, such as skin diseases, rabies, and roundworm; however, the incidence of this is very small. Dips and baths in solutions of medicines are often used to treat skin disease and external parasites. The best way to keep a dog or pup healthy is keep its area clean, remove droppings, and take it to the veterinarian for routine visits.

15-17. Medications are available for treating the skin problems of dogs, including ringworm, mange, fleas, and ticks.

Table 15-1. Puppy and Dog Vaccination Schedule	
Age	**Vaccination**
6- to 8-week-old puppy	distemper-measles-parainfluenza (CPI), Parvovirus
8- to 12-week-old puppy	DHLPP (distemper, hepatitis, leptospirosis, parainfluenza, parvovirus)
12 weeks	rabies
16 weeks	DHLPP
12 months	rabies
Boosters and others may be needed.	

Adapted from: Delbert G. Carlson, and James M. Giffin. *Dog Owner's Home Veterinary Handbook.* New York: Howell Book House, 1992.

FACILITIES AND EQUIPMENT

Dogs and puppies kept outside need shelter from the heat and cold. An outside dog requires a house that is warm in the winter and cool in the summer. At all times, it should be draft free, dry, easy to clean, and with an area for exercise. The house should be big enough that the dog can lay in the back and warm the house with its body heat at night and lay in front during the

15-18. A spacious pen with a house that provides for the well-being of the Alaska husky.

day with its head out to protect the area. Some owners even put a partition between the two areas.

The biggest problem is that houses are often too large and the animal's body heat cannot warm the area. A dog the size of a beagle needs about 2 × 2 feet (0.7 × 0.7 m) of floor space and a roof height no taller than 18 inches. A medium-sized dog of about 40 to 75 pounds (18 to 34 kg) requires floor space of 4 × 4 feet (1.4 × 1.4 m).

Fences are needed to keep dogs that are outside from straying. Wire, wood, and other materials may be used to build a "dog-proof" fence. In recent years, invisible sonic and radio fence containment systems have been used. The dog has a collar with a receiver that emits an audible signal followed by a mild electrical shock. With training, dogs soon learn where the boundaries of their yards are located.

An owner should also provide a collar, warm bedding, water and feed dishes, and a brush for the animal. An identification tag should be attached to the collar. This is useful if the dog runs away or is lost. The tag should have the owner's name, address, and telephone number.

15-19. A dog being trained to obey a radio containment system. (The small white flags mark the boundary of the yard for the dog. The flags are beneficial to the trainer. Once the dog is trained, the flags are no longer needed.)

TRAINING AND GROOMING

Training and grooming are used to have manageable and appealing dogs.

TRAINING

Training is the process of getting a dog to have the desired habits or qualities. With a puppy, it may be to not urinate or defecate on the floor. This is known as housebreaking.

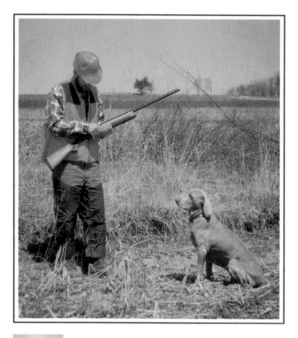

15-20. Dogs used in hunting receive special training.

Training includes obedience of verbal commands, such as "heel," "come," and "stay." Training often involves a collar with a leash. Walking the dog around and giving commands that are associated with movements of the leash are basic in training.

Some puppies attend obedience school. This is an organized program to train young dogs and their owners in a relationship of trust and following commands. Kindness and reward are fundamental in all training efforts. It is best to learn from an experienced trainer.

GROOMING

Grooming is washing, combing, brushing, trimming, and otherwise caring for the external appearance of a dog. Grooming may involve cleaning the hair, removing trash and parasites, cutting nails and hair, and treating minor wounds. Show animals must be groomed according to the

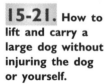

15-21. How to lift and carry a large dog without injuring the dog or yourself.

specifications of the breed and exhibit rules. Grooming can highlight the strong features of an appealing animal. It is best to learn grooming from an experienced dog groomer.

Hair Coat

The hair coat of dogs should be brushed regularly. Bathing is not a routine part of maintaining the hair coat of a dog. Bathing is done when the dog is especially dirty. Of course, a dog that has been running loose outside may need a bath before it can be brought inside. Too much bathing will remove oils from the coat and can cause skin irritation.

Long-haired dogs will require special care to remove plant burrs, excess hair, and hair mats (masses of hair usually found under the legs

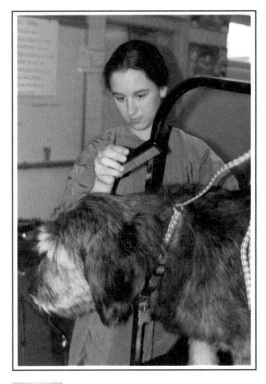

15-22. Using a dressing comb to remove dirt from a long-haired dog.

15-23. Using clippers to trim and shape the hair of a dog.

and behind the ears). Hair mats can often be removed by teasing them out but may occasionally need to be cut out. Care should be taken not to cut the skin when removing hair mats.

Nails

The need for nail trimming depends on the surface the dog is kept on. Inside dogs will need trimming more often than outside dogs. Be careful to trim the nail above the **nail bed** (the growing point of the nail). The nail bed is easy to spot on animals with light-colored nails, it is the darker area close to the paw. Use sharp clippers that will cut the nail, not crush it. A crushed nail can be painful and can cause bleeding.

Ears

A dog's ears should be cleaned monthly with a cotton swab and mineral oil or alcohol. Clean only the surface you can see and never put sharp objects in the dog's ear. Always check for parasites, such as ticks around the ears.

Eyes

Dogs may get irritants in their eyes. When this happens, flush the eye with an eyewash and watch it to see that it heals. Serious injuries and irritations that do not heal on their own should be called to the attention of your veterinarian. Check your dog's ears for mites. Dark grainy deposits inside of the ear are an indication of mites.

Teeth

The dog's teeth are generally resistant to decay. You may want to clean them regularly to prevent plaque and tartar. Use a small toothbrush with toothpaste, salt water, or a salt water/baking soda solution. Kits are available for cleaning the teeth of dogs. Clean the teeth from the gum line down toward the tip. Providing crunchy food

15-24. Brushing the teeth of a dog is easier with a ready-made brushing kit.

can help maintain healthy teeth. "Kibbled" food and dog biscuits that are hard help clean the teeth.

Note: Always be careful in grooming and performing other care on a dog. The dog may be surprised and become aggressive, growling, biting, and clawing.

SANITATION AND WELL-BEING PRACTICES

All animals deserve a clean and comfortable environment in which to live. A dog is not exception.

Animals that are in captivity (kept in cages and pens or within a fenced area) have special needs when it comes to sanitation and cleaning. Animals in the wild distribute their wastes away from areas in which they choose to live. Since much space cannot normally be provided, steps must be taken to ensure sanitary and clean conditions.

- Remove Body Wastes—Body waste (feces and urine) must be removed from the animal's quarters daily. This practice will keep the animal clean, keep the cage fresh, and prevent diseases and parasites.

- Provide Clean Bedding—Bedding materials should be changed on a regular schedule. Bedding provides warmth and comfort for the animal. Dirty bedding can promote parasites, make the animal dirty, and cause illnesses, like respiratory infections.

 15-25. A clean kennel cage is important to a young puppy.

15-26. Proper housing is needed. This plastic dog-house is for a large dog.

15-27. Regular care helps a puppy stay healthy.

■ Clean the Living Area—Clean animal pens and cages regularly with a good soap and disinfectant. The animal should be removed before cleaning and placed in a temporary holding area. Remove all old bedding and scrub the cage/pen with a brush. Remove all material and stains from the animal's living area.

Look after the dog's well-being. Sometimes, people fail to realize that what they do may not be in the best interest of an animal. Here are a few examples to overcome potential injury to a dog:

■ Never leave a dog in a hot car in the summer.

■ Always leave water available for a dog.

■ Provide secure facilities that offer protection from a harsh environment.

■ Never tie a dog in the hot sun, cold, or rain.

■ Provide appropriate food on a regular basis.

■ Neuter animals to prevent unwanted baby animals.

■ Provide proper care for newborn animals.

■ Have routine vaccinations, wormings, and other health care provided by a qualified individual.

■ Groom, trim, and otherwise help an animal maintain an attractive appearance.

■ Follow all local regulations with dogs and other companion animals. This may include paying tax assessments.

REVIEWING

MAIN IDEAS

Dogs are popular. They are versatile animals and may be used for many purposes—companionship, security, hunting, retrieving, herding, tracking, and guiding. The selection of a companion dog should be based on good reasoning. Getting a dog is a long-term responsibility. A dog needs feed, water, and exercise daily. Its living quarters must be cleaned on a daily basis and the animal's health must be routinely checked. All companion animals–including dogs–require a commitment of time and money. A pet could be a part of the family for up to 10 years, so make the decision a family decision.

It is important to know the use for the animal. If an owner wants an animal to do certain tasks, beyond that of being a companion, study the types of animals that fall into the proper category. Talk to people who own these breeds of dogs and find out the animals' strengths and weaknesses.

Dogs should be groomed regularly. Combing and brushing the hair of long-haired dogs is very important. Many animals are self-groomers and can indicate when they are not feeling well by their ungroomed appearance.

QUESTIONS

Answer the following questions using correct spelling and complete sentences.

1. What is the scientific name of the dog?

2. What are the classes of dogs? Briefly describe each class. List a breed in each class.

3. How does the fact that a dog is a carnivore affect its diet?

4. What is a mixed-breed dog? Assess the merits of a mixed-breed dog.

5. What factors should be considered in selecting a dog?

6. What is the general reproductive process of dogs?

7. What is an orphaned puppy? How is an orphaned puppy fed?

8. What is weaning? What should be the stage of development when puppies are weaned?

9. What vaccination program is followed with puppies and dogs?

10. What is involved in grooming a dog?

11. What are sanitation and cleaning? Why are they important?

12. What is involved in training a dog?

EVALUATING

Match the term with the correct definition. Write the letter by the term in the blank provided.

a. training   d. toy breed   g. stud dog
b. whelp   e. nail bed   h. herding dog
c. hound   f. bitch

_____ 1. A class of dog based on the use of the breeds for tracking.

_____ 2. A class of dog known for herding sheep and cattle.

_____ 3. A class of small dogs weighing between 4 and 16 pounds.

_____ 4. A bitch giving birth to puppies.

_____ 5. A female dog.

_____ 6. The growing point of a nail in a dog's foot.

_____ 7. A male dog used for breeding.

_____ 8. Getting a dog to have the desired traits or habits.

EXPLORING

1. Visit a nearby store that sells dogs. Make a list of the dog breeds in the store. For each breed, list your opinion of the advantages and disadvantages of the breed.

2. Volunteer to dog-sit or care for a dog for other people. This can become a profit-making enterprise by looking after pets while their owners are away. Be sure to visit with the owner about the needs of the dog and how it is managed. Develop a schedule of activities for the time you are responsible for the animal.

3. Tour a veterinary medical facility that treats small animals. Interview the veterinarian about the work. Determine the kinds of animals and the common veterinary medical procedures that are performed. Study how animals are restrained and moved. Draw a floor plan of the arrangement of the facility, including the animal boarding area. Note how the pens or cages are constructed and maintained.

16

CATS

Cute, playful kittens get the attention of everyone! People like to hold a kitten and listen to it purr. A tiny straw or short string can get the kitten moving, slapping, and jumping as the string teases the animal's instincts.

Kittens and cats have needs. To provide for their well-being, these needs—food, water, safety, and protection from the weather—must be met. Diseases need to be prevented and treated if they occur.

Cats make loving and playful pets once they are used to their home and owners. They have been domesticated and used for companion animals a long time, but not as long as dogs. The number of cats is less than the number of dogs. Most cats are smaller than dogs and are often well suited to living in an apartment or small home.

16-1. Cats are interesting animals.

OBJECTIVES

This chapter provides basic information on cats as companion animals. It has the following objectives:

1   Describe the biology of a cat

2   Name and explain the classes of cats

3   Assess important areas in selecting a cat

4   Describe management practices with cats

5   Identify training and grooming areas with cats

6   Describe cat reproduction

7   Explain the feeding and nutrition of cats

8   Describe important health areas with cats

9   Identify equipment and facility needs with cats

TERMS

declaw
hairball
kitten
kittening
litter
litter pan
long-haired cat

maternity box
pet carrier
queen
scratching post
short-haired cat
tomcat

CAT BIOLOGY

Cats carry out life processes similar to other animals. About 60 million cats are kept as companion animals in the United States. Some are also kept as rat cats in barns and warehouses. (A rat cat keeps down rodent pests that destroy grain and other products.)

Cats may live 12 to 18 years, with some living as long as 30 years. The female cat is known as a *queen*, while the male is known as a *tomcat* or tom. A baby cat is a *kitten*. Gestation is about 63 days. Queens often have litters of three to five kittens, though some have larger litters. The eyes and ears of newborn kittens are sealed; therefore, they cannot see or hear. Both eyes and ears are open at about two weeks of age. The kittens may nurse their mother for about six weeks.

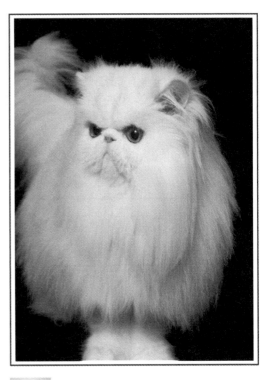

16-2. Cats have a wide range of physical appearances.

Though cats and dogs often do not get along with each other, they descended from the same ancient mammal, *Miacis*. Cats are carnivores and often will catch birds, mice, and other small animals as prey. The scientific name of the domestic cat is *Felis catus*. Many breeds of cats with unique features are used as companion animals.

SIZE AND SKELETON

Typical domestic cats weigh 6 to 15 pounds (2.7 to 7 kg), though some are larger and some are smaller. Cats with longer hair appear larger, but their weight may be no more than a cat that appears smaller.

Cats have about 250 bones. The exact number depends on the size of the cat and length of its tail. Their muscles, bones, and structure are designed for speed and quickness. Cats have about 500 muscles, with the largest muscles in the hind legs.

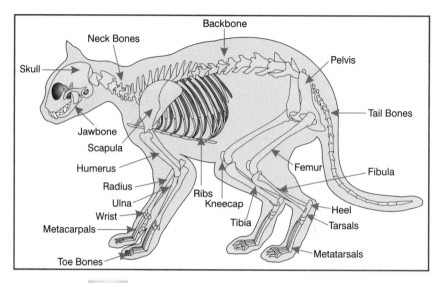

16-3. The skeleton of a cat has about 250 bones.

BODY SYSTEMS

Cats have similar body systems to other animals. As carnivorous animals, they require high-protein diets.

The sensory systems of cats are well developed. Their nose and paws are especially sensitive. Cats have extra smell and taste organs. These allow them to distinguish between foods.

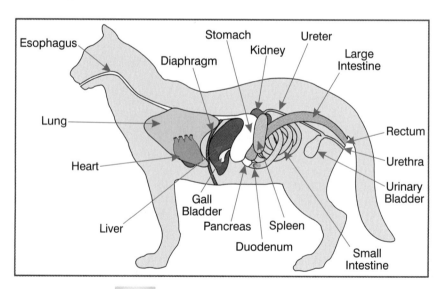

16-4. Gastrointestinal system of a cat.

The ears of cats are quite sensitive; they hear the slightest movements. Cats have three-dimensional vision as well as interesting eye color combinations. Their whiskers are tactile hairs; they are connected to sensitive nerves that help the cat protect itself and find its way in the darkness.

FEET

The feet of cats differ from those of other companion animals. Their foot structure is similar to that of exotic wild animals in the cat family, such as lions and tigers.

Cats' feet have claws, which allow them to climb trees. Pads on the bottoms of their feet serve as cushions and help the cat approach quietly when stalking prey.

The claws are dangerous to people and can damage furniture. Some cats are declawed. To ***declaw*** a cat is to surgically remove the claws from the cat's feet. The procedure is done by a veterinarian. A time of healing is needed for the cat to recover. A cat that has been declawed should never be turned outside; it is less able to protect itself.

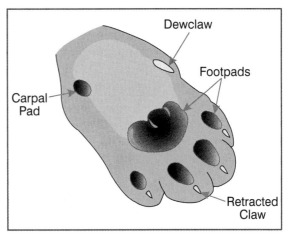

16-5. A bottom view of the foot of a cat shows the spring pads, which allow a cat to walk very quietly. The claws are retracted.

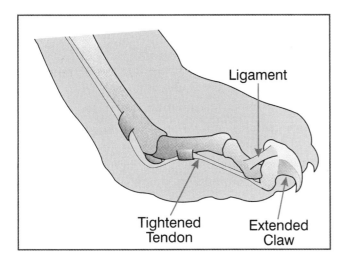

16-6. A side drawing of the foot of a cat shows the claws extended. Declawing is removing the claws at their base.

CLASSES AND BREEDS

The breed of cat selected by a person largely depends on personal choice. Basically, there are two choices: either the common house cat or a purebred cat.

The common house cat is not a specific breed, but a mix of various breeds. It is produced by chance and not by planning for a specific breed.

The purebred cat is bred by design. Careful planning and selection of bloodlines are used with purebred cats. There are two categories of purebred cats: short-haired and long-haired. Thirty-six breeds of cats are recognized in these categories in the United States. The American Cat Fanciers Association and the Cat Fanciers Association are the most recognized cat registries.

- Short-haired Breeds—A *short-haired cat* is low maintenance—it requires little or no brushing compared to a long-haired cat. Some of the most common short-haired breeds are the Abyssinian, American Shorthair, Burmese, Colorpoint Shorthair, Egyptian Mau, Exotic (Shorthair and Longhair), Havana Brown, Japanese Bobtail, Korat, Manx, Ocicat, Rex, Russian Blue, and Siamese.

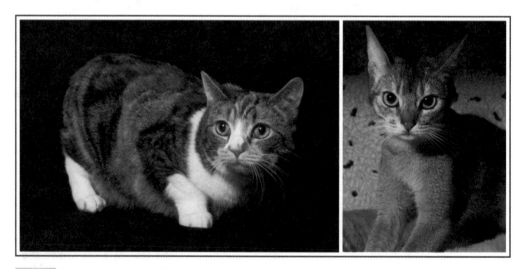

16-7. Two examples of short-haired cat breeds are the Abyssinian (left) and the Manx (right).

- Long-haired Breeds—A *long-haired cat* requires more attention to look its best and it is very pretty when groomed. Long-haired cats also get hairballs, which they must have removed. A *hairball* is a wad of hair that collects in the

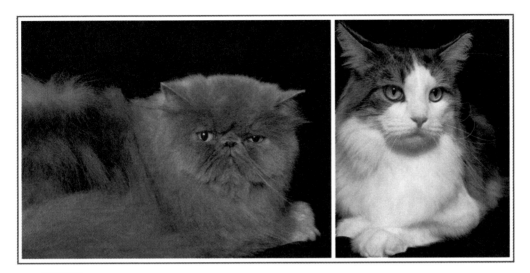

16-8. Two examples of long-haired cats are the Persian (left) and Main Coon (right).

animal's digestive tract. It can clog the movement of food in digestion and cause serious problems for the cat. There are products available that will prevent the forming of and the elimination of hairballs. Some of the most popular long-haired breeds are the Persian, Balinese, Birman, Himalayan, Maine Coon, and Turkish Angora.

SELECTING A CAT

Because cats require less care, they make better pets than dogs for some people. Cats are independent and self-sufficient. They are also more economical than a dog and some other companion animals. In selecting a cat, there are several considerations:

- Age—Kittens are usually cute and adorable, but they are also more work. They require special attention, care, feeding, training, and time to learn. Kittens are more active and playful than grown cats. Kittens adapt to a new home and other animals much quicker and with less stress than adult cats. It usually takes less than three weeks for the kitten to adapt—it does not have a very long memory of living in another place. Kittens change as they mature. The adult may have little resemblance to a tiny kitten.

- Gender—Male and female cats are much alike in temperament as kittens, but major behavioral differences develop at maturity. If the owner wants to raise kittens, then an obvious choice is a female. If the owner does not want kit-

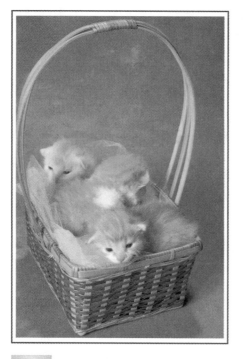

16-9. Kittens are cute when small but grow up to be cats!

tens, a cat of either sex could be chosen. Having the cat spayed or neutered at a young age should be considered. Female cats are spayed to eliminate the period of heat and unwanted pregnancy. Female cats may have strange behavior during heat. Male cats that are not neutered urinate frequently to mark their territory. This is annoying in a home. Urine stains can ruin a carpet. When let outside, tomcats may stray from home searching for a female or a male cat to fight.

■ Purebred or Mixed Breed—Most people enjoy mixed-breed cats. Some prefer purebred or pedigreed cats. If the owner plans on raising kittens or showing the cat, he or she may want to consider a pedigreed cat. A pedigree will allow the owner to trace the animal's family history. However, it is more expensive to buy pedigreed cats.

Connection

SAFETY IN RESTRAINING A CAT

A cat may be nervous about being handled. It generally doesn't like new things being done to it. It may hiss, bite, and scratch. A cat's claws can make severe skin wounds on the arm or hand of a person. The cat can also run, jump, and try to escape. Sometimes, a cat needs to be restrained even if it doesn't like it.

This shows one method of restraining a cat. The entire cat is placed in a specially designed zippered bag. Its head is out. Using the different arrangements of zippers, one foot at a time can be extended from the bag. The well-being of the cat is considered.

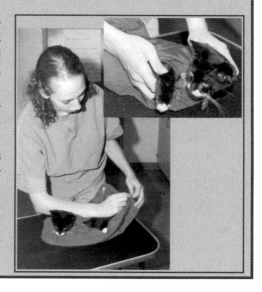

MANAGEMENT

Exercise care when handling a cat. To pick up a cat, slip one hand under the cat's chest and hold its front legs gently but firmly with your fingers. Put the other hand under the cat's hindquarters. Treat your cat carefully and make sure it feels secure. Kittens are sometimes picked up by the nape of the neck—this is how a queen moves her babies!

Owners should remember that every cat is different. Give a cat time to get used to its new home and owners. It is wise to keep the cat confined until it is used to the new owners and home. Give the cat time and space; do not force attention on the cat. While working with the cat, repeat its name often. This will reinforce the animal's learning. If the cat is frightened, it will fight and try to run away. Patience and time allow the cat to feel loved and comfortable in its new surroundings.

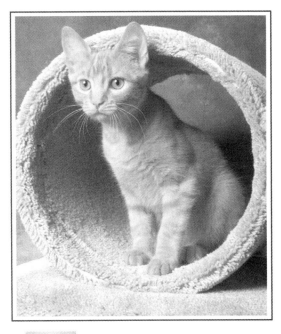

16-10. Cats are curious and enjoy playing.

Cats, if they are kept inside, require a bed, carrier, litter pans, a scratching post, and possibly some toys. The equipment does not have to be expensive, but for your cat to be happy and healthy the items are needed. If your cat stays outside, it needs protection from the elements and a safe place to eat where other animals do not get its food and water.

TRAINING AND GROOMING

Training and grooming are needed to ensure that the cat fits in well with its owners.

TRAINING

Most training focuses on toilet training. It is easier with cats than dogs. Queens usually help to train their kittens. A kitten can be trained with patience (and a few messy accidents). Put the kitten in the litter box. Use your hands to help it scratch with its front paws in the litter. Repeat this every time the kitten appears to be looking for a place to urinate or defecate.

Some furniture and other places are off limits. Providing a comfortable place for scratching and sleeping will help a cat adjust. Rewards for desired behavior work well. Firm voice commands are recognized by a cat when its behavior is undesirable.

GROOMING AND CARE

Grooming a cat is combing, brushing, trimming nails, and cleaning its ears, eyes, and teeth and gums. Most cats spend hours each day self-grooming.

16-11. A small, soft brush should be used in brushing a cat.

Hair coat

With few exceptions, the cat's hair coat should be brushed daily. Use a medium-bristle brush, along with a flea comb, on long-haired cats and a flea comb on short-haired breeds. Cats may occasionally need to be bathed. Do so carefully in a tub or sink with about four inches of warm water. Use a mild soap.

Nails

Cats that have a scratching post or spend time outside will probably not need to have their claws trimmed. If claws do require trimming, be careful not to trim below the nail bed. Use sharp clippers or a special cat clipper and file. Do not use scissors.

Ears

Check the ears during normal grooming. A discharge or stopped up ear canals require help. If ear mites are present, buy ear drops from the pet store and apply as directed. Outside cats may get ticks inside their ears. These should be carefully removed.

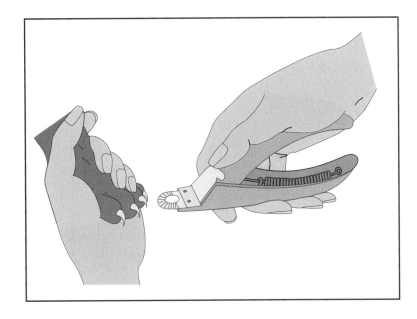

16-12. A pet nail clipper can be used to nip the tips of nails. Use a nail file to smooth the edges. Caution: Make small cuts to avoid cutting into the skin or a blood vessel.

Eyes

Regularly check the animal's eyes for any discharge, which indicates infection. You can apply eyewash to remove foreign materials. Serious problems need the attention of a veterinarian.

Teeth and gums

Teeth and gums should be checked regularly to detect any soreness. Use hard foods occasionally to promote clean teeth. Tartar and plaque should be removed by your veterinarian. In some cases, an owner can use a toothbrush kit designed for cats.

NUTRITION AND FEEDING

Cats will remain healthy and happy if they eat a well-balanced diet and always have water available. Many owners feed their cat from the table. This may sound like a good idea, but it does not provide the cat with a well-balanced diet. A cat should not be fed dog food. It does not contain enough protein.

Cats require a high-protein, high-fat diet. The best way to ensure a well-balanced diet is to feed a high-quality commercial cat food. The food is avail-

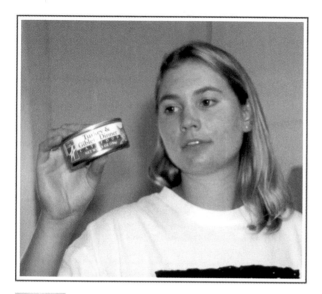

16-13. Read the nutrition information on cat food. This shows a can of turkey-based meat products.

16-14. Two separate bowls are used to provide a continual supply of food and water.

able in several forms: dry, semi-dry, canned specialty, and canned maintenance. Cats have different needs and different tastes; it is up to the owner to decide which food works best for their cat. Labels list the information needed to select the proper food for the cat.

If a cat is pregnant, she should continue to receive her same daily feed until the last 20 days of her pregnancy, then she should be allowed to eat all she wants. After the kittens arrive, the mother may not eat the first few days. Make sure she has both food and water available. During the time the mother is nursing the kittens, feed her all of the food she wants. Producing milk requires more nutrients.

The day the kittens are weaned, take all food from the mother. On the second day, increase feed to one-fourth of what the mother used to eat. On the third day, feed one-half the amount. On the fourth day, feed three-fourths, and on the fifth day, feed the normal amount. Limiting the mother's intake of food will help decrease her milk production.

REPRODUCTION

The female cat's heat period is usually easy to detect. She will growl, chew, maul, and call for the male. Gestation is 60 to 65 days. Giving birth is *kittening*. The female may continue her normal routine until the last week of pregnancy, when she should no longer be running up and down stairs and jumping on furniture.

During the last week of her pregnancy, the female will be restless as she looks for a place to have her young. A ***maternity box*** should be prepared for kittening. A maternity box is a quiet, draft-free, and secure place for parturition. Place the maternity box in a dark, quiet, warm, traffic-free location. All that is needed for the maternity box is a cardboard carton large enough for the mother to stretch out at full length on her side and have room left over. The top should be left attached to keep the area dark. A door should be cut with 3 to 5 inches of cardboard left at the bottom of the door so the kittens do not escape. The bottom should be lined with shredded paper and a towel. A temperature of 85° F is best for kittening. Cold, damp, drafty places result in more kitten deaths than any other condition.

The queen will give birth to a litter, which is usually three to four kittens. After the mother delivers the kittens, she should be left alone. No strangers should be allowed to see the mother or her kittens for at least three days. Most births are normal, with the front feet and head first through the birth canal. Occasionally a kitten will be presented wrong. Assistance of a qualified person is needed. The queen will lick the kittens to dry and clean them following birth. Kittens should nurse within an hour. Queens need more calories during lactation.

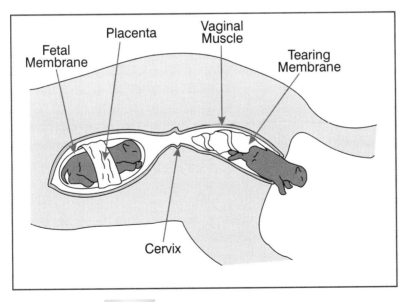

16-15. The birth process of kittens.

Kittens are ready to begin eating food by three weeks of age. Feed the kittens the same food as the mother. If it is dry food, moisten it with water to make it easier for the kittens to eat. Have water available. Some kittens are weaned at 25 days of age. Queens will often continue nursing for 6 to 7 weeks. At weaning, kittens should be eating adequate food for their nutritional needs. After weaning the kittens, do not overfeed the queen because it will cause her to gain weight and overproduce milk. This can lead to problems in the mammary system.

HEALTH CARE

Cats are generally healthy animals, especially if proper precautions are taken. A new owner of a cat should take it to a veterinarian to have it completely checked out. Vaccinations may be needed to prevent the animal from getting some diseases. The veterinarian will inform the owner about other shots, wormings, and office visits the animal needs to stay healthy.

Kittens should have a routine vaccination program. The diseases to protect against include feline infectious enteritis (panleukopenia), feline respiratory disease, rabies, feline leukemia, and feline infectious peritonitis. A suggested vaccination schedule is shown in Table 16-1.

If a cat does not feel well, it may show some of the following signs: coughing and sneezing; vomiting; runny nose; red, watery eyes; severe, prolonged diarrhea; loss of appetite for several days; dull coat; and a non-caring attitude. If a cat shows any of these signs, a visit with the veterinarian is in order.

Table 16-1. Possible Vaccination Schedule with Kittens and Cats	
Age	**Vaccine/Disease Controlled**
6–8 weeks	Panleukopenia (FPV), Phinotracheitis (FVR), Calicivirus (FCV
12 weeks	second FPV, FVR, and FCV; give first Feline Leukemia (FeLV) shot if negative to ELISA test
16 weeks	first rabies; second FeLV, and third FPV, FVR, and FCV
16 months and each year	FPV, FVR, VCV, FeLV, Rabies

Note: Get the assistance of a veterinarian in designing a vaccination program.

FACILITIES AND EQUIPMENT

The equipment for a cat does not have to be elaborate. A cat needs something to occupy its time and give it exercise. An outside cat needs protection from the weather and a safe place where other animals do not bother it.

A *pet carrier* is a carrying case for a cat or other animal. Since all cats must be taken to the veterinarian, they will, at some time, travel. The case should be large enough for the animal to turn around in, well ventilated, easy to clean, and lined with a towel in the bottom.

The bed for the cat does not have to be a fancy bed purchased from a pet store. It may be something as simple as a cardboard box with fairly high sides and the bottom lined with a cushion. Cats prefer quiet places to sleep that are secure, warm, and draft-free. To avoid drafts, it is a good idea to place the bed several inches off the floor.

16-16. Condominium poles provide a place for cats to play and exercise.

16-17. A pet carrier considers the well-being of a cat.

16-18. A warm, secure bed for a large cat.

16-19. A good litter box makes cleaning easier and serves the cat better.

A *litter pan* is a container designed for the cat to use in urinating and defecating. Most are made of plastic or baked enamel on metal, which will not rust—so they work best. The pan should be washed with soapy water when the litter is changed.

Litter is an absorptive material that helps prevent odor. Many types of litter are available, such as prepackaged sawdust, shredded newspaper, sand, clean soil, peat moss, or wood shavings, which are often used with a small amount of baking soda to help eliminate odor. If droppings are removed daily, changing the litter once a week should be sufficient. Use a slatted scoop to remove the soiled litter and place it in a plastic bag that is sealed. Soiled litter in a bag is typically disposed of with the household garbage.

Cats are curious and energetic animals. They need a scratching post for exercise—not to sharpen their claws as is commonly believed. A *scratching post* is a fabric-covered cylinder mounted in a vertical position on a wide base. Carpet or cloth may be used to cover the cylinder and a wide base will ensure stability. Owners can easily make a scratching post. Since it may be

16-20. Clean facilities are provided in this animal shelter.

hard for a scratching cat to know the difference between carpet on the post and carpet on the floor, set the post on a bare floor.

Cats also enjoy toys. However, avoid toys that are rubber, furry, or woolen, and can be torn apart and eaten. Watch the cat to see the types of toys it enjoys. Many toys can be made from things already around the house, such as a paper bag, shoe box, or a spool attached to a strong string and hung from a door.

REVIEWING

MAIN IDEAS

Cats make good companions for some people. They are less trouble and more economical than dogs. Providing proper care is essential for the well-being of the cat.

Mixed breed and purebred cats are chosen as pets. Purebred cats are classified as long-haired and short-haired. Several breeds are in each class. Most people keep mixed-breed cats. Purebred cats are needed for showing and raising kittens for sale.

Selecting a cat is important. Consider its age and gender. Female cats are typically better in homes. Toms (male adult cats) urinate on the floor and walls to mark their territory. Neutering can help prevent some problems.

Provide a good environment for a cat. Cats need a scratching pole. They need to be trained in how to use a litter box and in appropriate behavior in a home. Grooming is needed to keep them healthy, attractive, and to reduce cat hairs over the house and furniture. Particularly, cats should not be around food preparation areas and infants.

Cats need a proper diet. They need higher protein feed than dogs. Select a quality cat food designed for the age and condition of the cat. Read the label. Cats like meat, such as animal organs, fowl, and fish. The food can be dry or in cans. Some variety is good. Table scraps should usually not be fed.

A pregnant queen will kitten in 60 to 65 days following conception. A kittening box provides a safe, appropriate environment. Most queens have litters of three or four kittens.

All cats need regular health care. A vaccination program is recommended for kittens, beginning at six weeks of age. This is continued as an adult cat.

QUESTIONS

Answer the following questions using correct spelling and complete sentences.

1. How widely kept are cats as companion animals? How long do cats live?

2. What parts of the sensory system are especially well developed in cats?

3. What are the important features of the feet of cats?

4. What are the two classes of cats? Name two breeds in each class.

5. What should be considered in selecting a cat?

6. What are some considerations in managing a cat?

7. What training do cats need that are kept in a home?

8. What grooming may be used with cats?

9. What are the general nutrition needs of cats?

10. What is the reproductive process of cats?

11. What should be established to have good health with kittens?

12. What is a litter pan? How is a litter pan managed?

EVALUATION

Match the term with the correct definition. Write the letter of the term in the blank provided.

a. queen
b. kittening
c. hairball

d. declaw
e. maternity box
f. litter

g. pet carrier
h. scratching post

_____ 1. The process of a queen giving birth.

_____ 2. A female cat.

_____ 3. A carpet- or cloth-covered, exercise structure for the cat.

_____ 4. A container that is used to transport a cat.

_____ 5. The absorptive material placed in a litter box.

_____ 6. The surgical process of removing the claws from a cat.

_____ 7. A wad of hair that may form in the digestive system of a cat.

_____ 8. A box for kittening.

EXPLORING

1. Arrange to cat-sit for a neighbor or family member. Learn all the details so you can take good care of the cat. Keep an activity log each day you are responsible for cat-sitting.

2. For one day, job shadow a person in a pet store, grooming business, or veterinary medical clinic where cats are served. Write a report on your experience.

3. Prepare a poster or electronic presentation on proper cat care. Be sure to include all essential details for the well-being of the cat. Make a report to in class.

17

BIRDS, RODENTS, AND REPTILES

Many different animals are kept as companions. Sometimes, we fail to realize just how many there are! We forget, for example, that some people keep tarantulas as pets!

Regardless of the species kept, the well-being of the animals must be considered. There are right and wrong ways of caring for animals. Certainly, you would want to do the right thing! But, what is right?

Skills in animal care can be developed by learning and practice. Background information helps in understanding animals. If we know "why" an animal behaves as it does, we can learn "how" to care for it.

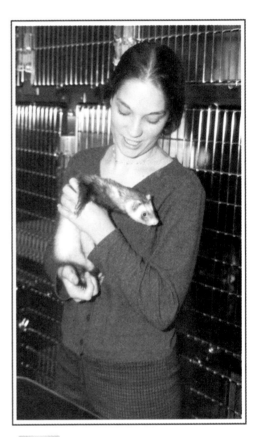

17-1. A ferret is being carefully handled.

457

OBJECTIVES

This chapter provides background information on several species of companion animals, including birds, rodents, and reptiles. It has the following objectives:

1 Describe the kinds of companion birds and their management

2 Describe the kinds of companion rodents and their management

3 Describe the kinds of companion reptiles and their management

4 Describe the kinds of rabbits and their management

5 Identify other companion animals and describe their keeping

TERMS

amphibian	grit
aviary	hutch
bird	nocturnal
bird cage	quill
birdkeeping	reptile
diurnal	rodent
fledgling	solitary
greenfood	

COMPANION BIRDS

Several kinds of birds make excellent pets in homes and apartments. Study the needs of each species and provide for its well-being. Special equipment and facilities will be needed to care for a bird. Birds must have food, water, and a good environment.

A *bird* is an animal with feathers. It is the only animal with feathers. Birds have four appendages: two wings and two legs. All birds have wings. Some birds can fly quite well; others cannot fly. Birds stand upright on two feet. Birds have bills instead of chewing mouths. Birds reproduce by laying eggs. All birds are in the class Aves. Scientists have further divided birds into 28 orders. For example, the parrot is in the order Psittaciformes.

Birdkeeping is raising birds in captivity. All areas of bird care are included—feeding, breeding, health care, and cleaning. Pet stores, discount stores, and other places sell equipment needed for birdkeeping. Many of these same stores sell birds.

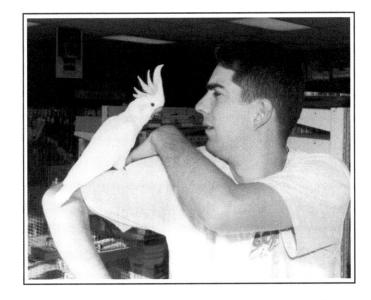

17-2. A lesser sulphur cockatiel is being trained by its owner.

GROUPS OF COMPANION BIRDS

Seven groups of birds are most popular with birdkeepers. Each group has several species and varieties. These have unique features that make them appealing to birdkeepers. Varieties represent different colors and other distinct features. The groups of companion birds:

17-3. Parakeets in a comfortable cage with food, water, and natural limb materials.

17-4. Lovebirds are colorful members of the parrot family.

■ Cockatiels and Parakeets—Cockatiels and parakeets are in the same order as the parrot. Cockatiels are easy to train and are lovable birds. Colorful yellows, oranges, and greys are present on the adult birds. They are good whistlers. Incubation is 25 to 26 days with fledgling occurring 55 to 56 days later. Cockatiels like cereal grains, fruit, and greenfood. Some are fed commercial parrot food. Length is 10 to 18 inches (20 to 45 cm) or more.

■ Finches—Finches are 4 to 5 inches long and come in several varieties and colors. Incubation period for eggs is 12 days. The fledgling stage begins at 18 to 21 days after hatching. A *fledgling* is a young bird that has left the nest but is not feeding independently. The males of some varieties have unique songs. Standard bird seed can be used as feed, especially small cereal grains, such as milo. Adults are typically 4 to 5 inches (10 to 12.5 cm) long.

■ Lovebirds—Lovebirds are short-tail parrots from Ethiopia. They are excellent cage birds and are quite colorful. Young lovebirds are easy to tame and show affection. Older birds are difficult to tame. Since lovebirds are in the parrot order, they often receive the same care as parrots. Lovebirds do not talk. Lovebirds like cereal grains, fruit, and greenfood. Adults are 5½ inches (14 cm) long.

■ Canaries—The canaries were introduced to the United States from the Canary Islands. Noted for bright colors, canaries are popular with birdkeepers. Some canaries are better singers than others, especially the males. Canaries generally prefer larger cages. Some varieties prefer outdoor aviaries year round. Incubation is 14 days. The fledgling stage is 14 days later. Canaries prefer a seed diet along with some greenfood. Adults are typically 5 to 7 inches (12 to 17 cm) long.

■ Budgerigars—Sometimes called budgies, budgerigars are native to Australia. They are friendly and talkative, with some knowing 600 words. Budgies often have health problems. They are colorful and appealing. Use a commercial budgerigar feed mix. Adults are typically 7 inches (18 cm) long.

■ Amazon Parrots—Amazon parrots are quite talkative and entertaining as adults. They are typically cuddly. Many colors and interesting plumage make

17-5. This Mexican redhead Amazon parrot is an interesting bird.

Table 17-1. Features of Selected Birds

Kind	Weight (grams)	Normal Life Span (years)	Considerations/Features
Finches	10–15	5–10	Easy to care for
Canaries	15–30	10–20	Easy to care for; males are singers
Budgerigars	30–35	10–20	Easy to care for
Cockatiels	80–90	10–20	Easy to care for and train
Lovebirds	40–50	10–20	Taming and training are more difficult
African greys	400–500	60–70	Can be trained to talk; enjoy talking
Macaws	1,000–1,400	60–70	Some can be trained to talk; interesting; colorful

them appealing to watch. Feed a commercial parrot food, fruit, and greenfood each day. Parrots are often kept in large cages. The length of an adult may be 15 inches (37.5 cm).

■ Macaws—Macaws are types of parrots. They are large, colorful, playful, and intelligent birds. They need more room than most birds and care is important. Macaws are hardy, but require protection from bad weather. Incubation is 25 days, with fledgling occurring 55 days later. They like seeds, fruit, and greenfoods. Some macaws prefer a little meat occasionally in their diets. The length of an adult is 13 to 36 inches (32.5 to 90 cm).

MANAGEMENT OF COMPANION BIRDS

As living organisms, birds have similar needs for life as other animals.

Reproduction

In general, birds reproduce by the females laying fertile eggs. The male has mated with the female several days prior to the laying of the fertile eggs. Eggs are incubated naturally or can be removed for artificial incubation. Nat-

Connection

NEVER AROUND YOUR NECK

Snakes like to wrap around things. They have strong bodies. Many obtain their food by wrapping around it and strangling it to death. Their strength can crush a small animal. Even small snakes are quite strong; larger snakes are even stronger. Some snakes can kill a pig or deer by tightening—known as constriction.

A snake draped over your shoulders or around your neck could suddenly tighten. If it is around your neck, you could quickly suffocate. The damage would be done before you could get it off or summon help. You would be found by someone but not before your were dead!

Make it a rule: Never drape a snake around your body!

ural incubation is preferred. Nesting facilities must be in the cage or aviary. Incubation periods vary with the species and variety of bird. Incubation typically ranges from two to four weeks.

Nutrition and Feeding

Use commercial feeds designed especially for the kind of bird. Some birds need supplemental fruit and greenfood. Apples are commonly fed fruits. *Greenfood* is non-dried vegetables with juice still present. It includes green peas, fresh corn, carrots, cabbage, and bean sprouts.

17-6. Commercial feeds are prepared for the needs of various kinds of birds, such as this parrot

Grit is sometimes needed to promote action of the gizzard. Grit is finely ground abrasive material similar to sand. Ground oyster shell or similar shell materials may be used. Small amounts of the ground shell are placed in the cage. Calcium and other minerals may be obtained from the grit.

17-7. Some stores sell loose feeds that can be purchased in the desired amounts.

Health Care

Birds are subject to a number of diseases. Good nutrition and protection from inclement weather help prevent disease. All new birds should be quarantined from existing birds for a period to determine if the new birds have any diseases. Sick birds lose their interest in food and water. They are inactive. Feathers become tattered. The services of a veterinarian may be needed as soon as a disease is suspected. Birds can get sick and die rather quickly from disease.

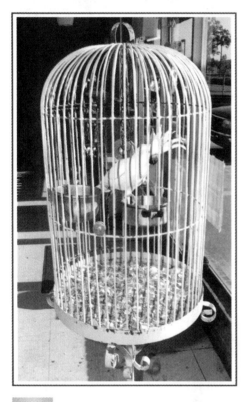

17-8. This cockatiel has a cage that promotes its well-being.

Facilities and Equipment

Birdkeeping is typically in one of two kinds of facilities: cage or aviary.

A *bird cage* is a box-like enclosure that keeps birds captive. Cages are made of wire; metal, wood, or plastic bars; and other materials. Some cages are made to be hung from stands. Other cages are boxes that sit on tables or the floor. Cages must have feeders and waterers as well as perches.

An *aviary* is a house-like enclosure for birdkeeping. Aviaries are large enough for people to walk inside. They allow plenty of room for the birds to move around and get exercise. Aviaries have roosts, nesting areas, feeders, and other equipment.

The size of the cage or aviary depends on the number and needs of the birds. Some birds need more room for flight than others. Obviously, many birds need more space than one or two birds.

RODENTS

Rodents have increased in popularity as pets. Several species have become more widely kept. A *rodent* is an animal with two large front teeth designed for gnawing. They are mammals. The scientific classification of rodents is phylum Chordata, class Mammalia, and order Rodentia. Several species of rodents, such as gerbils, hamsters, mice, and rats, are kept as pets. They require a clean cage, a well-balanced diet, clean water always available, and attention from their owner. The owner is repaid with an attentive, loving pet.

In the wild, gerbils, hamsters, guinea pigs, and mice and rats damage crops since they are rodents. It is important to make sure the animals do not escape. If they escape and multiply, they will cause serious crop damage. Gerbils are illegal to have as pets in California. The U.S. Department of Agriculture has warned about taking them into desert regions of the west.

GERBILS

Gerbils make fascinating pets because they are very quick and curious. The gerbil came to America in the 1950s for medical research. Unlike most rodents, gerbils sleep during the night and are awake during the day, they are *diurnal*.

The adult gerbil's body is 4 inches long. Its tail is another 4 inches long. A typical adult weighs about 3 ounces. Gerbils are dark brown on their backs, light brown on their sides, and light gray underneath. They have black and white whiskers.

Gerbils are shy animals. They must be handled gently. Scratching the ears and back of the gerbil will relax it. When holding a gerbil, keep your hand firmly around its entire body to make sure it does not fall or wriggle free.

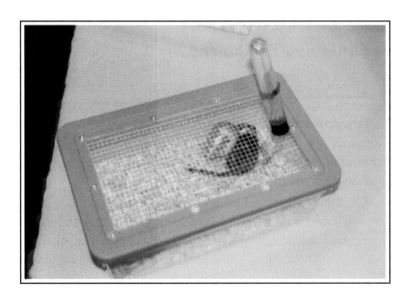

17-9. Gerbils in a comfortable cage with litter and waterer.

Reproduction

Gerbils are ready to mate at three months of age. A gerbil has only one mate. Potential mates should be kept separated from each other until they grow accustomed to each other's scent. The gestation period is 24 to 25 days. Most females have litters of about five babies. Both parents stay with the litter until weaning at six weeks of age. People should avoid contact with the young until their eyes are open.

Nutrition and Feeding

Gerbils eat a tablespoon of food each day. They only need to be fed once per day. Water should always be available. Gerbils eat grain, seeds, roots, grasses, sunflower seeds, corn, oats, wheat, watermelon seeds, bits of apple and lettuce, and may enjoy fresh grass as a treat. Using a commercial gerbil food is okay. Gerbils tend to enjoy different foods.

Health Care

Gerbils are fairly disease free. Gerbils will use one corner of their cage as their toilet, so it is important to clean it every one or two days. Gerbils do not like cold temperatures or drafty areas. Keep the temperature 65 to 80°F (18.3 to 26.6°C) and, if the gerbil is in the sunlight provide it with an area for shade. Gerbils also need hard, dry food or a block of wood to chew on. This will keep their teeth filed down, so they can chew properly.

Facilities and Equipment

Gerbils, because of their curiosity, prefer interesting homes, which may be purchased or made by the owner. A cage with two floors is best so the animal exercises by running up and down and all around. To create fun and excitement for the animal, the owner may want to create stairs, holes, and other play things for the gerbil. Make sure the cage has a top and keep it covered so the gerbil does not escape.

Gerbils work to make their cage a comfortable place. They need bedding in their cage. The bedding should be absorbent and clean. Cedar chips, sawdust, and small-animal litter work well when put into the cage about an inch thick. Gerbils build nests. Put a small piece of burlap, paper, or cardboard in the cage for use in nest building.

HAMSTERS

Hamsters are fun animals. The hamster came to America in 1938. It is *nocturnal*, which means that it sleeps during the day and is awake at night.

Three types of hamsters are used as pets. The Syrian or golden hamster has a golden body with white patches on its cheeks, front legs, and hind feet. The angora hamster has long, fuzzy hair, which comes in many colors, but is most common in tan, gray, or white. The Chinese hamster is solid black or gray and has short hair, which distinguishes it from the angora hamster.

Hamsters must be handled frequently to remain tame. It is best to stroke their backs until they get used to your touch. Pick up a hamster by placing your hand underneath for support. After a hamster trusts a person, it will crawl over him or her and ride in a shirt pocket. Always protect a hamster from dogs, cats, and anything else that may scare it, or it may try to escape. A good way to train a hamster is to give it a treat for doing a thing you like. Peanuts, cashews, and other nuts are good treats.

17-10. A mature hamster fits comfortably in a hand.

Reproduction

Male and female hamsters kept together in a cage will breed. If you do not want to raise hamsters, keep only one since they do not get lonely nor require the attention of another hamster.

Hamsters should not be bred until they are eight weeks old. Many keepers do not recommend breeding until three months old. After the female is one year old, it is very rare for her to produce offspring. Only well-developed, healthy animals should be used for breeding. Gestation is 16 days. You can have too many hamsters if you do not manage them carefully.

Start out the breeding process with hamsters by putting the cages of the male and female close to each other so they get used to one another. Place the female in the male's cage. If the male is put into the female's cage, she may kill him because she feels threatened. Wear gloves when handling the hamsters at this time. If the female is not in heat, they may fight and she will need to be removed. Put the female into the male's cage at night about 9 o'clock because most breeding takes place between 9:00 and 10:00 p.m.

Leave the female alone for nine days after she gives birth. Do not clean her cage for those days. A frightened mother may kill and eat her young. When the young reach two weeks of age, they are old enough to be weaned.

Nutrition and Feeding

Hamsters eat about ½ ounce of food each day. They should be fed at night when they are active. The natural food of hamsters is grains, seeds, and vegetables. Corn, oats, or wheat may be mixed with a prepared dog food. Commercial hamster food is likely the best way to feed hamsters.

As they eat, hamsters stuff their cheek pouches, hide the food, and return to do the same. In the wild, hamsters burrow into the ground and make one main tunnel with many side chambers in which they store food. As you clean the cage, you may find pockets of food. It is okay to put the food back where you found it if it is not spoiled.

17-11. Entertaining salt wheels may provide important minerals for rodents.

Health Care

Hamsters that are healthy have the following signs: soft, silken fur; a plump body; and prominent, bright eyes. A hamster that is long or skinny, has watery eyes, a runny nose, or a wet tail may be ill. If clumps of fur are missing, it may have a disease called mange.

Hamsters and guinea pigs may get diseases or parasites from wild mice or birds. If they do get fleas, lice, or other insects, it is okay to treat them with an insecticide for use on cats. Never use an insecticide for dogs on a hamster.

Facilities and Equipment

Many hamster owners buy bird cages or use old aquariums to house their animals. If a wooden cage is used, be sure to cover the wood on the inside so the hamster does not gnaw its way out.

Hamsters need a clean, dry cage away from drafts—55 to 80°F (12.8 to 26.7°C). If the temperature falls below 45°F (7.2°C), the hamster will go into hibernation. It may appear as though the animal is dead, but it is hibernating. Hold the animal to slowly warm it and feed it warm milk, one drop at a time, until it revives. When returning the animal to its cage, cover the cage with a heavy cloth. Hamsters also need a hard, dry food or even a block of wood. Chewing on a hard material will keep their teeth filed down so they can chew properly. Give the female scraps of cloth, tissue, paper, or cotton to line her nest. Hamsters use one corner of their cages for a toilet, much like gerbils.

MICE AND RATS

Mice and rats can be good company because they are friendly and curious. They can be trained to climb ladders, beg for food, do tricks, and ride on people's shoulders and in their pockets. They are used in psychological, biological, medical, and nutritional studies. Tamed mice and rats have gotten a bad reputation because of their wild relatives, which cause damage and are not loving and fun.

17-12. A hooded rat.

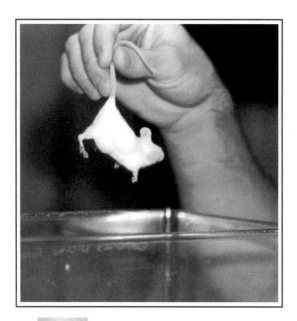

17-13. Pick up a white mouse by its tail.

Mice and rats come in various colors. Rats may be all white, black, brown, or hooded. Hooded coloring is white with brown or black around the head and shoulders. White rats are easier to tame and make nicer pets. Mice come in a variety of colors, such as white, black, tan, brown, or spotted. There is no difference in their personality based on their color.

Mice and rats are usually very tame animals, especially when the animal is acquired at a very young age. The younger the animal, the easier it is to tame, and the fastest way to tame it is with hand feeding. When picking up a mouse or rat, grab it by the tail near its body, place it in your other hand and then gently stroke its head and back. After the animal gets to know its owner, it will be at the door waiting to get out of its cage to crawl around on its owner. Treats may be used to encourage the animals. Since the animals are nocturnal, it is best to play with them in the early evening.

Reproduction

A female mouse may be bred at 8 to 10 weeks of age and a female rat at three months. It takes both animals about 21 days to give birth after they are bred. The male and female may be left together until the female gives birth and then the male must be removed until the babies are weaned at about three weeks old. Baby mice have a habit of jumping, so anyone holding the animals should be aware. Also, since the babies reach maturity so quickly, it is important to separate the males and females or more babies will be born. Since a litter consists of six to eight babies, it does not take long to be overwhelmed with young.

Nutrition and Feeding

Mice and rats have the same water requirements as all other animals. So, make sure they always have plenty of clean, fresh water. They eat dry dog food and that is enough to keep them healthy. Surprise them with new foods, such as seeds, nuts, rabbit food pellets, hard-boiled eggs, bread, breakfast cereals, rice, leafy foods, or raw potatoes. Feed dishes should be small enough so the animals cannot sleep in them and attached to the side of the cage above the floor. The food should stay clean and not get the animal's droppings in it.

Health Care

Mice and rats will live a healthy life if they are kept in good health. They need a block of wood to chew on to keep their teeth in good condition and a clean, dry cage. Like hamsters, it is okay to dust them with cat flea powder and disinfect their cage. However, never use a flea powder designed for dogs.

Facilities and Equipment

Mice and rats must be kept in wire or metal cages because they will gnaw through wooden cages. They need a hard, dry food or even a block of wood to chew on to keep their teeth in top condition. A solid upper platform should be provided for the animal to rest on and platforms and ramps should connect the floor and platform.

Since mice and rats are curious animals, toys, such as swings, perches, and exercise wheels should be included. Sawdust, cedar shavings, or cat litter should be used as bedding to keep the cage clean and dry. For nesting, cotton or shredded paper should be supplied.

17-14. White mice can be kept in old aquarium with litter, waterers, and feeders.

17-15. A silky guinea pig.

17-16. An abbey guinea pig does not have the smooth coat of a silky guinea pig.

GUINEA PIGS

Guinea pigs make friends and greet their owners with whistles. The guinea pig has a short, heavy body, short legs, and no tail. There are three strains of guinea pigs—the English, Peruvian, and Abyssinian. The English and Peruvian strains of guinea pigs both have short hair and are found in a wide variety of colors, but the Peruvian has rough hair that stands up in all directions, which distinguishes it from the English strain. The Abyssinian strain has long hair and has either solid or mixed colors.

Guinea pigs require attention and enjoy being talked to, especially when eating. When handling the guinea pig, even when it is small enough to fit into your hand, hold the animal with your thumb and forefinger just behind the animal's head and in front of its front legs. As the animal is lifted, use your other hand to support its back end. If the animal remains comfortable, it will remain calm, but if it feels uncomfortable, it may scratch you.

Reproduction

Female guinea pigs are called sows and male guinea pigs are called boars. Sows should not be exposed to the boar until they are three to five months old. The boar is always put into the sow's cage and she is never put into his cage. They are left together for three weeks. The sow will give birth between 63 and 72 days after she is bred. The sow will breed again within a few hours of giving birth to her litter. If she is not bred at that time, she will wait until the litter has been weaned. The litter should be weaned at three weeks old. The males and females in the litter should be separated to prevent breeding.

Nutrition and Feeding

Guinea pigs drink very little water because they eat so many greens. However, make sure they always have some clean water. Feed should be available at all times. At the end of the day, old food should be removed because the animal can get sick from rotten food. Favorite foods are alfalfa, apples, carrots, corn, dandelions, lettuce, cauliflower, clover, celery, lawn clippings, spinach, and tomatoes. Dry foods, such as rabbit food pellets, and a salt spool must also be available.

Guinea pigs, like humans, like different foods, so experiment to see what the animal prefers. Guinea pigs, monkeys, and humans are the only three animals that cannot make their own vitamin C, so make sure the guinea pig gets some of the foods mentioned above because they all contain some vitamin C.

Health Care

Guinea pigs are healthy animals. If they are not feeling well, the owner will be able to tell because the animal will sit perfectly still and hunched up. Its coat may be ruffled and messy. It will not eat and will lose weight. The droppings may be loose and watery. If any of these symptoms appear, the animal should be separated from other animals and taken to the veterinarian. If the animal picks up parasites, it may be treated the same way as the hamster.

Facilities and Equipment

Guinea pigs are not fighters and they do not jump or climb, so their cages may be small. Their cages should have some type of cover, even if it is just wire, to keep other animals away. A guinea pig needs at least 1 square foot of floor space, but many animals may be housed in one cage, as long as the cage is large enough. For example, a cage large enough for one male and three females would be 36 by 24 inches (0.9 by 0.6 m).

Guinea pigs need a dry floor of which at least half of it is solid and not made of wire. To keep the cage dry, a thin layer of shavings, sawdust, or straw may be put down. Do not use dusty material. Dust can make the animal sick. Guinea pigs will not grow if the temperature gets below 65°F (18.3°C) in the winter or if it is very hot in the summer.

REPTILES AND AMPHIBIANS

Reptiles and amphibians are similar in many regards. Both are kept by some people as companion animals.

REPTILES

A *reptile* is an ectothermic animal with dry, scaly skin and lungs for breathing. Ectothermic means that the animal's body temperature adjusts to its environment. Some people refer to reptiles as cold-blooded animals because they feel cold to the touch.

Two groups of reptiles are sometimes kept as pets: 1-lizards and snakes and 2-turtles. More species of lizards and snakes are kept than of turtles. Snakes have long tails but do not have legs, ear openings, nor eyelids. Most lizards have long tails, four legs, movable eyelids, and ear openings. Iguanas and skinks are types of lizards that are sometimes kept. Turtles have shells. They can pull their heads, legs, and tail inside the shell for protection. Turtles can live on land or water. Most reptiles reproduce by laying eggs.

People planning to use reptiles as pets need to carefully study the animal they propose to use. They need to know how to safely keep it and the require-

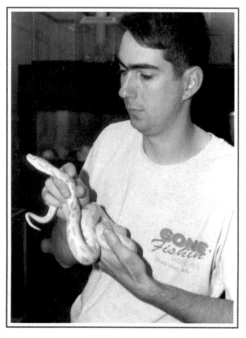

17-17. An Albino Burmese Python is a snake that may be kept.

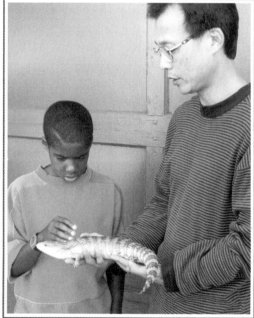

17-18. Keeping a blue-tongued skink requires special knowledge of its needs.

17-19. An iguana is a popular reptile pet.

17-20. Snakes prefer gravel and stones in their containers. This shows two red tail boa snakes in a securely covered aquarium.

ments for its well-being. Since many people are afraid of reptiles, owners must handle them carefully and keep the animal secure so it does not escape.

Keeping a reptile is a long-term commitment. They can live 10 to 20 years or more. Many are kept in aquaria with filters, heaters, and lights. They require special foods to meet their dietary needs. Snakes like small mice. Lizards and turtles may eat plants, small animals, and insects. Keepers usually buy a commercially prepared feed that meets the dietary needs of the species they are raising.

AMPHIBIANS

An *amphibian* is an animal whose body is covered with skin and lives part of its life in water and part on land. Common amphibians are frogs, toads, and salamanders. Frogs and toads are quite similar. As adults, they have four legs and no tail. Salamanders typically have long tails and four legs, though a few species have only two legs.

Amphibians are often kept in an aquarium. Part of the area is covered

17-21. A salamander.

with water and part is above the water line. The dry area is often of pebbles, stones, and wood. Food must be provided. Salamanders eat algae and insect larvae. Frogs and toads eat insects. Amphibians reproduce by laying eggs. The amphibians kept as companion animals are often captured from the wild or are from farms that specialize in the species.

RABBITS

Rabbits are widely kept as companions and for commercial food and pelt production. Information is presented on keeping rabbits as companions here. Detailed information on rabbit production is available in *Rabbit Production*, a book available from Interstate Publishers, Inc.

KINDS OF RABBITS

Rabbits are sometimes mistakenly referred to as rodents. Rabbits have four large, upper, front teeth, and this alone distinguishes them from rodents, which have two large teeth. Rabbits are classified in the order Lagomorpha and family Leporidae. This family includes rabbits and hares. Domestic rabbits descended from the European wild rabbit, *Oryctolagus cunniculus*. Hares are larger than rabbits and have long ears with black tips.

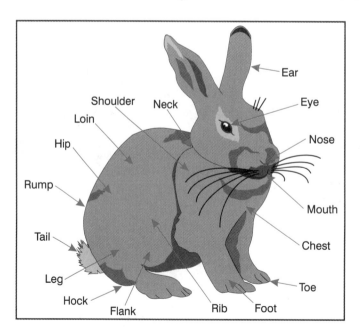

17-22. Major parts of a rabbit.

17-23. Dressed rabbit in a supermarket. (The rabbit is also known by its French name of lapin.)

Colors of rabbits range from solid white to solid black and includes browns, grays, and various spotted color patterns.

The 45 breeds of rabbits are placed in categories by weight. The weight categories and examples of breeds:

- Miniature—The miniature category is the smallest. It is sometimes known as the dwarf category. The Jersey Wooly and Holland Lop are two miniature breeds.

- Small—The breeds in the small category are slightly larger than those in the miniature category. Examples of breeds include Dutch and Mini Lop.

- Medium—The medium category is larger than the small category. Examples of breeds include English Angora and Sable.

- Large—The large category has breeds that are next to the largest category. Examples of breeds include the Californian and Satin.

- Giant—The giant category has the largest breeds. Examples are French Lops and Giant Angora.

17-24. Rabbits are in a range of appealing sizes and colors.

Reproduction

Rabbits are mammals. The females (does) may be bred for the first time at about six months of age. Gestation is 30 to 32 days. Females need to have materials (straw, paper, etc.) for nest building just before parturition. Litters average six babies. Baby rabbits are hairless and have closed eyes. They are weaned at six to eight weeks of age.

Nutrition and Feeding

Use a special pelleted commercial rabbit food. Select a feed based on the age and condition of the rabbit being fed. Changes in feed sometimes upset the digestive systems of rabbits. Fruit, vegetables, and greenfoods may be fed sparingly. Too much of these can also upset the digestive system. Do not overfeed rabbits.

Health Care

A good environment helps prevent disease problems. Avoid cool, drafty hutches. Health problems include colds, diarrhea, constipation, pot belly, eye infections, and sore hocks (feet). A change in the normal behavior of a rabbit is a sign of disease. Get the assistance of a veterinarian or other person qualified in rabbit diseases.

17-25. How to hold and carry a rabbit.

FACILITIES AND EQUIPMENT

Rabbits are often kept in cages known as hutches. A **hutch** is a cage specially designed for rabbits. Hutches should be covered to keep out rain. Walls may be needed to keep rabbits warm in cold weather.

The size of a hutch depends on size and number of rabbits to be kept. For example, a 10-lb. rabbit needs a hutch that is 18 inches high by 36 inches long and 24 inches wide. Wire attached to wooden frames is often used. Wire bottoms of hardware cloth are easy to clean and allow wastes to fall through and out of the cage. With large rabbits, wire can lead to sores on the feet; therefore, a wooden floor may be needed. Waterers and feeders should be in each hutch. Keep hutches clean.

OTHERS

Many other animals are sometimes used as pets. These include insects, ferrets, hedgehogs, chinchillas, deer, raccoons, and farm animals, such as horses, goats, pigs, and cattle. People who have wild animals near their homes often put out food and become friendly with them. Squirrels commonly live in trees in urban areas and become friendly to people. Some of these are wild animals that have not been domesticated. They can suddenly change behavior and attack people.

FERRETS

Ferrets are mammals. They are members of the weasel family and descended from the European polecat. Their scientific name is *Mustela putorius furo*.

Ferrets have long, slender bodies with long tails. A male is known as a hob and a female as a jill. They are primarily nocturnal. When well cared for, ferrets can live 11 years or so.

17-26. Ferrets like cages where they can exercise and play.

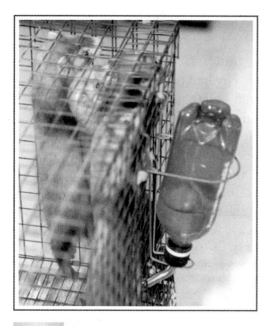

17-27. Water container attached to the outside of a cage.

Ferrets are kept in cages. Most cages should be 24 inches wide by 24 inches long by 14 inches high. Use special ferret or cat food. Water should be available. Keep cages clean. Ferrets sometimes appear to create more filth than some species.

Female ferrets breed in the spring after they are 10 months old. Gestation is about 42 days. Litter size is six to eight babies. Eyes open in three to four weeks. Babies begin eating bread or water-softened dry cat food when their eyes open. Young can be weaned after six weeks. They reach nearly full size in 14 weeks.

Some health care and grooming are needed. Ferrets are susceptible to several diseases, with canine distemper usually fatal. It is best to establish a health care program with a veterinarian when the ferret is young. Nails need to be trimmed regularly. Use care and only remove a small amount of the tip. The trimmers used with dogs and cats work well with ferrets.

Ferrets are sometimes removed from their cages. Always keep them under careful watch and never outside where they can escape. Some incidents of ferrets attacking infants have been reported.

17-28. Trimming the nails of a ferret.

HEDGEHOGS

Hedgehogs have increased in popularity as companion animals in recent years. A hedgehog somewhat resembles a miniature porcupine because of the quills that cover its body. A *quill* is a modified hollow hair that is stiff and has a point. Quills on hedgehogs are ½ to 1 inch (1.3 to 2.5 cm) long. The quills on a hedgehog are not particularly dangerous when the hedgehog is handled properly. When frightened, hedgehogs roll into a ball. Their muscles cause the quills to stick outward and prevent attack. Hedgehogs also roll into balls for sleeping.

Hedgehogs are native to Europe and Africa. Colors vary but many resemble "salt and pepper," with others being chocolate brown. The scientific name of the hedgehog is *Erinaceus europaeus*. An adult European hedgehog is about 12 inches (30 cm) long. They may weigh as much as 3 pounds. An African hedgehog is smaller and shorter. African hedgehogs are most popular in the United States.

In the wild, hedgehogs move quite slowly except for quick darts. Many eat dead animals, worms, and other easy-prey foods. Hedgehogs are kept in cages or unfilled aquaria. A small plastic "swimming" pool as used with children can be arranged nicely. Hedgehogs will not get over the sides of the pool and escape unless equipment placed in the pool allows them to do so. Exercise

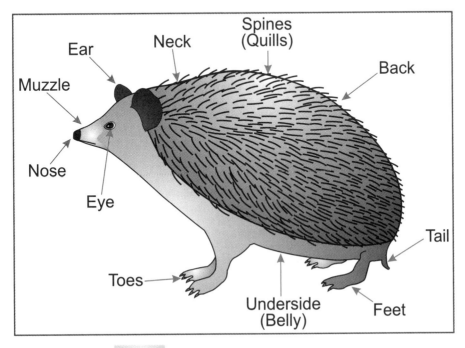

17-29. Major external parts of a hedgehog.

17-30. An African hedgehog.

17-31. Mealworms for feeding an African hedgehog.

equipment similar to that used with hamsters should be in their home. Use wood shavings or other litter on the bottom of the cage or container.

Hedgehog health is evident by clear, bright eyes. Their nose should be dry and free of discharge. Ears should be erect and clean. Their body covering should be soft and appear in good condition. Areas around the anus and genitals should be clean and free of feces.

Clean water should always be available. Hedgehogs are meat-eaters and are often fed live mealworms. A small amount of vegetable or fruit may be fed, such as two peas or kernels of corn. Commercial catfood may also be used with hedgehogs.

17-32. Tarantulas may be kept as pets. This shows a rose hair tarantula.

Hedgehogs of either sex may be kept. Normal social behavior is solitary. *Solitary* means that hedgehogs do not need companions. They typically get as far away from each other as possible. Males in particular move far apart as adults. Two hedgehogs are not needed. Of course, a male and female are needed if breeding is to be done. Most hedgehogs prefer separate cages unless the male and female are to be bred. Hedgehogs can breed after about seven weeks of age. After a short gestation, the female gives birth to a small litter. The male will eat the babies if present in the cage. Females nurse the young for about five weeks, at which time they can be weaned.

REVIEWING

*M*AIN IDEAS

Birds, rodents, reptiles, and other species may be successfully kept as companion animals. Study and learn all that you can before attempting to keep any companion animal.

Birdkeepers often have favorite birds. Companion birds are in seven groups: finches, canaries, budgerigars, cockatiels and parakeets, lovebirds, Amazon parrots, and macaws. Good management is essential for the well-being of the bird. This includes a cage or aviary, food and water, and providing a good environment.

Several species of rodents are used for companion animals. Gerbils, hamsters, mice and rats, and guinea pigs are most widely kept. Proper equipment is needed to assure their well-being.

Reptiles and amphibians are kept by some people. Other people are frightened by them. Snakes, lizards, and turtles are the reptiles that are kept. Amphibians include salamanders and frogs. Amphibians must have an environment of water and land.

Rabbits are widely kept as companions as well as animals for food and fur. Hutches are needed to provide for their well-being. Rabbits are typically provided a pelleted commercial food.

Ferrets and hedgehogs are other species that may be kept. Both are mammals. A good environment must be provided. Proper food, water, and nesting materials are needed.

*Q*UESTIONS

Answer the following questions using correct spelling and complete sentences.

1. What groups of birds are kept? Briefly describe each group.
2. What are the distinguishing features of birds as organisms?
3. What is the role of grit in bird nutrition?

4. What two major kinds of facilities are used with birds? Distinguish between the two.

5. What is a rodent?

6. What species of rodents are kept as companion animals? Select one species and briefly describe how it is kept.

7. What is a reptile? Amphibian? Name one example of each that is kept.

8. How are rabbits kept?

9. How are ferrets kept?

10. What social behavior of hedgehogs is important in keeping them?

EVALUATING

Match the term with the correct definition. Write the letter by the term in the blank provided.

a. birdkeeping   d. aviary   g. solitary
b. fledgling   e. diurnal   h. nocturnal
c. greenfood   f. hutch

_____ 1. An animal that is active at night.

_____ 2. A type of facility used for keeping rabbits.

_____ 3. Raising birds in captivity.

_____ 4. Cabbage, peas, and similar foods.

_____ 5. A young bird that has left the nest but is not capable of providing food for itself.

_____ 6. An animal that is active in the daylight.

_____ 7. Behavior of an animal that lives alone.

_____ 8. A facility for keeping birds.

EXPLORING

1. Make a tour to a birdkeeping facility. Observe the cages and aviaries that are in use. Interview the manager about the kinds of birds that are kept, how they are managed, and the problems that are encountered. Prepare a written report on your observations.

2. Job shadow a worker in a companion animal store for a day. Observe the species of animals and how they are managed. Prepare a report that includes detailed information on the care provided different species.

18

ORNAMENTAL FISH

Keeping an aquarium is an interesting venture. People express themselves in the design of the aquatic environment. Proper care is essential. Without proper care, fish will die and the aquarium will not be appealing.

Ornamental fish form a large industry in the United States. These fish are more valuable than food fish based on weight. People want to buy new fish and all the supplies. This means that ornamental fish must be raised. Some people find ornamental fish a profitable crop to grow.

People enjoy watching fish. They like to see them swim and go about life in an aquarium or open pool in a courtyard. Watching fish even relieves stress and helps people lead healthier lives, according to research.

18-1. Ornamental fish need regular care.

OBJECTIVES

This chapter is an introduction to keeping ornamental fish. It has the following objectives:

1 Explain ornamental species and list examples

2 Explain the equipment and facilities needed for ornamental species

3 Describe the water environments for ornamental species

4 Describe how to care for ornamental species

TERMS

aeration
aging
aquarium
aquarium maintenance
  schedule (AMS)
biological filtration
biological oxygenation
chemical filtration
companion fish
egg-laying fish
filtration
freshwater ornamental fish
habitat

livebearing fish
mechanical filtration
nitrogen cycle
ornamental fish
oxygenation
salinity
saltwater ornamental fish
shoal
synthetic seawater
tapwater
thermostat
tropical fish
vat

ORNAMENTAL SPECIES

Ornamental fish are fish kept for their appearance and personal appeal to people. Some ornamental fish have bright colors and large, fancy fins. These fish are not typically used for food.

Tropical fish form a subgroup of the ornamental species. A *tropical fish* is a small, brightly colored fish that is popular in home and office aquaria. These fish thrive in warm water, breed rapidly, and tolerate conditions in home aquaria. Tropical fish range from 1 to 12 inches in length.

Companion fish are kept in homes for human companionship or as pets. An aquarium with fish may provide an important use of leisure time. Companion fish amuse, entertain, and provide enjoyment. Companion fish may be tropical fish or they may be other species of fish, as well as other aquatic animals, such as snails or turtles, and aquatic plants.

18-2. An attractive freshwater aquarium.

Sources of Fish

Most ornamental fish are not native species. They have been brought into North America from another place. Releasing these fish into streams is often illegal.

Ornamental fish are captured from the wild or raised in confinement. In the past, most ornamental fish were caught from streams and lakes. People have found cultured ornamental fish to survive better in aquaria. A few people raise ornamental fish for stores and pet shops. Florida is the leading state in the production of ornamental fish.

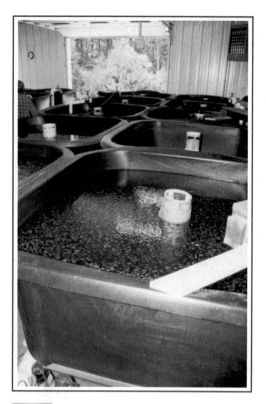

18-3. A volume producer of ornamental fish in Maryland uses large tanks.

Wild ornamental fish are typically caught in the tropical waters of the Pacific Islands, Asia, and South America. Proper treatment of captured fish can rid them of disease and help them survive in captivity.

FRESHWATER ORNAMENTAL FISH

Freshwater ornamental fish are most popular. These are species that grow in freshwater. They will tolerate only a very small amount of salt in the water.

The freshwater ornamental fish reproduce in two ways: laying eggs and giving birth to live young.

Egg-Laying Fish

Egg-laying fish reproduce by the female fish releasing eggs that are fertilized by sperm from the male fish. Incubation usually lasts only a few days.

Popular egg-laying ornamental fish include koi, goldfish, gouramis, tetras, barbs, and catfish.

- Koi—The koi is a variety of common carp (*Cyprinus carpio*) that was developed by selecting fish with distinctive coloring. The colors may be bright red, black, gold, or white. Koi, a freshwater fish covered with scales, may reach a length of 37.5 inches (95 cm) and weigh 3 to 10 pounds or more. Most koi grow in pools or large tanks. Koi clubs are found in some places. The members raise and enter koi in various competitive events.

- Goldfish—People have been keeping and raising goldfish (*Carassius auratus*) for more than 2,000 years. In the koi family and sometimes known as golden carp, goldfish are smaller than koi. They are adapted to a wide range of environments. New varieties of goldfish with bright colors, fancy fins, and interesting eyes have been developed. Goldfish are hardy and easy to keep. Their size depends on the size of the container in which they are grown and their nutrition. They range from a couple of inches to more than 2 feet in length. A

18-4. A large koi.

goldfish must have at least 2 gallons of water to grow to 2 inches (5 cm) in length. The goldfish is a good species for the person who is just beginning with ornamental fish. However, the water must be cleaned often because goldfish are messy. Goldfish will live up to 15 years.

- Gouramis—Several species are known as gouramis. The most interesting is the kissing gourami (*Helostoma temmincki*) because of the unique kissing behavior of pairs of the species. Other popular gouramis include the three-spot gourami (*Trichogaster trichopterus trichopterus*), blue gourami (*Trichogaster trichopterus*), and the pearl gourami (*Trichogaster leeri*). Gouramis are freshwater fish that may reach lengths of 12 inches, depending on the species. Most gouramis prefer water that is in the range of 75°F (24°C).

18-5. Goldfish are often kept in small bowls.

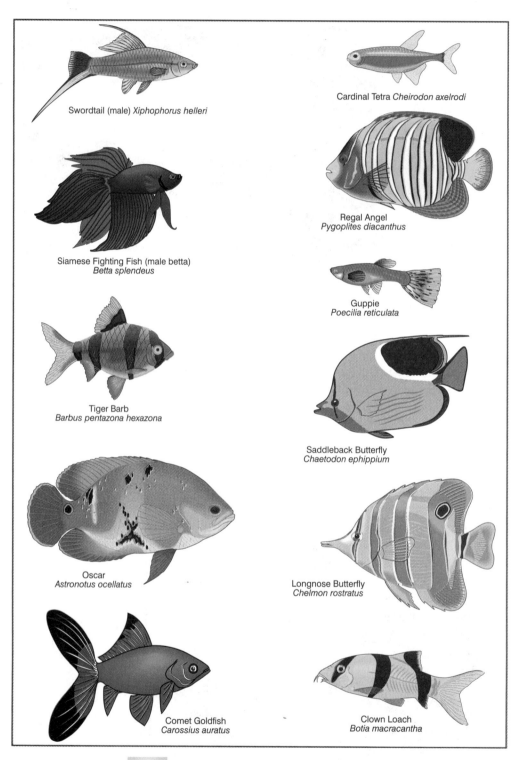

18-6. Common ornamental fish kept in aquaria.

■ Tetras—The tetras include several species that are easy- to medium-care ornamental fish. The neon tetra (*Paracheirodon innesi*) has an electric-blue stripe the entire length of its body. The cardinal tetra (*Cheirodon axelrodi*) has a brilliant reddish color. Several other tetras are often raised. Most tetras grow to a maximum length of 1.5 to 3 inches (3.75 to 7.5 cm). Tetras prefer water that is 72 to 85°F (22 to 28°C).

■ Barbs—Barbs are popular. They are easy to raise and are in the same family as the goldfish. Barbs reach a mature size of 2 to 4 inches (5 to 10 cm) in length. They prefer freshwater that is 70 to 80°F (21 to 27°C). Most barbs like aquaria with plenty of light. Three common barbs are the spotted barb (*Barbus binotatus*), rosy barb (*Barbus conchonius*), and tinfoil barb (*Barbus schwanenfeldi*). Some barbs eat their eggs after spawning.

■ Catfish—Some species of America's most popular foodfish are grown as ornamentals. The upside-down catfish (*Synodontis angelicus*) is interesting to watch in an aquarium because of its behavior in swimming upside-down. The glass catfish (*Kryptoptereus bicirrhis*) and electric catfish (*Malapterurus electricus*) are also interesting species. The glass catfish has a clear-like body

Connection

GETTING BACTERIA

Water does not always have the bacteria needed with fish tanks. Well water and water from municipal systems would be objectionable for drinking if it had large numbers of bacteria. Bacteria are needed in the filtration system, on pipes, tank areas, and other places.

Getting the needed bacteria may involve inoculating the water. Bacteria are bought and added to the water. This assures that the proper bacteria are present and increasing in numbers. Without an inoculant, several days or weeks may be needed for a system to grow the needed bacteria.

These photographs show the clean filtration system that needs bacteria and a small bottle containing many millions of bacteria. Just a few drops into the water gives all the bacteria a system needs to get started.

that may appear as a rainbow. The electric catfish can give off an electric shock that will kill smaller fish. Catfish prefer water that is 70 to 80°F (21 to 27°C). Most catfish prefer the darker areas of an aquarium.

Livebearing Fish

Livebearing fish give birth to live young. They like to live in a group of five or more, known as a *shoal*. Rather than releasing sperm over the eggs, as with egg-laying fish, the males deposit sperm inside the female fish. The fertilized eggs develop into nearly fully formed fish. Females of some species store sperm from the male until it is needed for reproduction. Four common livebearers are guppies, swordtails, mollies, and platys.

- Guppies—The guppy (*Poecilia reticulata*) is the most popular livebearer. Many varieties have been developed, with the primary difference between them being the shapes of their fins and tails. Guppies like plenty of food and will eat different foods. Most producers feed a dried commercial food. Guppies may reach lengths of 2.5 inches (6.35 cm). Females may give birth to 200 baby guppies at a time, though the average is about 50 babies. They prefer water that is 68 to 75°F (20 to 24°C). Guppies reproduce profusely, but the adults may eat the young.

- Swordtails—Similar to guppies, a major distinction of the swordtail (*Xiphophorus helleri*) is the long, sword-like appearance of the caudal fin. Many different colorings are found on swordtails. They may reach lengths of 3 to 4.75 inches (8 to 12 cm) and are easily kept with other livebearers. Swordtails often eat their young when they are born, so producers use special traps through which the young can escape. Swordtails, like guppies, eat live and dried food. Their environmental requirements are similar to the guppy.

- Mollies—Most mollies are of two species: small fin (*Poecilia sphenops*) and large fin (*Poecilia latipinna*). With some mollies, the fins get so large that swimming is nearly impossible. Most mollies are black. Other colors are increasingly in demand, such as the white molly. Mollies prefer water that is 72 to 82°F (22 to 28° C). Mollies like to eat vegetable foods, such as algae and cooked spinach. They like to form large schools in tanks with other mollies. It is best to keep several mollies together in an aquarium.

- Platys—Platys are popular in aquaria and have been bred to achieve certain desired characteristics. The variatus platy (*Xiphophororus variatus hybrid*) is especially popular. It has a yellow color with a distinctive orange-red caudal fin and dark-lined vertical markings. Some platys may be brownish-yellow in color. They grow to about 2.5 inches long (6 cm) and prefer water that is 68 to 77°F (20 to 25°C). They are hearty eaters, being especially fond of live and dried food. Platys are easy to keep with other livebearers.

SALTWATER ORNAMENTAL FISH

Saltwater ornamental fish are those that live in saltwater. The water is a complex mixture of various minerals, with salt (sodium chloride) being one of the substances that is found in the largest amount. Other minerals in seawater include magnesium chloride, magnesium sulfate, and calcium carbonate. Very small amounts of several other minerals are found in seawater, such as zinc and molybdenum. Most saltwater ornamental fish require natural or synthetic seawater. A saltwater aquarium is a marine aquarium. All of the species included here are egg-layers.

- Angelfish—There are many different species of angelfish; some can live in freshwater. Most angelfish appear delicate, but they are hardy and capable of living a long time in a well-managed aquarium. A popular angelfish is the French angelfish (*Pomacanthus paru*), which has a black body and vertical yellow stripes. The coral beauty angelfish (*Centropyge bispinosus*) has a wide range of colors, including orange and black. Most angelfish prefer a water temperature of 77 to 86°F (25 to 30°C). Angelfish may be 6 inches (15 cm) long and 10 inches (26 cm) high (tip of fin to tip of fin). Angelfish eat a wide range of live food in natural environments, but will eat commercial food in captivity. Their eggs hatch in about 36 hours. As part of the incubation process, the eggs are carried in their parents' mouths and placed on the leaves of water plants and later in the sand.

- Butterfly fish—Butterfly fish are beautiful and popular marine aquarium fish. Most of the butterfly fish need to have ample space in an aquarium. The long-nosed butterfly fish (*Forcipiger flavissimus*) has beautiful bright yellow and black colors, with the snout and throat being green. This fish will grow to lengths of 6 inches (16 cm). Butterfly fish prefer temperatures of 75 to 82°F (24 to 28°C). Butterfly fish are territorial and are preferably isolated from other butterfly species. Some people keep only one butterfly fish in a tank.

- Basslets—Basslets are small colorful fish that are popular in marine aquaria. The royal gramma (*Gramma loreto*) is among the best-known basslet. The royal gramma is a hardy fish and particularly good for the person who is just beginning a marine aquarium. Royal gramma are very aggressive toward others of their species. They are, therefore, usually isolated from other royal gramma. They can be mixed with other species of fish. Another popular basslet is the bicolor basslet (*Pseudochromis paccagnellae*), which has a yellow and violet color. The bicolor basslet is also an aggressive fish. Basslets prefer tiny, living food, such as brine shrimp. They like water that is 79 to 82°F (26 to 28°C).

EQUIPMENT AND FACILITIES

Ornamental fish need a good environment to survive and grow. This environment is artificially created by the equipment and facilities that are used. Needs vary with the species being grown. For example, goldfish can exist in simple systems, but other fish, such as angelfish, require more elaborate equipment.

The basic equipment begins with a water container. Other equipment makes the environment better for fish. Water containers are often decorated with colorful gravel, aquatic plants, and other items.

WATER CONTAINERS

Several different kinds of water containers are used. Some are used to display fish; others are used to reproduce the fish and keep them in good health.

Tanks (Aquaria)

Tanks vary widely. Some are for production; others for enjoyment and often have clear walls for good viewing.

An *aquarium* is a container used to hold water. Sometimes the containers are known as "fish tanks" or "tanks." An aquarium should be watertight.

18-7. The basic equipment for an aquarium.

Begin with small aquaria and expand into larger sizes. Several smaller aquaria also have advantages over one larger aquarium, especially in disease control and water management.

A simple aquarium is the goldfish bowl. This is a rounded glass bowl with a water capacity of about 1 gallon (4 liters), but it may be smaller or larger. It is used with species that are hardy and can survive in water that may get fouled and is not regularly aerated.

The most common aquaria are rectangular, though some are square or spherical. Specially built aquaria are available, but are often expensive. Aquaria range in size from 10 gallons (42 l) to 30 gallons (127 l), 50 gallons (212 l), 100 gallons (424 l), or larger.

Aquaria of all-glass materials are preferred, especially with saltwater. Aquaria that have metal corners or other parts will corrode and are more likely to leak. All-glass aquaria are easier to clean. The glass in an aquarium should be at least $1/4$ inch (64 mm) thick for small aquaria and $3/8$ inch (96 mm) thick for larger aquaria. A disadvantage of all-glass construction is that the larger aquaria are very heavy—thicker glass is used in making them. Only high-strength glass should be used in building an aquarium.

Aquaria need to be on stands that will support their weight. The stands should be level and located where they will not need to be moved often. Moving an aquarium is a big job.

Pools and Fountains

Decorative pools and fountains may be used for goldfish, koi, and similar species that are hardy. These structures are made of plastic, fiberglass, concrete, or other materials. The construction should be watertight and easy to

18-8. Koi in a decorative interior water garden.

clean. The arrangement of the facilities should make it easy to manage the water and empty the container, should the need arise.

Vats

A *vat* is a large tank made of concrete, fiberglass, or a similar material. Vats are in a fixed location and cannot be moved about. These structures are used to raise and reproduce fish and not usually for displaying fish. Vats are rather large and may hold hundreds or thousands of gallons of water. The structures should be designed for ease of management. Water circulation, aeration, drainage, and other features should be designed into vats. Vats should be watertight and designed to meet the needs of the species that will be produced in them.

Earthen Ponds

Adapted ornamental species may be produced in small ponds made of earth, known as earthen ponds. These are frequently used with koi, goldfish, sunfish, and other species that are adapted to the local climate. The pond's construction should prevent seepage or overflow from nearby streams. Select a location that keeps predators and undesirable species out of the ponds. Grassy areas are needed around the edges to keep the water from getting muddy and to reduce erosion.

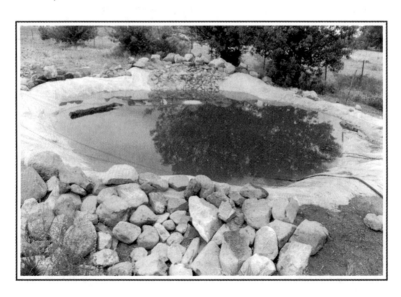

18-9. A small, plastic-lined, earthen pond used for keeping koi.

WATER QUALITY EQUIPMENT

Equipment is needed to keep the water in a fish tank appropriate for the growth of the species.

Oxygenation

Oxygenation is the process of keeping adequate dissolved oxygen (DO) in the water. Oxygen that has been dissolved in the water is removed by the gills of the fish and used for life processes. Water that has little movement in a tank takes on little new oxygen. Fish can quickly use up the DO and die.

Several methods are available to keep dissolved oxygen in the water. Some are much more practical than others.

Aeration is the process of bubbling air into water or splashing water in the air. This uses mechanical methods of moving the water and air. Aeration promotes the gas transfer between the water and the air. The supply of oxygen in the water is replenished by aeration.

With aquaria, air or oxygen may be injected into the water. An air pump may be used to force air through a plastic tube into the water at the bottom of an aquarium. With vats, aerators may be suspended on the surface of the water and fans used to splash the water in the air. Splashing increases the rate of diffusion of oxygen in the water.

Biological oxygenation is using phytoplankton (tiny plants) and other aquatic plants to replenish oxygen. These plants carry on photosynthesis to produce food. In the process, oxygen is released into the water.

Filtration

Fish create solid wastes in the water. These wastes foul the water and make it unfit for fish. *Filtration* is the process of removing solid particles and gases from the water to keep a good environment for the fish. Three kinds of filtration are used: biological, mechanical, and chemical.

18-10. Underwater gravel filters are sometimes used in aquaria.

Biological filtration is using bacteria and other living organisms to convert harmful materials into forms that are less harmful. The bacteria feed on the uneaten feed, feces, and gases, such as nitrogen, in the water. Snails are sometimes used in aquaria to remove organic matter from the water. Crawfish may be used to scavenge food particles at the bottom of the aquarium.

Bacterial action changes nitrogen compounds into less harmful forms and replenishes the DO in the water. Wastes may form ammonia, a compound containing nitrogen. This material undergoes denitrification by bacteria in the nitrogen cycle. The bacteria are primarily found in the filter bed of the aquarium.

Mechanical filtration is using various kinds of filtration devices to remove particles from the water and keep the water clear. It involves flowing water over and through filters made of gravel, charcoal, and fibrous materials (known as floss). Filtering materials must be cleaned often; otherwise, they will become clogged with wastes and will not work very well. Many mechanical filtration systems are tied in with oxygenation systems.

Chemical filtration is using chemical processes to filter water. Special kinds of chemical filters are used. Ozone and ultraviolet irradiation may be used. Chemical filtration often uses activated charcoal to help keep the water clear and prevent water yellowing.

Aquaria often use combinations of biological, mechanical, and chemical filtration. Devices are added to a fish tank so filtration can take place.

18-11. A canister filter located in a secure position near the aquarium.

- Undergravel filters—Undergravel filters are placed at the bottom of the fish tank. The water is pulled through the bed of gravel. The gravel bed is both a mechanical and biological filtration system. The gravel particles screen solid materials out of the water and provide a place for beneficial bacteria to grow.

- Canister filters—Canister filters are filtration systems outside the aquarium. Pumps and a system of tubes move the water through several layers of filtering material, such as activated charcoal and filter floss. The material removes solid wastes and provides a location for the bacteria to act. The material must be cleaned or replaced every few weeks, depending on the extent of fouling of the water.

■ Outside filters—Outside power filters hang on the back of the aquarium. Water is moved through the filtration material much as a canister filter. These are common on small aquaria.

■ Nitrifier/denitrifier filters— Nitrifier/denitrifier filters supplement the action of bacteria in the nitrogen cycle. These filters are expensive and used only in special situations in aquaria.

18-12. An outside filter attached to an aquarium.

Thermometers and Heaters

Thermometers measure the temperature of the water. They often have thermostats attached to them. A ***thermostat*** is a control device that turns on the heater when the water temperature gets below a certain level.

Heaters help keep the water warm enough for fish. Most heaters are glass-enclosed, electric, heating elements that extend into the water. The thermostat in the heater turns the heating element on when the water is cool and off when the water has been warmed sufficiently.

Thermometers may be independent of thermostats and heaters. Each species of fish has its own temperature requirements. Most of the ornamental fish in aquaria prefer temperatures in the 70 to 80°F (21 to 27°C) range.

COVERS AND LIGHTING EQUIPMENT

Most fish tanks should be kept covered and properly lighted. Covers keep the fish from jumping out of the container and prevent predators or objects

18-13. A combination cover and light for an aquarium.

from falling into the fish tank. Covers are designed for easy removal or opening to feed the fish and remove or add fish.

Lighting may be built into the cover. Lighting is needed so people can see the fish as well as to encourage biological processes. Water plants need light to live and grow. Algae and other organisms may need light to grow. Lights often have reflective hoods that allow better lighting. Caution: Be very careful when using electricity around the water in a fish tank.

OTHER EQUIPMENT

A wide range of other equipment may be useful. Examples include dip nets, transfer containers, containers for aging water, decorative items for a fish tank, and hydrometers to determine the salt content of the water in a marine aquarium.

THE WATER ENVIRONMENT

Keeping ornamental fish requires a good water environment. Freshwater and saltwater species usually cannot be mixed. Some species also prefer brackish water, which is a mixture of freshwater and saltwater. Managing the systems varies with freshwater and saltwater; however, many of the principles are the same.

ARTIFICIAL HABITAT

Setting up an aquarium is establishing a desired habitat. A *habitat* is a place where a plant or animal naturally lives. The aquarium is set up to duplicate the environment that nature provides. Knowing what to include requires careful study of natural habitats. In some cases, aquaria are designed for a particular climate, such as the Amazon River.

Various plants, animals, minerals, rocks, and other materials may be used in trying to get balance in the artificial habitat. The items are arranged to provide a pleasing appearance as well as a good place for the fish to live and grow.

WATER SOURCE

A supply of quality water is needed. The water must be appropriate for the species that are being grown.

Freshwater

Freshwater varies in quality. Water quality depends on its source and the substances in it. The dissolved chemicals in water can make it unfit for fish.

Water for fish tanks comes from three major sources: tapwater, rainwater, and wells.

Tapwater is the water at the faucets in homes, offices, and businesses. Tapwater is readily available but usually cannot be used as it comes from the tap. Municipal water systems often add chlorine or other chemicals to water. These substances keep the water safe for humans to use, but make it dangerous to fish. Tapwater can often be readied by "aging." *Aging* is collecting the water in an open container and allowing it to stand for

18-14. Water for a small aquarium can be collected in a jar for aging.

several days. Chlorine and other substances are released into the atmosphere. Of course, a large aquarium will require a lot of "aged" water. Containers in which water is collected and stored must be clean. Dirty containers pollute the water. Tablets that prepare water for use can be purchased at a pet supply store and put into the tapwater.

Rainwater can be used in aquaria, but must be collected so it is not polluted. Acid rain and other substances in rainwater may make it unfit for fish tanks.

Large-scale producers use wells. The water should be tested to determine its mineral content and other substances. In some cases, well water may be unfit for fish tanks. Gases in water can be released by storing the water in an open container for a few days.

Saltwater

Saltwater may be from seawater or made using a synthetic salt mix. Natural seawater is often used near oceans or lakes with saltwater. Seawater is a complex solution of many chemical compounds. Natural seawater must be collected away from sources of pollution. This may involve going out into the ocean, away from inflowing streams, and collecting the seawater in a large tank that is brought back to shore.

Synthetic seawater is made by mixing freshwater with a saltwater mix. These mixes can be bought at pet supply or aquaculture stores. Follow the instructions that are included with the mix for preparing the seawater. In most cases, it is easier and safer to make synthetic seawater than to use natural seawater.

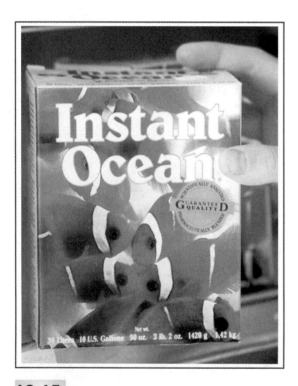

18-15. Synthetic seawater can be made using a specially prepared mixture of salt and minerals.

WATER QUALITY

Water quality is the suitability of the water for use in a fish tank.

Water has to be prepared before placing fish in it. This section is about maintaining water quality.

Temperature

Fish thrive best if the temperature of the water is within an appropriate range. The requirements for each species should be determined.

In some cases, the water will need to be warmed. In moving fish, the water into which they are placed should vary only a few degrees from the water where they have been. Sudden changes in water temperature cause stress.

Most ornamental species prefer water that is 70 to 80°F (21 to 26°C). Saltwater species often prefer water that is slightly warmer than the freshwater species. Heaters may be needed to keep the water at a uniform temperature.

pH

pH is the acidity or alkalinity of the water. It is measured on a scale of 0 to 14, with 7.0 being neutral. pH readings below 7.0 are acidic; the lower the number, the higher the acidity. pH readings are increasingly alkaline as the number goes above 7.0. Ornamental species prefer water with a pH of 7.8 to 8.3.

Changes in pH result from reactions in the water. Many things in an aquarium influence pH. For example, the materials used in a filter are continually being dissolved into the water. Those made from calcium-based ma-

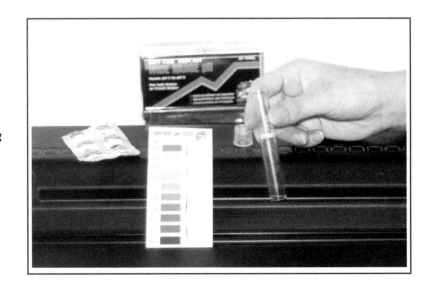

18-16. Using a kit to test pH.

terials can result in higher pH readings. The nitrification process and carbon dioxide in the water can lower the pH. Buffering solutions and powders can be bought at pet supply and aquaculture stores to help maintain water within the right pH range.

Nitrogen Cycle

The **nitrogen cycle** is the process of converting animal wastes into ammonia, ammonia into nitrites, and nitrites into nitrates. It is a natural process that occurs in water.

Ammonia is toxic to fish. Ammonia can sometimes be noticed as an odor above the water. It is more likely to be a problem when fish are fed too much or when solid wastes are not filtered out of the water. Commercial test kits can be used to test for ammonia. They must be used quickly and properly because, as a gas, ammonia can quickly go into the air. Overfeeding and overstocking the fish tank contribute to ammonia problems.

Dissolved Oxygen

Dissolved oxygen (DO) is gaseous oxygen in the water. Fish and other aquatic life remove the DO from the water. Most fish need at least 5.0 ppm of DO. Below 3.0 ppm DO level, some species show stress and die. Most fish die in a few minutes when DO is as low as 1.0 ppm. Oxygen meters and test kits can be used to measure DO.

Salinity

Salinity is the salt content of water. Freshwater species do not want and cannot survive in water with much salt. Saltwater species require sufficient salt in the water. Salinity is measured as specific gravity using a hydrometer. Water has a specific gravity of 1.0. Saltwa-

18-17. Using a hydrometer to check salinity. (Water in a marine aquarium should have a specific gravity of 1.020 to 1.025.)

ter has a slightly higher specific gravity. Most saltwater species want water with a specific gravity of 1.020 to 1.024 at a water temperature of 77 to 80°F (25 to 26°C). Salinity can be increased by using commercially prepared sea salts.

CHANGING THE WATER

Water in fish tanks should be changed periodically. Since this is a big job with large tanks, careful management should be followed to prevent water fouling. Water should be obtained and aged, as with a new aquarium, before it is used. Fish will need to be temporarily moved to another water tank. Care should be used to avoid stressing the fish at this time.

18-18. Use a sponge brush to clean the inside walls of an aquarium.

18-19. Using a siphoning tube to remove solid materials from the bottom of an aquarium.

CARING FOR ORNAMENTAL FISH

Ornamental fish have needs that must be met. How long fish live and their rate of reproduction is often related to the care they receive.

FEEDING

Fish need the right amounts as well as the proper nutrients in their feed. The amount a fish is fed each day is its ration. The ration should provide a diet that meets the nutrient needs of the fish. The amount of feed that fish eat is related to the temperature of the water as well as the kind of fish and their stage of life. As the water temperature rises, the use of food by the fish increases; therefore, they should be fed more.

The amount to feed is no more than the fish will eat in a few minutes. This involves careful observation of the behavior of the fish. Overfeeding fouls the water.

Most ornamental fish growers feed a commercial food. These feeds are specially formulated for different fish. Some include animal protein, such as shrimp. Others include grain and related ingredients. All feed containers should have labels that list the ingredients and nutrients. Most feeds are recommended for certain species.

18-20. Properly feed and care for fish each day.

HEALTH

A good environment and proper feeding help to keep fish healthy. Minimizing stress also reduces the chance of a disease outbreak. Stress occurs when fish are handled or when there is a sudden change in their environment. Regularly and carefully observing fish helps people know the behavior of healthy fish. A disease should be suspected when one or more of the following are observed:

- Fish scratching on objects

- Fish failing to eat

- Increased respiration (gill movement) and gasping at the surface of the water

- Clamped or folded fins

- Cloudy eyes

- Bloody spots or fuzzy patches on the body

- White spots on the body

- Frayed fins

- Color changes

- Protruding eyes or abdomen

- Material hanging from the anus of the fish (could be a worm parasite)

Sick fish should be removed from the tank and kept in a separate tank. Disease treatments can be given to the diseased fish as well as the fish in the healthy tank. All medications should be used according to instructions.

18-21. Signs of good health are important in selecting fish.

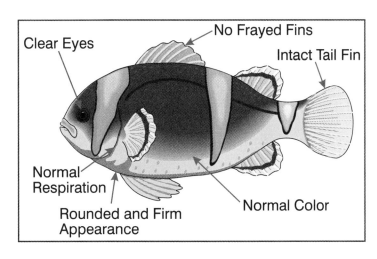

Clear Eyes

No Frayed Fins

Intact Tail Fin

Normal Respiration

Rounded and Firm Appearance

Normal Color

When selecting new fish, observe the fish for signs of disease. If a fish has the signs, do not get the fish. Also, do not get healthy fish from an aquarium where other fish are sick. You will likely be bringing disease to your aquarium!

MAINTENANCE SCHEDULE

Following a regular maintenance schedule can be valuable in keeping an aquarium in good condition. An *aquarium maintenance schedule (AMS)* is a listing of important activities and the frequency with which they should be performed in keeping a good aquarium environment. Some activities must be done each day, while others can be done weekly, monthly, quarterly, or annually.

- Daily—provide feed; check the heater, temperature, aeration, and filtration; remove any dead fish; and observe the fish for unusual behavior.

- Weekly—check the water level and pH; add water and chemicals as appropriate.

- Monthly—change and add water; siphon off dead material on the bottom; remove algae; and tend to any plants that may be in the aquarium.

- Quarterly—clean the filter; check electrical connections; check hoses and pump.

- Annually—clean the tank thoroughly; rinse the bottom gravel; and replace light bulbs/tubes.

REVIEWING

MAIN IDEAS

Ornamental fish are kept for beauty and personal appeal. Some ornamental fish are captured from the wild; others are produced in fish tanks. Many different species of ornamental fish are kept. Some are egg-layers; others are livebearers. Some live in freshwater; others live in saltwater.

Producing ornamental fish involves establishing and maintaining a good environment. The needed facilities include water containers, water-quality equipment, and tank covers and lighting.

Water source is important to the success of ornamental fish production. Freshwater can be obtained from drilled wells, municipal water systems, and rain. Saltwater can be obtained from the sea or synthetically made using commercially available salts. In nearly all cases, the water must be conditioned or aged before it is used.

Water quality—the suitability of water for fish—must be maintained. It includes temperature, pH, nitrogen cycle, dissolved oxygen, and salinity. The solid materials must also be filtered out of the water.

Care of ornamental fish includes feeding, controlling disease, and performing maintenance activities on daily, weekly, monthly, quarterly, and annual bases.

QUESTIONS

Answer the following questions using correct spelling and complete sentences.

1. What are the sources of ornamental fish?

2. What are examples of ornamental fish? Name two examples each of freshwater egg-laying fish, freshwater livebearing fish, and saltwater fish. Briefly describe each species.

3. What equipment is needed to produce ornamental fish?

4. What is oxygenation? Why is it important?

5. What kinds of filtration are used? Distinguish between the three common kinds.

6. What should be considered in selecting a source of water?

7. What are the important areas of water quality with ornamental fish?

8. What is the purpose of feeding? What guidelines should be followed in feeding?

9. Name three fish behaviors that indicate a possible disease problem.

10. Describe a maintenance schedule for ornamental fish production.

EVALUATING

Match the term with the correct definition. Write the letter by the term in the space provided.

a. egg-laying fish   e. nitrogen cycle   i. salinity
b. aquarium   f. shoal   j. dissolved oxygen
c. oxygenation   g. guppy
d. filtration   h. habitat

_____ 1. A container used to hold water for ornamental fish.

_____ 2. Fish that reproduce by the female laying eggs.

_____ 3. The process of removing solids and gases from water.

_____ 4. Gaseous oxygen in water.

_____ 5. Salt content of water.

_____ 6. The process of keeping adequate DO in the water.

_____ 7. The process of converting wastes into ammonia, nitrites, and nitrates.

_____ 8. Place where a plant or animal naturally lives.

_____ 9. Group of fish living together.

_____ 10. Popular ornamental livebearing fish.

EXPLORING

1. Set up two fish tanks in the classroom. Establish a saltwater environment in one tank and freshwater in the other. Select the appropriate equipment and install the oxygenation, filtration, and lighting systems. Select the appropriate species and establish the tank. Establish an artificial habitat that is pleasing to view as well as appropriate for the species selected. Set up a routine aquarium maintenance schedule (AMS) for the tanks. (Additional resources may be needed. A visit to a local pet store will provide access to materials and information on the species to use.)

2. Take a tour of a large aquarium and study the different aquatic environments that have been installed. Ask the manager to let you observe the water filtration systems and other aspects of managing the aquarium.

19

WILDLIFE ANIMALS

Wildlife is enjoyed in many ways. Watching animals is a favorite pastime with some people. Others enjoy hunting. Many animals are taken by sporting enthusiasts each year. These are for trophies, food, and other uses.

People differ in how they view the use of wildlife. Some feel that legal hunting is a good sport and helps keep wildlife populations controlled; others disagree. People enjoy wildlife in many ways. What is your favorite wild animal?

As we have grown to appreciate and enjoy wildlife, we have also come to realize that a healthy population of wildlife demands planned management. As human populations grow and the amount of land available to wildlife decreases, the management task becomes more difficult.

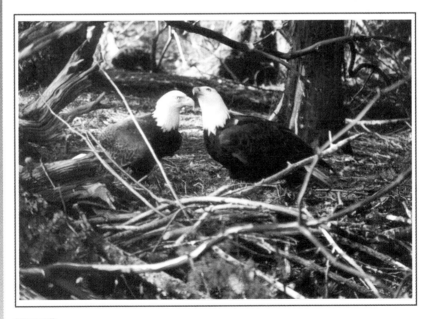

19-1. Once endangered animals, bald eagles are among favorite birds to watch in some places.

511

OBJECTIVES

This chapter is an introduction to animal wildlife and management. It has the following objectives:

1 Explain the importance of wildlife animals

2 Classify the major game animals

3 Explain species endangerment and practices to prevent endangerment

4 Explain important practices in wildlife management

5 List important considerations for sports enthusiasts

6 Describe ways to enjoy wildlife

TERMS

animal wildlife
biodiversity
birding
carrying capacity
endangered species
game
hunting

limiting factor
niche
non-game animal
pelt
territory
trapping
urbanization

THE IMPORTANCE OF WILDLIFE

Animal wildlife is any animal that has not been domesticated. This includes insects, birds, fish, rodents, and many others, such as deer, elk, and bear. These animals benefit humans in many ways. Those hunted for food and other uses are known as *game*.

Why should we be concerned with wildlife populations? Our society has evolved to a point where we can go to the grocery store and buy any kind of meat or fish we could possibly want. There is no longer a need to hunt or fish for our food.

This argument would be true if the only reason we wanted to have fish and wildlife was to provide food. There are many other reasons to be concerned with wildlife and game fish populations. Several in this chapter include economic, beauty, recreation, and biodiversity.

*E*CONOMIC VALUE OF WILDLIFE

Large sums of money are spent on wildlife in North America each year. Millions of dollars are made every year in department stores that sell hunting and fishing supplies. The communities around popular hunting, fishing, and wildlife areas and parks depend on the income from nature lovers, hunters, and fishers to operate the services of the community.

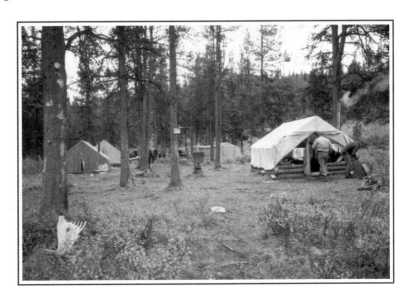

19-2. Hunters spend a great deal of money on guns, travel, and other things, such as camping.

Hunting is taking animals with guns or other weapons. The animals are used for meat, hides, or other products and to make trophies for home or office decorations.

Trapping is capturing animals for their products. The animals are usually caught in traps and remain alive until someone checks the traps. Most of the animals are valued for their hides. Some trapped animals have other products, such as people in local areas may use the meat of the raccoon for food. Some animals are trapped to move them to other places. People who trap animals are trappers.

A *pelt* is the whole hide with fur attached. Pelts are used to make valuable fur coats and other products. Some animals used for pelts include fox, rabbit, squirrel, raccoon, and mink.

19-3. Pelts in a Sitka, Alaska, store were taken by trappers.

BEAUTY AND AESTHETICS

People stop to look at a deer grazing in a field or a young family of mallard ducks swimming on a pond. We enjoy the show wildlife animals provide. Many people want to vacation in

19-4. The beauty of a bull elk in Yellowstone National Park helps us appreciate wildlife.

the country or at a park in hopes of seeing wildlife animals. The beauty of wildlife captures our attention and imagination.

Chances are that you have visited a zoo or been a part of a wildlife officer's program that let you get close to a wild animal. You may even be in a club formed to enjoy wildlife. Bird watching clubs, conservation clubs, and organizations at local parks have become a popular way to enjoy wildlife.

Recreation

Every year, hundreds of thousands of people enjoy fish and wildlife through hunting, fishing, photography, scouting, bird watching, and other recreational activities. Adults and children can enjoy nature through recreational activities.

Although hunting for food is no longer required for us to survive, it gives hunters a chance to relive what it must have been like for our ancestors. Most hunters are true sports-people who eat what they kill. Hunting is one of the oldest sports in the world.

Connection

BAG LIMITS

Bag limit is the maximum number of game that a hunter can take. The number may be on a daily, seasonal, or other basis. The limits may apply to any game—fish, fowl, deer, or others that are covered by the regulations. Non-game species are not protected by bag limits.

The purpose of bag limits is to protect wildlife and have a harvest that maintains the population. If one hunter takes more than his or her share, wildlife is threatened and other hunters may not have animals to take.

The number in a bag limit is based on game populations. Changes in wildlife populations may result in bag limits being raised or lowered depending on what is needed to protect the wildlife. In some cases, the bag limit is only one per season for a hunter, such as this hunter has with the turkey.

Fishing with a hook and line has been called relaxing, tense, laid back, and exciting—all are true! Many people who fish today still fish for food. Many others fish for sport, releasing most of their catch. We are fortunate to live where we are never far from a place to fish.

Photography, scouting for wildlife, bird watching, and other recreation activities are enjoyed by people who live in the country, suburbs, and in the city.

BIODIVERSITY

North America has a variety of wildlife animals. From the desert southwest to the maple-aspen forests of the northeast, many unique environments and many unique species of wildlife are found.

Having a variety of wildlife species adds to the *biodiversity* of our planet. Biodiverse populations are less likely to be harmed by disease or illness. One concern is the growing number of wildlife animals that are *endangered species*. Endangered species are animals that are close to becoming extinct. Every time an animal reaches extinction, the biodiversity of our planet decreases.

19-5. A black-tailed jackrabbit in the Arizona desert represents biodiversity from wildlife found in other regions.

MAJOR TYPES OF WILDLIFE

It is impossible to list all of the wildlife in North America. A few of the major types of fish and wildlife animals follow.

GAME ANIMALS

Animals hunted for sport, meat, and other products are game animals. The kinds of animals used for game varies with where you live. Game animals can be large, like the white-tailed deer and the antelope. They can also be small animals, like the rabbit or squirrel.

In most places, game animals can be harvested by hunters during a regulated hunting or trapping season. Personnel in state departments of game and fisheries decide when the season occurs and how many animals can be harvested. A license may be needed to hunt. Hunter education is expected before a person begins hunting with a gun.

Popular species of large game animals include the white-tailed deer, mule deer, elk, antelope, bighorn sheep, and alligator. Small game species include the cottontail rabbit, squirrel, fox, mink, raccoon, and muskrat.

19-6. Squirrels are popular game animals.

NON-GAME ANIMALS

More non-game animal species than game animal species exist. These animals can be found deep in the forest or in urban areas, or just about any place in between! The diversity of non-game species is amazing.

A ***non-game animal*** is one that does not provide a consumable product like meat or furs. It provides beauty and adds to the biodiversity of our planet. Some non-game animals serve as a food source for predators. Others are scavengers and help nature recycle by eating dead animals. Non-game animals are an important part of nature.

19-7. Bison have been hunted but are now protected as non-game animals.

Most people do not think of wildlife in their neighborhood. Many non-game animals can be found in urban and rural areas across the country. Mice, rats, opossum, and others are comfortable in an urban setting.

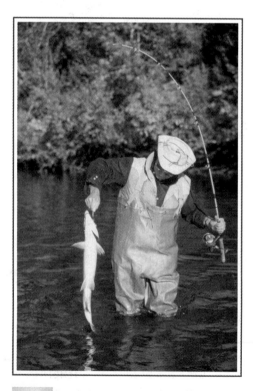

19-8. A sport fisher with a 14-pound coho salmon.

GAME FISH

Game fish are sought to provide food or sport to the fisher. Game fish exist in both fresh and salt water and come in a variety of sizes. Most game fish species are caught with a hook and line.

Freshwater game fish can be caught in every state. Again, the types of game fish will differ from state to state. Major types of freshwater game fish include black bass (largemouth, smallmouth, and spotted), sunfish, walleye, trout, pike, muskellunge, salmon, striped bass, crappie, and others. Popular saltwater species include groupers, sea trout, cobia, snapper, flounder, redfish, mackerel, tuna, sailfish, dolphin, and many more.

NON-GAME FISH

Just as non-game animals outnumber game animals, non-game fish outnumber game fish. Non-game species can be found in small freshwater creeks, ponds, lakes, and in saltwater habitats. Many species of non-game fish are too small to be caught on a hook and line, but some non-game species are very large. For example, the sturgeon can grow to over 2,000 pounds!

GAME BIRDS

Game birds are hunted for food and sport. They can be divided into three basic categories that include migratory (non-waterfowl), waterfowl, and upland game birds.

The woodcock and the crow are examples of non-waterfowl game birds that are migratory. Game waterfowl include many species of ducks and geese. Upland game birds include the pheasant, grouse, and quail.

Harvest of game birds is regulated by state and federal law. Since migratory birds can travel to other countries, laws governing their harvest often have international input.

BIRDS OF PREY

Eagles, osprey, owls, and hawks are all examples of birds of prey. Most birds of prey are protected by laws that make shooting them or attempting to catch or harm them illegal. These birds are hunters. They eat small rodents, fish, snakes, reptiles, amphibians, and small mammals.

19-9. A red tail hawk is a bird of prey.

NON-GAME BIRDS

Non-game birds include all other birds. The sparrow, blue jay, robin, and pelican are all examples of non-game wild birds. Some are known as song birds and are found in areas where people have homes. Non-game birds live in

19-10. An egret is an interesting water bird.

a variety of environments. You can find these animals in wetlands, plains, uplands, and anywhere in between.

ENDANGERED SPECIES

Unfortunately, people have had a negative impact on many wild animals. Through **urbanization** (building cities, suburbs, and all of the support services needed by people in these areas), agricultural growth, pollution, and neglect, many species of wildlife are extinct or endangered.

With work, the fate of some endangered species can be changed. The American alligator was once close to extinction. Today, the alligator is doing very well. In fact, the population is in such good shape that alligators can be legally harvested in some places in the southeastern United States! Other examples could be given, such as bald eagles and bison.

WILDLIFE MANAGEMENT PRACTICES

Think for a moment about where you live. How many people live in your home? Does everyone have a bed and a chair at the table? People need space in which to live. You would not be very comfortable if 10 people shared your bedroom, or if 50 of your classmates came to your house for dinner. There are limits to how many people your home can accommodate. Wildlife species are limited by the same factors.

HABITAT

Habitat is the place where animal wildlife live. It has four components: food, space, water, and cover. Think of your home as a habitat. It is a place

19-11. A grizzly has unique needs for survival.

where your basic survival needs are met. You have cover in a house or apartment. You also have food, water, and space to live.

Wildlife must have the proper habitat to meet their needs. All four components must be in place to support a wildlife species. If one component is missing or lacking, it is called a *limiting factor*. A deer needs to have browse. If there is no browse, the deer cannot survive in that habitat—food is the limiting factor. Pheasants need tall grass to nest. If there is no tall grass in the habitat, there will be no pheasants—cover is the limiting factor.

Even if a habitat is perfect for a particular species, there are limits to how many animals the habitat can sustain. The habitat's *carrying capacity* is the number of animals it can support. Carrying capacity is usually low in the winter and high in the summer. The population in a given habitat is directly related to the carrying capacity.

INCREASING WILDLIFE POPULATIONS

The best way to increase wildlife populations is to meet their basic needs in the habitat. While this is true, there are other considerations.

Niches

Every animal has a *niche* or a special place in a habitat where it belongs. The animal fills an important role in consuming and producing for the overall habitat. For example, white-tail deer and squirrels both live in the same

19-12. A jaguar enjoys a special niche in its territory.

habitat. The deer eats small branches and likes to bed in fields of high grass—this is its niche. The squirrel lives in the trees and eats nuts, berries, etc. The squirrel fills a different niche. Some animals may share similar niches, but they are seldom the same.

Interspersion

When things are interspersed, they are mixed. For example, in an 8-acre plot of land, 2 acres could be dedicated to food, 2 acres to water, and the remaining 4 acres to cover. If these areas are arranged in large blocks, there would be very few places where animals could have easy access to all three areas. Placing these resources in smaller blocks would allow quicker access to all three and increase the number of animals in the habitat.

Territory

Some animals claim an area, or **territory**, which they protect and keep other animals from entering. Bears will mark their territory to warn others from invading. Space can quickly become a limiting factor for territorial animals.

CONSIDERATIONS FOR WILDLIFE SPORTS ENTHUSIASTS

Wildlife sports are important traditions in some communities. Sports include traditional hunting and trapping, as well as watching (observation), attracting wildlife, and photography. Sports enthusiasts need to keep some simple rules in mind when enjoying wildlife.

EXERCISE CAUTION

Hunting and trapping are sports that involve deadly force. Guns can kill or injure people as well as animals. Hunter and trapper safety courses are required in many states before a license can be obtained. These courses will teach you how to use traps and firearms and to respect their power.

Another potential area of danger involves human/animal interaction. Wild animals are wild! They are not pets or zoo animals. They are driven by instinct and will try to get away from you and out of danger any way they can. Often, this includes biting, scratching, goring (puncturing with horns), or attacking, using any means available to them. When hunting, trapping, observing, or photographing, remember to stay at a safe distance.

TAKE ONLY WHAT YOU NEED

Wild animals are a renewable resource. Game harvests are closely regulated by state and federal agencies. These agencies set limits on the number of animals that can be taken. To maintain wildlife populations, it is necessary for sports enthusiasts to obey those regulations.

There are times when a fisher or hunter may be having an excellent season and can harvest limits of fish or game on a daily basis. The sports enthusiast needs to ask, "How much will I use?" If the harvest will be wasted, should you take the animal even though you are within the legal limits? A conscientious sportsperson would answer, "No."

RESPECT THE HOME OF ANIMALS

How would you like to have visitors in your home leave their trash on the floor? Would you be upset if someone came to your house and broke all of the furniture? You probably would not be happy with either situation.

People are often in the homes of animals. While we are in their habitat, we need to respect where they live. Leaving paper, cans, and bottles, or damaging trees and animal homes, is not the way a responsible sports enthusiast uses the resource.

WAYS TO ENJOY WILDLIFE

Most people enjoy wildlife in one way or another. Someone who lives in the city can enjoy many kinds of birds and small animals that have adapted to city life. Some ways to enjoy wildlife without hunting follow.

BIRD WATCHING

Many people enjoy watching birds in the wild or even in their own backyard! Watching and studying wild birds is *birding*. We can attract birds with feeders or with plants that they need for food. It is easy to attract birds to your home with a simple birdfeeder. You can purchase a feeder or make one that will hold seeds and grains the birds eat. Hanging flowering plants like fuchsia will even attract hummingbirds! If you start "birding" be aware that it can be very addicting! Some people will keep logs of the kinds of birds they spot. Others will even take trips to faraway places in hopes of adding new birds to their list!

PHOTOGRAPHY

Taking pictures of wildlife animals can be very exciting! Capturing an animal on film is not as easy as it sounds. Often the photographer must plan carefully and wait for long periods of time to get their shot! In many ways, the photographer uses the same skills that a hunter uses, but the animals are left in the wild and the trophy is the photograph.

19-13. Birding provides wonderful insights into the lives to birds. This shows a flycatcher with four young.

ANIMAL WATCHING

Some people simply enjoy watching wildlife animals. It is special to see a doe and her fawn feeding in a field or even watching a family of skunks scurry around the park. There are probably several places in your community where you can go to watch animals. From bat houses to public parks there are usually many options for you to choose from. Some people will set up animal feeders or plant crops that will provide the wildlife with food in areas where they can watch them.

REVIEWING

MAIN IDEAS

Animal wildlife exists in large numbers across the country. These animals are important to us economically. They add beauty to our world. They provide recreation. In addition, they add to the biodiversity of our planet. Major types of wildlife species include game and non-game animals, game and non-game fish, game and non-game birds, birds of prey, and endangered species.

Managers of wildlife must be concerned with providing food, water, cover, and space for the animals. Wildlife populations can be increased by providing niches for animals, practicing interspersion, and meeting territorial needs of animals.

Sports enthusiasts should exercise caution when dealing with wild animals. Those that are conscientious take wildlife that they will use. All people who use wildlife need to respect the animal's habitat and home.

There are many ways to enjoy wildlife animals including birding, photography, and animal watching. For more information on wildlife, refer to *Wildlife Management: Science and Technology* available from Interstate Publishers, Inc.

QUESTIONS

Answer the following questions using correct spelling and complete sentences.

1. What are the four reasons that animal wildlife is important? Briefly explain each reason.

2. Distinguish between animal wildlife and game animals.

3. What is an endangered species? Give an example of an animal that was once endangered, but is no longer.

526 INTRODUCTION TO LIVESTOCK AND COMPANION ANIMALS

4. What is habitat? What are the four components of a habitat?

5. What can be done to increase wildlife populations?

6. How should sports enthusiasts respond to animal wildlife?

7. What is your favorite way to enjoy wildlife?

EVALUATING

Match the term with the correct definition. Write the letter by the term in the blank provided.

a. hunting   d. game   g. carrying capacity
b. niche   e. animal wildlife   h. pelt
c. endangered species   f. territory   i. birding

_____ 1. The natural skin covering an animal used in making clothing and other products.
_____ 2. Animal close to becoming extinct.
_____ 3. The number of animals a habitat can support.
_____ 4. The place and role of an animal in its habitat.
_____ 5. Taking animals with guns and other weapons.
_____ 6. An animal harvested for food or other products.
_____ 7. The area an animal will defend as its own.
_____ 8. Animals that have not been domesticated.
_____ 9. Watching and studying wild birds.

EXPLORING

1. Develop a personal library of information on game animals in your area. Collect brochures and other materials in your library. These are available from a local conservation officer, game and fish commission, or office of the Cooperative Extension Service.

2. Tour a wildlife refuge. Learn the kinds of animals in the refuge and the practices followed in improving habitat for the animals. Prepare a written or oral report on your findings.

3. Select a species of wildlife found in your area. Prepare a report that describes the animal, its habitat requirements, and other conditions that provide for its well-being.

20

SERVICE AND SAFETY ANIMALS

Animals help people in many ways! Think about the people you know who have animals. What kinds of animals do these people have? How do their animals help them?

Some animals are used in ways that do not provide food or pleasure. These animals help people live better lives. In some cases, they help people live independently. Close your eyes for a moment and think about what it would be like to be unable to see. Can you find your way to school with your eyes closed? A trained leader dog for the blind helps visually impaired people live independently.

Many different animal species are used for service and safety. Training is usually required so the animals will know what to do and how to do it. The training usually includes close bonding between the animal and its human owner.

20-1. A trained Border Collie is herding sheep.

OBJECTIVES

This chapter has information on using animals for service and safety. It has the following objectives:

1   Explain service and safety animals and give examples of each

2   Describe practices with service and safety dogs

3   Describe practices with service and safety donkeys

4   Describe practices with service and safety llamas

5   Describe practices with service and safety geese

TERMS

cria	K-9
guard dog	multipurpose dog
handler	safety animal
jack	service animal
jenny	single-purpose dog

KINDS OF SERVICE AND SAFETY ANIMALS

Service and safety animals help people in many ways. A *service animal* is an animal that gives assistance in some way. These animals often provide an important service. A *safety animal* is an animal that helps protect people and property. Service and safety are not widely thought of as uses for animals, but they are very important uses.

There are many kinds of and possibilities with service and safety animals. Animals can do various tasks if they have the correct instincts and are properly trained. Examples of service and safety animals are dogs, donkeys, llamas, and geese.

20-2. A specially trained dog helps its owner to be active and independent.

Service and safety animals help and protect. Dogs and geese in a person's yard warn of intruders. Llamas, donkeys, and dogs put out to pasture with sheep and goats protect them from predators. Today, many ranchers are using other animals to control the aggressive predators and protect their animals.

DOGS

Dogs are common service and safety animals. Many homes have dogs. Dogs are used not only as companions, but to warn the occupants of the home when someone arrives. Dogs help with law enforcement, assist people who have visual or mobility problems, guard property, and herd livestock.

An example of how dogs are helpful is the police dog. The *K-9* is a dog trained to assist with law enforcement work. The K-9 is a highly versatile, dedicated member of a police department. When properly trained, the police K-9 can perform a variety of services.

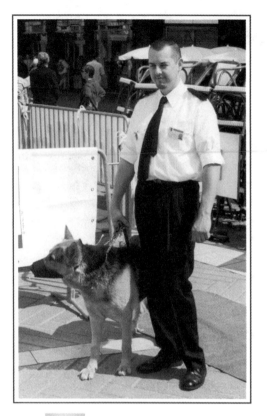

20-3. A German Shepherd K-9.

20-4. A Rottweiler is a massive, powerful dog that is often used for human protection. (Rottweilers can attack strangers and cause severe injury or death.)

K-9 unit dogs are commonly the German Shepherd breed because they are easily trained, strong, aggressive, and good tempered. An aggressive dog is needed for police work; however, a mild-tempered dog is best with the public. Some law enforcement agencies use standard poodles and Rottweilers.

Farmers, ranchers, and business owners use trained dogs as guards. A **guard dog** is one that is trained to protect property. It may bark, bite, and attack potential danger. With sheep, a guard dog stays with a flock of sheep and protects them. With businesses, it stays inside a building or fenced area. Its presence deters thefts and vandalism.

The breeds most common for guard dogs in the United States are large dogs weighing 80 to 120 pounds (36 to 54 kg). They stand at least 25 inches (63 cm) at the shoulder and have flat ears covered with hair, short and blunt muzzles, and long tails. They are usually fawn, gray, or all white with dark muzzles. The most common breeds used for guard dogs are Great Pyrenees from France, Komondor and Kuvasz from Hungary, Maremma Sheepdog from Italy, Cão de Castro Laboreiro from Portugal, Shar Planinetz from Yugoslavia, Tibetan Mastiff from Tibet, and Akbash Dog and Anatolian Shepherd Dog, both from Turkey. The most successful breeds are the Akbash and Pyrenees.

Some ranchers use dogs for herding. A trained herding dog can move sheep, cattle, and other animals from one pasture to another or into a corral. Border Collies are well known as herding dogs.

TRAINING

K-9 units require special management practices. First, a handler is selected. A **handler** is the person who works solely with a dog. They go through training together and will work together later.

Dogs are trained at one to three years of age. They go through 12 weeks of training with their handlers. The dogs are trained to respond to commands and to bite and hold rather than to maul. Dogs trained in specialty areas, such as explosives and narcotics, require at least another 8 to 10 weeks of training. After that, the dog still goes through two days of training per month. K-9 units may be used for searching buildings for burglary suspects, tracking suspects of crimes, locating lost persons, locating explosive or incendiary devices, searching out hidden narcotics, and protecting their officer partner.

A dog is used in law enforcement until it is eight or nine years old. This depends largely on the area of work. The harder the job, the earlier the dog is retired. The handler usually has the opportunity to adopt the retiring dog.

Dogs may be trained as single-purpose or multipurpose animals. A **single-purpose dog** is trained and used for just one activity, such as patrols, tracking, narcotics, or explosives. A **multipurpose dog** is trained and used for more than one activity. The choice of training depends on how the dog is to be used in law enforcement. An agency that is small and has only one dog may want a multipurpose dog. Large agencies may have several dogs with specialized training.

20-5. A K-9 in active training for law enforcement work.

Training and care are needed. Guard dogs are not like donkeys or llamas who need very little training and care. Training a guard dog takes time. First, the dog must be acquired when it is young, about six to eight weeks old. It should be intelligent, alert, and confident. There is no difference between the performance of male and female pups. However, if a female pup is selected, it is recommended that she be neutered.

Puppies that will work sheep must be reared with them. They are given little human contact from six to eight weeks of age. The pup must be monitored and bad behaviors must be corrected immediately. The training for the pup will take up to a year. Patience is essential. Give the pup the opportunity to learn with the flock. Start the pup out in small areas and keep working with it.

When a pup is weaned from the litter and ready for training, the socialization with the sheep begins. The pup is placed in an area of about 150 square feet (14 m²) with three to six lambs. Orphan lambs work well because all of the animals are lonely and they are more likely to establish bonds. The pup should have an area it can escape to that the lambs cannot get to, which should contain the pup's food. The water dish for the pup and lambs should be shared to force bonding between the two. The pen should be checked several times the first few days to make sure all animals are healthy and adjusting to each other.

Gentle play between the pup and the sheep is okay. Be alert if the play gets too rough. A lamb that dominates the pup should be removed and another lamb put in its place. The reason to put the pup alone with the lambs is to build confidence and to bond with the sheep. The pup may be petted by the owner each day when it is checked. Excessive handling must be avoided.

The animals may have to be moved to bigger pens periodically if the pup grows too big for the pen. After the pup is 16 weeks of age, it and the group it was raised with can all be put out with the flock. The pup should not be put into an area where there are predators until it is ready to take on the guarding role. The age varies with dogs, but it is sometimes as young as 4½ months. The dog will begin patrolling and taking on other activities of guarding. It should be large enough to defend itself from predators.

It takes pups up to two years or more to mature. Puppy foolishness and guard dog genetics make them independent. Also, keep the goal in mind. This pup is not a herding dog, but a guard dog. Its job is to stay with or near the sheep and protect them from predators. Herding dogs, such as Border Collies, Australian Shepherds, and

20-6. A Border Collie pup lives next to lambs as part of his training.

Connection

BEWARE OF DOG

Some dogs are dangerous. They will attack and bite people. Bites can cause severe injuries and deaths, especially with children. This can happen quickly when the owner is not around.

The owner of a dog is responsible for its behavior. Dogs trained for law enforcement and retired may retain their aggressive instinct. These dogs must be controlled at all times. They must be penned and on leashes. A dog that may attack should never be allowed to run loose. In some cases, euthanasia is appropriate. Posting a sign is informative but not adequate if a dog runs free.

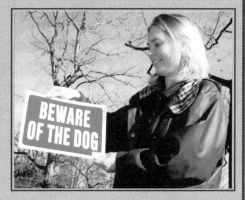

A dog should never run free in residential areas. All dogs cause damage to property. Some can cause human injury, pain, and death.

Australian Kelpies are used to move sheep from area to area by nipping, chasing, and barking at the sheep.

One-fourth of the dogs trained as guards will not be successful. Failure may be due to starting training of the pup when it was too old, or improper rearing by the owner. Some dogs also will harass, injure, or kill stock, and roam away from the flock. Always post warning signs to protect unknowing people from a guard dog. Never allow a guard dog to roam where it can attack innocent people.

The number of guard dogs needed varies. It depends on how old the dog is and how much area the dog can handle. The sheep may graze close together or scatter. The rule is to start with one dog and let it get established before adding others. A first-time guard dog user should start with one pup.

Three behaviors for an effective guard dog are

■ be trustworthy to not injure the stock;
■ be attentive and stay close to the flock; and
■ be protective whenever a predator shows up.

A guard dog that is not on duty (with the sheep) is not where it is supposed to be!

FEEDING AND NUTRITION

A dog used for service and safety work must receive proper food if it is to perform effectively. Each breed varies somewhat in requirements. In general, a dog should receive needed energy to do its work and other needed nutrients for it to stay healthy. Commercial feeds selected for the age, condition, and work activity are likely the best to use.

Throughout the work day, dogs should receive plenty of good water. Never let them drink from a puddle in the street or another dog's bowl if the dog is a disease suspect. Information on the feeding of K-9 unit dogs and guard dogs can be found in Chapter 15.

HEALTH CARE

A service and safety dog should be properly vaccinated, wormed, and otherwise receive health care. Those that work around people will also need grooming to maintain an attractive appearance. K-9s are usually under routine care of a veterinarian.

Guard dogs do not usually live to a very old age. It is easy for a rancher to forget to worm and vaccinate the dog since it is out with the sheep. Besides disease, some dogs just disappear, others are poisoned, shot, or hit by vehicles. Guard dogs may only live to be about three years old, so it is important to start training a new pup. Ideally, the training will occur while the older dog is working well with the sheep.

20-7. Maremma pups are eating highly nutritious goat's milk. (Courtesy, Luck-E-G Nubians, Idaho)

20-8. A trained Maremma male dog herds goats on an Idaho farm. (Courtesy, Luck-E-G Nubians, Idaho)

*F*ACILITIES AND EQUIPMENT

Proper facilities and equipment are needed to care for the well-being of service and safety dogs.

Using a K-9 unit is a 24-hour-per-day, 365-day-per-year decision. The K-9 unit dogs have varying facilities. Some law enforcement departments mandate the dog go home with its handler. The dog is with the handler 24 hours per day. The dog rides with the handler, does duty with the handler, and lives in the home with the handler and his or her family. Some officers say they spend more time with their dog than their spouse or children. Other K-9 unit dogs are kept in kennels at the police department or in a kennel at the handler's home.

Sheep guard dogs spend all of their time with the sheep. Special facilities, separate from the sheep, are not needed, except, possibly, in the winter if the sheep are penned.

Leashes, food and water bowls, grooming equipment, and other items may be needed. These vary with the nature of the dog and its work.

20-9. The skills of a German Shepherd are demonstrated at a dog show.

DONKEYS

Some farms and ranches use donkeys for service and safety. Ranchers may use them to protect sheep and goats. Donkeys may be used to teach a show calf to be halter broken. Donkeys were first used to guard sheep in Montana and Utah.

Donkeys are classified as either a *jenny*, which is a female, or a *jack*, which is a breeding male. A gelding is a castrated male. When selecting a donkey, know that jacks and geldings may be too aggressive with the sheep and kick or bite them. Geldings are less likely to be aggressive. They are selected over jennies because they are stronger. Jennies work well with the sheep because they are less aggressive, unless they have newborn babies of their own. This may be remedied by removing the jenny and her newborn for a week or ten days, then return the two to the sheep flock.

Donkeys have a natural hate for all members of the dog family such as fox and wolves. Donkeys do not kill predators, but they chase them with their teeth bared, ears back, and striking at their feet. Guarding is a natural part of the donkey's temperament. Donkeys are replacing dogs in some places because they are more economical to buy and feed than dogs.

TRAINING AND MANAGEMENT

The key to managing donkeys as security animals with sheep is to make sure they bond with the sheep. Donkeys are gregarious, so they want to be around other animals. If there are no donkeys to be around, they will assume the role of a sheep and fit in with the crowd. Bonding takes time. The donkey needs to get familiar with the flock. One donkey is needed for every 150 sheep.

20-10. Young donkeys.

Donkeys make good guards with no training. Not every donkey bonds with sheep. It is important that a donkey is tested with sheep before it is released with them. Often, it is best for the producer is to buy a young female donkey raised on a sheep farm. If more than one donkey is with the sheep, they will bond and spend time with each other rather than with the sheep.

Donkeys have limitations; though, they can kill dogs, coyotes, and foxes. A donkey can be killed by bears, mountain lions, and wolves. Donkeys know this and are afraid of these animals. It is important to know what sheep need protection from to ensure that donkeys are the correct animals for the job.

Besides knowing the limitations of the donkey, it is important to understand the techniques the donkey uses to protect the sheep. The donkey will bray loudly and stamp its feet at night if it hears or sees something unusual. They will chase dogs and coyotes that get too near.

20-11. The head of a mature donkey.

FEEDING AND NUTRITION

Donkeys usually graze and feed along with the sheep flock. If additional feed is provided, the donkey should not be fed in a separate area. This will lessen the bond between the donkey and the sheep. The donkey may even drive the sheep away so it can eat, especially early in the relationship. For a strong bond to exist, the animals must graze, sleep, and live together and establish the same routine.

HEALTH CARE

Donkeys are healthy animals. They do not require very much veterinary care, because they are hardy. Donkeys may be susceptible to most of the diseases of horses. Injuries may also occur when fighting off attacks by predators. Chapter 13 has more information on health care with donkeys.

20-12. A guard donkey has bonded with this flock of sheep.

*F*ACILITIES AND EQUIPMENT

Donkeys do not need special facilities and equipment. They should be kept with the sheep. They should sleep with the sheep since this is when the predators attack. If the donkey is not there during the hours it is needed, it cannot protect the sheep. It is important that the donkey is kept with the sheep so it bonds with the sheep. Donkeys can survive in the same climate as sheep, so no special housing is required.

LLAMAS

Llamas are members of the camel family. They are from South America. Llamas have been used as pack animals for many years. Their usefulness as guard animals has come about in recent years.

Llamas are used on some farms as protection for other animals. Llamas were first brought to the United States as novelty animals. Initially, llamas were put with sheep because of convenience. Producers quickly realized that they were losing fewer sheep to predators when the llamas were with the sheep.

Llamas that encounter predators get very aggressive and protective. They are proud animals that remain calm and do not get scared and run from intruders. They take an aggressive role. When they encounter coyotes, dogs, foxes, or bears, they give an alarm call, walk or run toward the intruder, chase it, kick it, paw it, and sometimes kill it.

The female lays down to be bred. Gestation is 350 days. Most give birth without assistance. The young are called *cria* and are usually born in the morning. The female, if approached by the male after she has been bred, will spit at him.

*T*RAINING AND MANAGEMENT

Unlike the donkey, llamas quickly bond to sheep. The two species get along well. The llama is also very instinctive about what to do with the sheep. No training is needed for a llama. They take over and protect sheep as their companions.

20-13. A llama can become aggressive when danger approaches.

FEEDING AND NUTRITION

Similar to donkeys, llamas work well as guard animals with sheep because they eat the same food as sheep and sleep with them at night. This helps them know when threats occur to the flock's safety. Llamas have the same lifestyle as sheep, so they fit naturally alongside them.

Llamas may be fed with the sheep because they will stand aside and let the sheep eat while they look on. They eat the same foods as sheep and are very satisfied with a bale of hay and water.

HEALTH CARE

Llamas, much like donkeys, do not have many health threats. Also, they may last as long as 20 years as a guard animal with the sheep. In the end, they cost less money than a guard dog, which initially costs less, but requires more in care. A dog lasts only three years, of which, the first is only training. The llama is automatically trained because it is all instinct.

FACILITIES AND EQUIPMENT

Llamas can die of heat stress and require shade in hot weather. However, they do not have other special requirements. They can live happily on ⅛ of an acre with only a three-sided windbreak for shelter. The "toilet habits" of the llama are clean and disease-free because they select one corner of their pen for defecation and do not deviate from that area.

GEESE

Geese are used to protect homes and their owners from intruders. A goose can be a very mean and ruthless animal. When an intruder arrives, the goose begins honking and then chases the intruder. If the intruder is caught, the goose will hold onto the person with its beak and beat the person repeatedly with its strong wings. This may not seem too threatening, but wait until a goose comes to attack you!

Another use of geese is in weeding crops, especially cotton. This use has declined in importance. Large gaggles of geese could be placed in a field to remove the weeds and grass. Geese do not readily eat the cotton plants.

TRAINING AND MANAGEMENT

Geese used as protection should be obtained as goslings or baby geese. They should arrive in the late spring or early summer. Caring for the birds is easier when the weather conditions are more favorable.

Geese normally do not need training. They need to know the boundaries of the property where they live. Fences usually help in this regard.

FEEDING AND NUTRITION

Geese need to receive feed that meets their needs. Grains and other supplements are used.

Goslings (young geese) should be fed all they can eat and drink. A 20- to 24-percent protein commercial feed should be used. A chick starter or a duck or waterfowl formula will be fine if the goslings are allowed outside to forage for food on their own. Goslings grow faster than chicks, up to 10 pounds in ten weeks. They need food in addition to the starter. Goslings will eat grass and insects to get the necessary food to balance their ration. Goslings need water at all times. If weather permits, goslings should be outside during the day and inside at night.

Adult geese can be fed whole corn for feed and then left to forage on their own. Many times, geese are used to weed crops, such as strawberries, asparagus, sugar beets, and orchards. The geese will eat the weeds and not harm the plants. However, the geese should be removed from the strawberries when they start to ripen so they do not eat or ruin the berries.

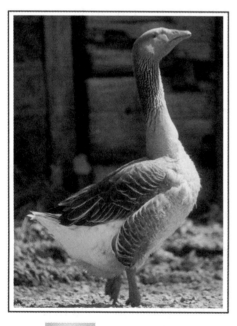

20-14. A Gray Lag goose.

HEALTH CARE

Goslings and geese are healthy animals. They need ample water to drink and to use to stay cool in the warm weather. Goslings, when they are very young, should not be outside in the rain or they become waterlogged—their down soaks up all the moisture it will hold. Goslings have no feathers to repel the water. Their tiny legs cannot support the weight of the water. Geese are known for being disease-resistant, so little has to be done to maintain their health.

FACILITIES AND EQUIPMENT

Adult geese may need shelter from inclement weather. A shed where they can go to escape predators and roost is beneficial. The feeder and waterer may be in this area.

Goslings require a warm, dry place for the first four weeks of their lives. If, in late May or June the weather is still cold, the goslings must be kept dry, warm, and away from drafts. A 250-watt heat lamp may be installed 15 to 18 inches above the floor to keep the goslings warm. If the goslings are piling up on each other, it is too cold and the weaker birds may get smothered. If the goslings are comfortable, they will sit evenly distributed on the floor under the heat lamp.

Goslings require litter on the floor of their pen. Litter, such as straw, sawdust, shavings, or crushed corn cobs, will work. The material must be dry and absorbent since goslings are very messy.

REVIEWING

MAIN IDEAS

Several species of animals are used for service and safety. Dogs, donkeys, llamas, and geese are three examples.

The K-9 unit has become important in law enforcement. When properly used, the trained police dog is an asset that cannot be overlooked. It increases the chances of capture and the safety of the officer.

Guard dogs are used effectively as protectors for sheep. They require more work than either llamas or donkeys. The guard dog is a highly trained animal that works effectively with the sheep if it is reared correctly.

Donkeys make good protection providers for sheep. Not all donkeys perform as they should. A donkey should be tested for how well it bonds with sheep.

Llamas are proving to be successful protection animals because of their instinct to bond with and protect sheep. The viciousness of a llama can be provoked by a predator. Their popularity as companion animals is also growing.

Geese are animals that may not be used by everyone, but for those that choose the birds, they work out well. The birds are vocal and very aggressive, but also easy to care for and disease resistant.

When selecting and considering types of animal protection, it is important to note the time commitment to train the animal, the efficiency of the animal, the length of time the animal can perform the job, the type of protection needed, and which animals are best suited for that type of protection.

QUESTIONS

Answer the following questions using correct spelling and complete sentences.

1. What animals are commonly used for service and safety?
2. What are the major considerations in using dogs as guard animals?
3. What are the major considerations in using donkeys as guard animals?
4. What are the uses of llamas as guard animals?
5. What are the uses of geese as service and safety animals?
6. Why are donkeys and dogs not used together?

EVALUATING

Match the term with the correct definition. Write the letter by the term in the blank provided.

a. handler   d. K-9   g. jack
b. service animal e. jenny   h. single-purpose dog
c. safety animal   f. cria

_____ 1. A male donkey.

_____ 2. A baby llama.

_____ 3. A female donkey.

_____ 4. A dog used for one safety or service activity.

_____ 5. The person who directs a K-9.

_____ 6. Animals that give assistance.

_____ 7. A trained police dog.

_____ 8. Animals used to protect people and property.

EXPLORING

1. Contact local law enforcement officials and learn the use they make of K-9 dogs. Interview the handler and determine how the dog is trained, what it does, and how it is managed. Prepare a written or oral report on your findings.

2. Prepare a report on using llamas, geese, or donkeys as service and safety animals. Interview owners of such animals and use library resources for information. Give an oral report in class on your findings.

21

SCIENTIFIC AND LABORATORY ANIMALS

Did you know that research with animals made it possible for you to avoid having polio? Polio was once a dreaded disease. A simple vaccine now keeps people from getting it. Wonderful! The same methods are used in other areas.

Animals had a major role in the development of antibiotics, insulin for diabetics, anesthetics, radiation and chemotherapy for cancer, open-heart surgery, joint replacements, and hundreds of other medical advancements. People need to understand just how important animals are in making scientific advances.

Opinions differ on the use of animals. What do you think about using animals in research? Regardless, animal well-being is always important. People live better and longer thanks to animals.

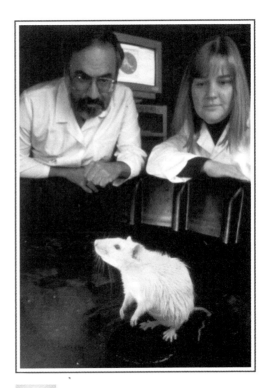

21-1. Scientists in a laboratory research program use a rat to test memory capacity. (Courtesy, Agricultural Research Service, USDA)

OBJECTIVES

The purpose of this chapter is to introduce the use of laboratory research animals. It has the following objectives:

1 Describe types and methods of animal research

2 Identify the kinds of animals used in laboratory research

3 Discuss animal models used in modern research

4 Explain important management practices used in animal research

5 Describe environmental considerations with laboratory animals

TERMS

animal by-product tissue
animal tissue culture
applied research
basic research
cavy
clinical research
laboratory animal
living research animal

mathematical and computer
  model
models for animal research
nonhuman primate
nonliving research system
primate
research

LABORATORY RESEARCH METHODS

Scientists use research to gain information. **Research** is a systematic attempt to answer questions. The research involves using procedures that are carefully planned and designed to provide accurate findings. Animals are sometimes used in the research.

ANIMALS IN RESEARCH

Answers to some questions can only be found by using animals. Agricultural researchers have long used animals in research. How would researchers know how animals respond to new feeds or methods of care without studying the animals?

A *laboratory animal* is an animal raised and/or used in research laboratories. They are carefully tended to assure their well-being. Laboratory animals are often kept in controlled environments to assure accurate research.

Animal by-product tissues are also used in research. An *animal by-product tissue* is a tissue that has been taken from an animal when it was used for other purposes. The tissues, organs, or other parts of animals slaughtered for human food may be used in research. These materials are collected in slaughter plants and carefully preserved.

Animal products are also used in research and to produce medicines and similar products. A good example is the chicken egg. Eggs contain albumin,

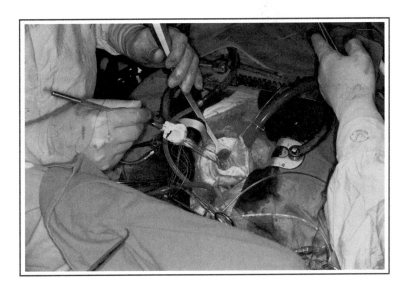

21-2. A replacement valve is being sutured into place in a human patient. Pigs are now being used to grow valves for use in humans.

21-3. Cattle by-product pituitary glands will be used to synthesize and artificially culture enzymes in a laboratory at Georgia Tech. (This package contains 100 pituitary glands.) (Courtesy, Corinne Mounier-Lee, Georgia Tech)

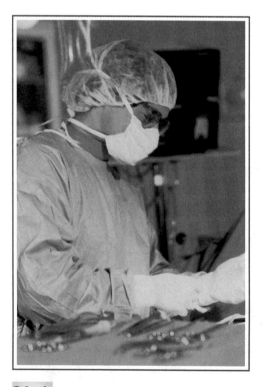

21-4. Clinical research is carried out in a medical facility or clinical setting.

which is rich in nutrition and an excellent medium for the production of many vaccines.

TYPES OF RESEARCH

Animals are used in three types of research: basic, applied, and clinical. Each type of research helps answer important questions.

Basic research is carried out in a laboratory, with a computer, or in nature. According to the American Medical Association, the goal of basic research is to increase the knowledge and understanding of life processes and disease. Basic research does not have a predetermined focus. It uses observing, describing, measuring, and experimenting to provide data that are often used in the other types of research.

Scientists may have specific purposes in mind for their research. A good example is developing a new vaccine or a medical procedure that builds upon existing knowledge. This is known as **applied research**. Animals are used in applied research, but this type of research could involve computers, nonanimal substitutes, or even people.

Research that takes place in a medical facility or clinical setting is called **clinical research**. Clinical research focuses on a specific human or animal problem. It often involves the animal species that the research is intended to help. Clinical research is supported by knowledge gained in basic and applied research.

Scientific and laboratory animals are used in basic and applied research for human- and animal-related research objectives. They are used in clinical research when their species is the subject being studied. For example, research on chickens involves using chickens in carefully controlled settings.

ANIMALS USED IN RESEARCH

Some animals are better than others for certain kinds of research. A few selected species of animals are used for basic and applied research. Clinical research animals involve any species that is being studied.

Most basic and applied animal research takes place in a laboratory setting. Scientists need animals with certain traits that make a particular species useful for research purposes. Most research facilities have a limited amount of space. Animals that require small amounts of space are most desirable. Larger animals also eat more food and create more waste. It is not surprising to learn that small animals are the animals of choice! In fact, more than 90 percent of the animals used in biomedical research are mice,

Connection

SENSING INFORMATION

Animals provide research information about themselves. Sometimes, devices are attached to animals to help in collecting the information.

This shows a sensor being attached to the neck of the steer. The sensor provides information about the grazing habits of the steer. This lets scientists know the amount of time a steer spends each day grazing and other details.

Is this information useful? Does collecting the information injure the steer? The well-being of the animal is a top priority in this research. (Courtesy, Agricultural Research Service, USDA)

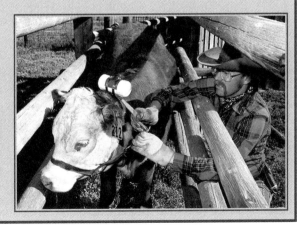

rats, and other rodents. Recent human research with tissue has focused on pigs. Important similarities exist between pigs and humans.

Another consideration is the similarity between the animal and the benefactor of the research. For example, a strain of mice has been developed that have immune systems that closely resemble humans. Such mice could be used in HIV and AIDS research.

Mice

Mice (*Mus musculus*) are the most popular animals for scientific and laboratory use. They are small, relatively inexpensive animals that need little space and food. Mice have a mild temperament and are easy to handle and work with.

Because they are efficient breeders, they are especially suited to research that requires large numbers of animals. Mice are most often used in genetic, immunology, and infectious disease research.

There are more than 100 documented bloodlines of mice available to researchers. The bloodlines are designated by a system of letters and numbers. Examples are C3H, C57BL/6, and CBA.

Mice, like most other rodents, are nocturnal. They spend much time grooming themselves. This is useful in medical research because an early sign of disease is an ungroomed appearance. Like all rodents, mice's teeth grow constantly. This is not usually a problem in laboratory settings, as the food provided is hard and keeps the teeth trimmed.

21-5. White mice are often used in laboratory research.

Rats

A larger cousin of the mouse is the laboratory rat (*Rattus norvegicus*). It was developed from the wild brown Norwegian rat. Rats are commonly used in behavioral and nutritional research.

Five strains of rats account for most of the animals used in modern research. Examples of the strains are Wistar (WI), Long Evans (LE), F344, and LEW.

Rats are also nocturnal rodents. They are usually docile and easily handled. They are larger and stronger than mice and greater effort is required to restrain them. Since they seldom fight, male rats can be kept in the same cage together.

Hamsters

The golden Syrian hamster (*Mesocricetus auratus*) is the most commonly used hamster for research. It has a stout body, short tail, and reddish-gold fur on the head, back, and sides, with grey-white fur on the underside. All hamsters have cheek pouches, which they use to transport and temporarily store food.

Hamsters are aggressive and will bite when startled. They should be handled carefully and gently. When they are handled properly, they will become easier to work with. Females usually dominate the males and larger animals dominate smaller ones. Animals kept in the same cage develop a "pecking order" can live peacefully.

Guinea Pigs

A Guinea pig (*Cavia porcellus*) is also known as a **cavy**. Guinea pigs differ socially and anatomically in several ways from other rodents. Important research strains include strain 13, strain 2, Dunkin Harlet (DH), and Shorthair (SH).

The cavy has a stocky body and short legs. It cannot climb as well as most other rodents and lacks agility. These characteristics, along with their docile nature, make cavies easier to handle than most laboratory rodents.

One nutritional difference between guinea pigs and other rodents is their inability to produce vitamin C. Like humans, guinea pigs must supplement their diet with this vitamin. Vitamin C deficiency causes a fatal disease known as scurvy.

21-6. Guinea pigs are used in human nutrition studies.

Rabbits

Rabbits (*Oryctolagus cuniculus*) were first domesticated in the 16th century. They serve as models for human and animal diseases, produce serum antibodies, and are used for drug testing. The New Zealand white rabbit is the most commonly used research animal.

Rabbits are used in areas of human research where rabbits and humans suffer from similar diseases and conditions. For example, rabbits have atherosclerosis (hardening of the arteries), emphysema, spina bifida, and cleft

21-7. Rabbits are used in antibody research and drug testing.

pallet. Laboratory rabbit research helps understand these conditions in humans.

Rabbits are curious animals that enjoy exploring their surroundings. They are gentle, usually docile, and easy to handle if not excited. Rabbits can be frightened by loud or sudden noise and often show fear by stomping their hind feet. A frightened rabbit may try to bite or scratch.

CATS

The cat (*Felis catus*) has been living with humans for more than 5,000 years. The domestic cat is of particular importance in brain research. The brain of a cat seems to be at a development stage between primates and lower mammals. Much of the present human and animal brain research is with cats. The American Shorthair is the dominant research breed.

Healthy cats are curious and aware of their surroundings. Their ears are erect and eyes are bright and alert. Most cats enjoy being petted and purr when handled. Some cats are raised specifically for laboratory use. Other lab cats are caught as abandoned cats or taken from local animal shelters.

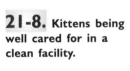

21-8. Kittens being well cared for in a clean facility.

DOGS

Dogs (*Canis familiaris*) have been used for experiments for more than 300 years. The most common research with dogs includes physiology, pharmacology, and surgical studies. Beagles are the breed of choice for most nonsurgical research.

21-9. A beagle represents the breed that is used in lab research involving dogs.

21-10. Chimpanzees are nonhuman, higher-order primates that are used in researching areas of human behavior.

Daily contact improves the dog's physical and psychological well-being. A healthy dog greets its handler with bright eyes, alert ears, and a wagging tail.

Good facilities for dogs include a large cage with waterer, feeder, and rest area. Indoor and outdoor runs are used to ensure that the animal is able to exercise.

NONHUMAN PRIMATES

A *primate* is an animal with an opposable thumb used for grasping things. The human being is a primate. Look at your thumb. How important is it to you?

Several species of nonhuman primates are used in a broad range of scientific and laboratory settings. A **nonhuman primate** is an animal very similar to humans, with thumb-forefinger opposition being a big similarity. Examples include chimpanzees and monkeys.

Monkeys can be divided into two basic categories: old world and new world. Old world monkeys come primarily from Africa and Asia. They have close-set nostrils that open downward and cheek pouches capable of carrying food. Old world monkeys also have pads on their buttocks.

New world monkeys have nostrils that open to the side, and do not have buttocks' pads or cheek pouches. New world monkeys also have long tails used for grasping and climbing. A unique characteristic of nonhuman primates that separates them from other scientific and laboratory animals is their social behavior. Communication through verbal and nonverbal means is of special interest to researchers.

Most research is with conditioned animals that are taken from the wild. Very few nonhuman primates are bred domestically for scientific and laboratory use.

MODELS FOR SCIENTIFIC AND LABORATORY INQUIRY

The use of animals in research is strictly controlled and monitored. Scientists have developed a set of acceptable *models for animal research*. These ensure that scientific and laboratory animals will contribute to knowledge and not be misused.

Four basic models exist for the use of animals in research.

1. Living animals

2. Living animal systems (tissue culture)

3. Nonliving systems

4. Mathematical or computer models

Using these models, scientists can better understand the human body, human behavior, diseases, treatments, and the effects of drugs on the body.

Living research animals (including humans) provide the best vehicles for psychological and human research. A *living research animal* is an animal that is alive and responds to treatments. They can react to their environment and are readily examined. The trend is away from using living animals and toward other models.

Animal tissue cultures can tell scientists a great deal about how body systems will react to stimuli. An *animal tissue culture* is the use of living tissues cultured in a lab. The tissues are taken from animals, includ-

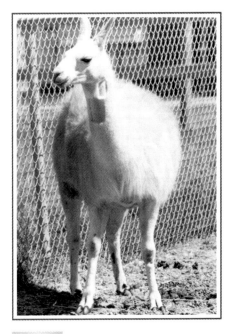

21-11. This llama, a large laboratory animal, is being used in veterinary medical research. (The shaved area on its neck does not represent anything that would threaten its well-being.)

ing humans, and grown in a laboratory. A great deal of the research in cancer treatment, through radiation and chemotherapy, is done using tissue cultures.

Another model is with nonliving systems. A *nonliving research system* is using mechanical models developed by scientists that mirror animal activity. They are used in research that is related to movement or effects of injury on living animal systems. An example related to humans is the development of the artificial hip. Many models had to be developed in the laboratory before a working design could be developed.

Computers have opened new approaches. The *mathematical and computer model* is a model that mimics animal behaviors and systems. It was developed after many hours of basic research. The models do not involve animal subjects in research. Scientists can manipulate these models on the computer and see results of their manipulation in seconds; conducting the experiment with live animals in the laboratory would take much longer. Computer and mathematical models are limited to what we know about a particular animal or situation and, therefore, are not perfect.

MANAGEMENT IN ANIMAL RESEARCH

Research with animals must meet acceptable standards. These standards focus on working with living beings. The well-being of animals is a top priority.

REGULATION OF ANIMAL RESEARCH

Animal research is heavily regulated. Researchers themselves are much involved. Animal research also has other regulations.

Universities doing animal research must follow strict regulations. Research institutions are members of the Institutional Animal Care and Use Committee (IACUC). All animal research conducted must be approved by this committee. This committee looks at the research proposal to make sure that protocols of animal treatment are followed. For example, the committee will look at medical care provided for the animals, the qualifications of the personnel conducting the research, and the use of anesthetics.

External review of the use of animals in research is based on the Animal Welfare Act of 1985. This federal legislation helps assure the proper use of animals in research. The U.S. Department of Agriculture is the federal agency responsible for ensuring animal welfare in research. This agency reg-

21-12. Cultured enzymes from by-product pituitary glands are being used in a study of ways to alleviate human pain. (Courtesy, Georgia Tech)

ulates and inspects animal research facilities through the Animal and Plant Health Inspection Service (APHIS).

In addition, the Public Health Service (PHS), the Environmental Protection Agency, and the Food and Drug Administration have animal welfare regulations. Most states also have regulations that are intended to protect the welfare of animals.

LAB ANIMAL NUTRITION

It is important that the animals used in research are fed nutritious, balanced diets. If an animal is malnourished, the results of the research may be clouded by the effect of nutrition and not by the effect of the experimental treatment! Care is needed on the part of researchers and laboratory technicians to give animals involved in research a proper diet.

Animal diets will vary greatly depending on the type of animal, the animal's size, maturity level, and activity level, and the environment. The animal's diet includes water, protein, carbohydrates, fats, vitamins, and minerals. These elements provide energy and nutrients that help the animal grow and maintain itself.

LAB ANIMAL ENVIRONMENT

How would you feel if you were reading this book in a room with poor lighting, loud noise in the background, and a temperature below freezing? You probably would not be concentrating on what is written here! Just as humans have comfort levels in their environment, animals also have ranges in

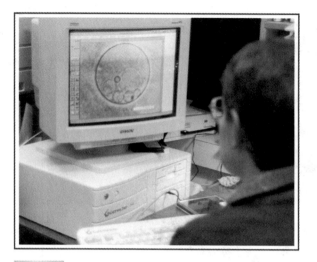

21-13. Synthetic cell structures are being studied with computer systems—no animals are needed.

which they are comfortable. The only life experience most scientific and research animals have is in the laboratory environment. Making their existence comfortable is a major concern of animal researchers.

Five areas related to the environment are included: space, temperature, light, ventilation, and noise.

■ Space—Probably the biggest concern for animal welfare in scientific and laboratory settings is the animal's living space. An animal must be given enough space to eat, sleep, and live out its life in relative comfort. Space requirements will vary with the type and size of the animal. Suggested minimums range from 6 square inches (15 cm²) for a small mouse to 144 square feet (4.4 m²) for horses. Space requirements are available from the National Institutes of Health in publication 86-23.

■ Temperature—Animals are usually most comfortable at a temperature between 65 and 85°F (16.6° to 29.4° C). Relative humidity of the room should fall in the 30 to 70 percent range for most animals.

■ Light—Lighting in the facility should be uniformly distributed and be bright enough for researchers to see and effectively work with the animals. Animals require a daily period of darkness to help them rest. The light and dark periods should remain on a set schedule to avoid stressing the animals.

■ Ventilation—Fresh air provided through adequate ventilation eliminates odors and reduces the number of airborne diseases. Fresh air is a necessity in maintaining a healthy environment.

■ Noise—Although it is impossible to eliminate noise from the environment, the animals should not be exposed to loud noises or areas of high activity (loading docks, building entrances). Noise can cause stress in animals and even lead to health and behavioral problems.

LAB ANIMAL HYGIENE

Hygiene is a primary concern for researchers in a laboratory setting. Diseases result from poor sanitation. Diseases can ruin experiments by causing animal sickness and even death.

Sterilization of equipment and materials that come in contact with animals is a fact of life in the laboratory. Bedding that is used to line the cages, instruments used for examination, even the cages used to house the animals must be sterilized to avoid spreading disease.

Disinfectants are used to kill or inhibit the growth of bacteria and viruses. These chemicals are usually too strong to use directly on animals and are applied to cages and equipment used by animals.

Sanitation is the practice of controlling bacteria and other organisms. This may be accomplished by removing contaminated objects with which animals come in contact. For example, removing animal waste, washing the cages, and cleaning examination tables are common sanitation practices. Sanitation includes changing the animal's bedding or cleaning the cage floor.

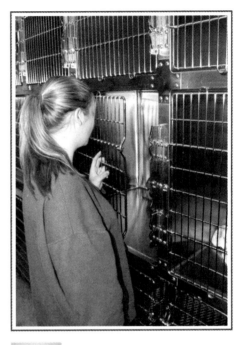

21-14. Sanitation includes clean cages, feeders, and waterers.

REVIEWING

*M*AIN IDEAS

Animals contribute to the welfare of people and of other animals. Scientific and laboratory animals are used in basic and applied research and to a lesser extent in clinical research.

There are many types of scientific and laboratory animals, but the most popular animals are the rodents. More than 90 percent of the animals in biomedical research are rodents. Other popular research animals include rabbits, cats, dogs, and nonhuman primates.

The advancements made in animal research have allowed scientists to develop four basic models that utilize animals in research. Only one model, the living animal model, actually conducts research with live animals.

Management in the laboratory or research facility is a top priority for scientists. Animal welfare is the goal of animal research scientists. Scientific and laboratory animal research is strictly regulated, internally and externally.

The living environment is regulated to provide a healthy, comfortable life for the animals. Space requirements, temperature, humidity, noise, and lighting are all components of the animals' environment. Keeping the animals healthy through hygiene is accomplished through sterilization, disinfection, and sanitation.

QUESTIONS

Answer the following questions using correct spelling and complete sentences.

1. Distinguish between using animals and using animal by-products in research.
2. What are three kinds of research? Briefly define each.
3. What animals are used in laboratory research? Which is most popular?
4. What animal models are used in scientific and laboratory research?
5. How is laboratory animal research regulated?
6. How is the well-being of animals provided for in a laboratory environment?

EVALUATING

Match the term with the correct definition. Write the letter by the term in the blank provided.

a. cavy
b. laboratory animal
c. basic research
d. animal by-product tissue

e. nonhuman primate
f. living animal model
g. Animal Welfare Act
h. clinical research

_____ 1. A federal law that assures proper use of animals in research.
_____ 2. A guinea pig.
_____ 3. Research animals that are kept alive for study.
_____ 4. Animals similar to humans.
_____ 5. Research in a medical facility.
_____ 6. An animal raised for and/or used in laboratory research.
_____ 7. Tissues from an animal used primarily for purposes other than research.
_____ 8. Research that increases knowledge to help understand animals.

EXPLORING

1. Prepare a report on the use of laboratory animals to alleviate a human disease. Be sure to investigate the disease and its effect on humans. Do you feel that animals should have been used for this research?

2. Organize a class debate or discussion to explore the use of animals in research. What is the conclusion drawn from the information that is shared? Do you agree with the positions taken by different individuals in the debate or discussion? Why?

22

EXOTIC ANIMALS

Animals are special in many ways! What would a circus be without elephants, or a zoo without a zebra? We look forward to the entertainment animals can provide. Just watching them go about life is fascinating.

People enjoy working with animals. Some people have jobs caring for them; others own and raise them. New and interesting kinds of animals are occasionally grown. The people who grow them are pioneers. They learn by trial and error. They do not have the benefit of research and written information to help them.

Special animals often need a little help from humans to survive. The animals may need to be cared for in protected areas or receive a little extra attention. There may be at least one of these that is highly interesting to you.

22-1. A Dama Gazelle is a sophisticated exotic animal.

OBJECTIVES

This chapter introduces exotic animals. It has the following objectives:

1 Identify and describe major species of exotic animals

2 Explain how to assess opportunities in exotic animal production

3 Explain important management practices with exotic animals

TERMS

animal refuge
biologist
circus
exotic animal
exotic animal dealer
exotic animal investment
  species

feral animal
performing animal
rare breed
zoo
zoo curator

MAJOR KINDS OF EXOTIC ANIMALS

An *exotic animal* is an animal that is not native to where it is being raised or is an animal that is rarely seen. Exotic animals are often new or are used in new ways. In some cases, they need extra care to survive. Exotic animals are used for a wide range of purposes. Often, formal marketing structures are not in place for exotic animals or their products.

Exotic animals are relatively newly domesticated animals or animals that are beginning to be used for food and fiber production. They may also be used as investment animals and zoo and performing animals.

EXOTIC FOOD AND FIBER ANIMALS

Some exotic animals are used for human food and other products. In a broad sense, all current domesticated livestock started as an exotic species. Exotic animals used for food are new agricultural animals or animals being used for new purposes. A few species in the United States that are now considered exotic are the llama, alligator, elk, and bison. Many other animal species are raised to produce agricultural commodities. Some exotic animals are harvested from the wild.

Llama

Llamas are mammals that have the potential of being raised commercially for their coat. Llama meat has a low fat content and is delicious. At this time, few llamas are eaten. Breeding animals can sell for more than $10,000 and to eat one would make for a very expensive meal. Some llamas are used as guard animals with sheep and goats. In remote parts of the world, llamas are used as pack animals.

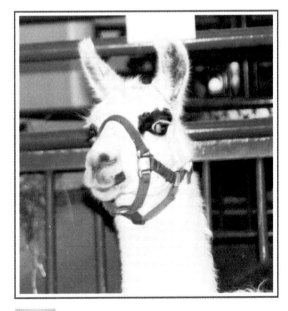

22-2. A llama has a proud and alert appearance.

Alligator

The American alligator, once on the brink of extinction, is now con-

22-3. The skeleton of an alligator is used to learn more about the species.

sidered an exotic food animal. Alligators are raised on farms in many southern states. Their meat is light in color, has a firm texture, and a distinct flavor. Their skin is used for making expensive leather products. Alligators grow and reproduce well in captivity.

Bison

The American bison (sometimes called buffalo) is being used to produce an alternative to beef. Although few purebred bison are grown for meat, the genetics of the bison are used in crossbreeding programs with domesticated beef animals. This cross, often called a beefalo, yields meat similar to beef with a lower fat content and less cholesterol.

Elk

Elk are typically thought of as wild animals. Several individuals are now producing elk much as other livestock

22-4. A bison calf in a herd of bison. (Courtesy, American Bison Association)

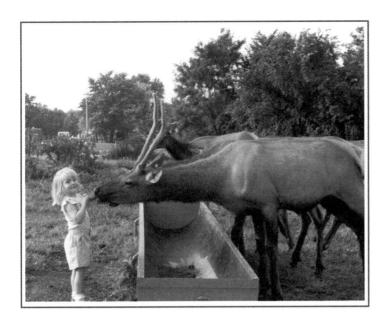

22-5. An elk undergoing domestication is tame. (Courtesy, North American Elk Breeders Association)

are raised. The producers are learning the requirements for successfully reproducing and caring for elk. Little information has been available on how to raise elk. Practical experience is important.

Rare Breeds

Breeds of livestock and poultry that were common just a few years ago are now rare. A *rare breed* is a breed of domesticated animal that was once popular and is now disappearing. Producers often take a special interest in a breed and strive to continue its existence. Producing these breeds is an alternative similar to keeping other exotic animals. Effort is made to keep the bloodline of the rare breeds pure for future breeding and research. This includes breeds of hogs, chickens, cattle, and other farm animals that were popular 50 or more years ago.

Other Exotic Food and Fiber Animals

Several species of animals are raised for food, fur, and fiber that are

22-6. The Barred Plymouth Rock is becoming a rare poultry breed.

considered exotic. Eel, turtles, mink, and upland game birds, deer, and game animals are a few examples.

EXOTIC SPECIES HARVESTED IN THE WILD

Some businesses have started to market products from exotic wild animals. Meat from the antelope, deer, elk, and small game is sold throughout the country. Other wild products include snake meat, shark steaks, and skins, feathers, hides, and hair from a variety of birds and animals.

EXOTIC INVESTMENT ANIMALS

Since exotic animals are new animals, or animals being used for a new purpose, most people are afraid to invest the money needed to start an exotic animal business. Fortunately, there are some entrepreneurs who take the risk. Investing in exotic animals can pay off well or it could leave you broke.

Connection

ELEPHANTS ARE LARGE IN MANY WAYS

Elephants are favorite exotic animals. As the largest of all terrestrial animals, their size makes them impressive. They can weigh 12,000 pounds (5,400 kg) and stand over 12 feet (4 m) tall. And, they never completely stop growing as long as they live!

An animal the size of a mature elephant takes a lot of food. Elephants eat up to 300 pounds (136 kg) of food a day when on the range. They like grasses and tree fruits, leaves, bark, small branches, and roots. Large elephants can push over 30-foot (9 m) tall trees to get food. An elephant can drink up to 40 gallons (150 L) of water each day.

When kept in captivity, elephants need open areas enclosed with strong fences. They can live 60 or more years if they receive good care. Natural death is typically when their last teeth wear away and they die of malnutrition.

An ***exotic animal investment species*** is an animal raised in hopes of making a large sum of money. If successful, they are more profitable than cattle. There is a high degree of risk and many people fail to earn big profits.

Early investors in the pot-bellied pig market paid thousands of dollars for the first breeding animals that came to the United States. They hoped to make their money back selling these animals as pets. Although there was interest in the animals, the American public did not buy many pot-bellied pigs. Some investors were stuck with expensive pets.

Other investors in exotic animals have fared better. A trend for sports enthusiasts has been hunting in animal preserves. With the areas available to hunters decreasing, animal preserves have become a good alternative. Investors are realizing a boom in business, while sports enthusiasts are provided with a place to practice their sport.

Investing in Agriculture

One area of investment is in agriculture. Several species are available for an entrepreneur to choose to invest in today. Good business people keep their eyes open for new investment opportunities. No one knows for sure what the next big exotic species opportunity will be!

Investing in Pets

There will always be people who want to own unique pets. New companion animals could be a good investment for the investor. Roadblocks for the new companion animals include state and federal laws restricting new ani-

22-7. In the 1980s, there was high interest in the pot-bellied pig.

mals, international law, and the compatibility of the animal with the environment and other animals.

Investing in Research

Many animals, especially strains of mice and rats, are used in research. A potential investor could fund development of a new strain or improvement of an existing strain. New research animals could also be added to assist with research.

Exotic zoo and performing animals

Zoo and performing animals are raised in captivity or captured from the wild. Training often begins at a young age to prepare animals that will be performers.

Zoo

A *zoo* is a zoological garden. It is a parklike area where animals are kept for viewing. Many of the species are exotic animals that are not native to the area where the zoo is located. Sometimes, the animals will breed in captivity. Environments are constructed that resemble the wild environments for the animals. Probably the best-known zoo in North America is the San Diego Zoo in California, which has a total of 3,200 animals representing about 800 species. The Smithsonian Institute Zoo in Washington, D.C., is well known for its panda bears. Many cities have one or more zoological parks.

Zoos have four roles: education, entertainment, research, and animal conservation. Many people with special training in zoology work in a zoo. A *zoo curator* is a person trained in exotic animal zoology who looks after the animals.

Zoos get animals in several ways. Some get animals by breeding and raising them. In some cases, zoos exchange animals. For example, one that has raised a young giraffe might exchange it with another zoo for an animal of equal value. Many zoos get their animals from animal dealers. An *exotic animal dealer* buys and sells exotic animals. They help meet the needs of zoos for certain species of animals.

Animal well-being is a high priority. Each animal is fed according to its needs. Health care is provided to keep the animals disease-free. Occasionally, the animals must be protected from people who are diseased. Some species get diseases from humans. An example is the closure of a popular and attrac-

22-8. Four animals commonly found in a zoo are (clockwise from top left) sable antelope, tiger, grizzly bear, and zebra. (Note the long nails on the grizzly!)

tive monkey exhibit at a major zoo because the animals contracted tuberculosis. They got the disease from eating food scraps thrown into their area that had been infected by humans!

Some zoos have trained animals and provide performances. Others limit their work to caring for the well-being of the animals.

Performing Animals

A *performing animal* is one taught to go through a routine of unusual activity. Performing animals may be domesticated or exotic species. Dogs and horses are often taught to perform tricks. Exotic animals, such as elephants, tigers, and monkeys, are also taught similar activities.

Many circuses have trained animals as part of the show. A *circus* is a variety show with animal and human performances. Humans are closely involved with the animals.

Who do you think breeds and raises circus animals? The people who do so have skills in agriscience! It is true that some animals come from the wild, but many are bred and raised in captivity.

ANIMAL REFUGES

An *animal refuge* is a protected place where animals are cared for to assure their well-being. Some animal refuges are for wildlife; others include exotic animals. A bird sanctuary is a kind of refuge.

There are about 475 National Wildlife Refuges in the United States. These range from small areas to protect one species of bird to large land areas for the protection of several species. Private individuals also have refuges that cater to exotic animals.

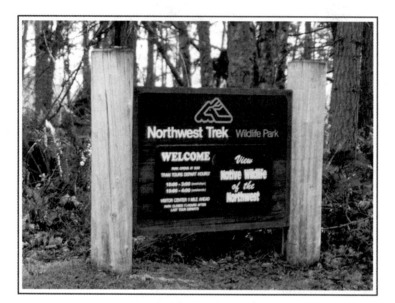

22-9. The Northwest Trek (near Seattle, Washington) is a wildlife park featuring animals of the northwestern United States.

22-10. People can view animals, such as this river otter that is sunning on a limb, at the Northwest Trek.

22-11. Visitors to the Chincoteague National Wildlife Refuge can often see the feral ponies.

Some animal refuges are homes to feral species. A *feral animal* is one that was once domesticated by now lives in the wild. Horses, dogs, hogs, and others live as feral animals in various areas of the United States.

Animal refuges study animals and develop new ways of meeting their needs. They often try to maintain and reestablish a threatened species. Biologists often work with the animals. A *biologist* is a person trained in areas dealing with living organisms. Wildlife biologists have specific training in the biology of wildlife.

OPPORTUNITIES IN EXOTIC ANIMAL PRODUCTION

Many opportunities exist for the entrepreneur to get into the exotic animal business. But, before you get started, you should examine the requirements to start an exotic animal production business.

PROFITABILITY

Making a profit is the goal of all businesses. The exotic animal business is no different. If you decide to enter this part of livestock and poultry production, you need to know the costs, advantages, and regulations that will affect your business.

Start-up Costs

Start-up cost depends greatly on the species you choose to raise. As mentioned earlier, llama production would be a very expensive enterprise. Investing in a species of tropical fish, on the other hand, would be much less expensive. One must consider start-up costs of breeding stock, facilities, and equipment.

Operating Costs

Operating costs include feed, medicine, labor, marketing, utilities, permits, insurance, and any other cost of being in business. You can get information to estimate these costs from extension agents, feed salespeople, or others in the same business.

CHOOSING A SPECIES TO RAISE

Suppose you have decided to raise an exotic animal. How will you decide which species to raise? Here are areas to consider.

Location Limitations

Many exotic species are environment-specific. For example, alligators would not grow well in Ohio. They need to be in a subtropical region to grow and reproduce. Fortunately, there are other species that are not as environmentally sensitive. Also, some species (tropical fish for example) can be kept indoors. You need to be aware of limitations you have because of your location.

Comparative Advantages

You should also compare advantages and disadvantages between species. If you are considering several alternatives, look at ease of production, costs, marketing, and ease of leaving the industry.

Regulations

Ask the proper agencies in your state and community about regulations for exotic animals. You must find out if you are allowed to possess the species or if you need a license to produce the animals.

Regulations on some exotic animals are strict. States like Florida, Louisiana, and Mississippi have experienced many problems with exotic animals invading the wild. One example is the nutria. This large rodent is not native to the United States. It was accidentally released and has now become a major problem. The animal is a vegetarian with a big appetite! These animals are destroying thousands of acres of wetland vegetation every year.

Regulations are also in place to prevent the spread of disease, protect people and property, and ensure that animals are treated humanely. Housing, zoning, shipping, and purchasing regulations must be explored before starting your business.

Market

You could be an excellent producer and go out of business if there is no market for your product. Make sure there is a market before making an investment. You should decide if you are going to sell directly to consumers or sell to a dealer or processor. You will likely need to develop a market before beginning production.

EXOTIC ANIMAL MANAGEMENT PRACTICES

Although there are many similarities between exotic and other agricultural animal species, there are important differences. To raise an animal is to be responsible for its well-being. Responsible producers will know the needs of animals.

SIMILARITIES WITH OTHER ANIMALS

The basic rules for animal production apply to producing exotic animals. They must have a balanced ration. They must be given adequate space to live. Their living area should be kept sanitary and safe. The environment must be comfortable. They must receive preventive care and timely treatment for illness and injury. In these ways, raising exotic animals is very much like raising hogs or sheep.

DIFFERENCES WITH OTHER ANIMALS

While there are many similarities, the differences are key to being a successful exotic animal producer. Each species has its own characteristics that make it a unique animal. Major areas of difference between traditional domestic livestock and exotic species are nutritional requirements, health needs, housing and living space, and reproduction.

Nutritional Requirements

Although every animal needs a balanced ration, every animal differs in its nutritional requirements. Domestic livestock rations have been carefully studied. We know with a great degree of certainty what they require in their diet. Information on most exotic animals is sketchy at best.

What would you feed alligators? How can you find out what they need? Fortunately, alligator diets have been researched by universities and alligator producers. This information would be easy to find from these sources. Most other exotic animals are the subject of research and you can learn about their requirements from universities or the Cooperative Extension Service.

Health Needs

Health care may be needed for exotic animals. Some veterinarians and others in animal health care are familiar with the health needs of domestic animals. Unfortunately, they may not know about particular needs of exotic animals. The veterinarians trained to work with exotic, zoo, circus, and performing animals are usually at colleges of veterinary medicine or work for agencies that have exotic animals.

22-12. A trained panda performs in a circus.

Housing and Space

Housing needs will vary with the species. You need to find out whether animals can be kept together, how much space they need, what temperature they prefer, and the type of housing they need.

The exotic animal owner will also need to know how to restrain their animals. For example, keeping an ostrich in confinement requires strong fences!

Reproduction

Animal production requires reproduction. Some species will reproduce well in captivity, while others need to be given great care to encourage reproduction.

REVIEWING

*M*AIN IDEAS

An exotic animal is any new species of animal that is being used for agricultural purposes, or an existing animal that is being used for a new purpose. The major types of exotic animals include food and fiber animals, investment animals, and zoo and performing animals.

Opportunities exist for entrepreneurs in the exotic animal industry. One must consider the profitability of the animal, the costs of production, and choose a species to raise.

Exotic animals and domestic animals have many characteristics in common. They all need proper nutrition, health care, sanitary living conditions, and space to live. They also have differences that must be attended to. To find out about these differences, you should seek help from universities, the Cooperative Extension Service, or veterinarians that specialize in exotic animal care.

*Q*UESTIONS

Answer the following questions using correct spelling and complete sentences.

1. What is an exotic animal? How are they different from other domesticated animals?

2. What are three species of exotic animals used for food and fiber?

3. Why did investors often lose money with the pot-bellied pig?

4. How does a zoo get its animals?

5. Give an example where zoo animals contracted disease from humans.

6. What are the major questions to answer in trying to determine if an exotic animal business will be profitable?

7. What should you consider in deciding on the species of exotic animal to raise?

8. What are the major practices in managing exotic animals?

EVALUATING

Match the term with the correct definition. Write the letter by the term in the blank provided.

a. zoo curator
b. performing animal
c. zoo
d. animal refuge

e. exotic animal
f. exotic animal dealer
g. biologist
h. exotic animal investment species

_____ 1. An animal that is raised in hopes of having a large profit.

_____ 2. Person who buys and sells exotic animals.

_____ 3. Animal that is not native to the area where it is raised and is being used for new purposes.

_____ 4. Parklike area where animals are kept for viewing.

_____ 5. Animal taught to perform a routine of unusual activities.

_____ 6. Person trained in understanding living organisms.

_____ 7. Person trained in exotic animal zoology.

_____ 8. Place where animals are protected to assure their well-being.

EXPLORING

1. Visit a zoo or animal refuge. Develop a list of the animals that you see. Study the kind of environment provided. Interview a zoo curator about feeding and other management practices with the animals. Prepare a report on what you observed.

2. Select an exotic animal. Use various reference materials and prepare a report on the origin of the animal and the kind of environment it needs to survive. Report in class.

23

ENTREPRENEURSHIP IN THE ANIMAL INDUSTRY

The animal industry offers many opportunities. Some people start their own ranch or other animal business. Others work at jobs in the animal industry. People who like to work with animals often go into animal production. What are your preferences?

People can work for themselves and reap the benefits of their hard work. The animal industry is large. It accounts for nearly one-half of the money from agricultural sales. People need to get the right education and experience to be successful.

Begin planning for a future in the animal industry while in high school. Take courses in agriculture, science, and business. Get on-the-job experience by working for other people. This will help you understand what is involved in having an animal enterprise.

23-1. Producers must often tend to the health needs of animals. (Courtesy, College of Veterinary Medicine, Texas A&M University)

OBJECTIVES

This chapter introduces the opportunities for entrepreneurs in the animal industry. It has the following objectives:

1 Explain consumers and the role of consumption in animal production

2 Describe the scope of animal production in the United States

3 Explain entrepreneurship in livestock and poultry production and list available opportunities

4 Identify important areas of management in successful entrepreneurship

TERMS

animal producer
beef
business plan
consumer
consumption
dairy product

decision-making
entrepreneur
management
part-time animal producer
per capita consumption
risk

THE ROLE OF CONSUMERS

The animal industry is set up to meet the needs of people. It provides a supply of wholesome products in the forms that people want. And people want many products!

A *consumer* is a person who eats, wears, and otherwise uses goods and services. All people are consumers! Consumers want products that satisfy needs. They use or consume products.

What consumers want is important. They have strong ideas about what they will buy. They want good value for their money. Every time a person goes to the store to buy food or other items they are consumers. They make decisions based on price, the money they have, and the desirability of the products.

POPULATION

The world's population is nearing 6 billion people. It takes an extraordinary amount of food and clothing for them. The number of people will increase sharply in the future. Some forecasters say that the number will reach 10 to 12 billion by the year 2030. Of course, animal products are more important in some parts of the world than others.

In North America, people like animal products. Canada, Mexico, and the United States have a combined population of 450 million people. The United States has about 266 million people, with California, New York, Florida, and Texas being the states with the most population.

23-2. Consumers make choices every time they make a purchase. (This is the International Market Place in Honolulu.)

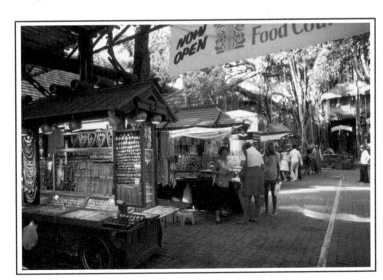

Consumption

Consumption is the use of animal products and other goods and services that have value. Every time you drink milk, consumption has occurred. Overall, Americans have a high consumption of animal products.

Information is compiled on the amounts that people consume. This is often reported as the amount for one person. *Per capita consumption* is the average amount consumed per year by a person. It may be reported in weight or another measure.

The total value of the U.S. livestock and poultry industry was approximately $99 billion in 1997. This value is distributed over 1.9 million farms. Cattle make up 41 percent of the livestock and poultry industry, while poultry comprise 23 percent, dairy makes up 19 percent, and swine contributes 14 percent of the total industry. Other animals make up the final 3 percent of the livestock and poultry industry.

The scope of the animal food industry is amazing! For example, in 1996, U.S. farmers produced 15 million metric tons of poultry meat products that generated almost $14 billion of income. Dairy producers put 70 million metric tons of milk into the food industry. In 1998, cattle producers had 99.5 million animals with an estimated value of about $60 billion.

For a typical U.S. citizen, that translates into about 112 pounds of red meat, 15 pounds of fish, 30 pounds of eggs, 46 pounds of pork, and 65 pounds of poultry per year. In addition, the average American consumes nearly 580 pounds of dairy products per year! We get a lot of our food from animals.

SCOPE OF ANIMAL PRODUCTION

The scope of the five main areas of food animal production is briefly described here.

Dairy

Nationwide, milk production has been profitable over a number of years. Milk is sold by the hundredweight (100 pounds or about 12.5 gallons of milk). Many dairy products are made from milk. A *dairy product* is milk and any products in which milk is the major ingredient, such as cheese and ice cream.

The annual per capita consumption of dairy products is about 575 pounds. In the last 25 years, dairy product consumption has varied up and down. The peak consumption was 601.2 pounds per capita and the low has been 535 pounds per capita.

For years, Wisconsin has been the top dairy production state. In the early 1990s, California moved ahead with 1.4 million cows. Other important dairy states are New York, Pennsylvania, and Minnesota. The nature of dairy farming varies. Some places have small family-owned dairies. Other places have large corporate-owned dairies.

BEEF

Beef is the meat produced by cattle. More beef has been eaten than any other meat in the United States for many years. However, the per capita consumption has declined in recent years. There were about 100 million beef animals in the United States in 1998. Beef cattle numbers declined slightly in the 1990s.

There are 1.7 million beef producers in the United States. These producers sell nearly 54 billion pounds of beef per year. Texas has been the number one beef cattle state for many years. Texas had 14.3 million beef animals in a recent year. Nebraska ranked second, with 5.9 million beef animals. Rounding out the top five states were Kansas with 5.89 million, Oklahoma with 5.3 million, and Missouri with 4.6 million beef cattle.

POULTRY

The poultry industry is composed of two distinct segments: meat and eggs. Chickens are the primary producers of eggs. Chickens, turkeys, ducks, and a few other species are used for meat.

23-3. Owning a rabbit is a start for an entrepreneur.

23-4. A poultry producer using a computer system to monitor and control the conditions in poultry houses from a central location.

The poultry meat industry has expanded in the last decade. Production is now more than 31 billion pounds per year. Many chickens are marketed as broilers. Most broilers grow fast and are marketed at seven weeks of age. People feel that eating broilers has health benefits.

Egg production has had only a slight increase in recent years. Current annual production is about 77 billion eggs. Concern about cholesterol in the human diet has resulted in some drop in egg consumption. More than 280 million hens are used to produce eggs in the United States.

California is the leading state in egg production. Ohio is second, followed by Pennsylvania and Indiana.

In addition to chickens, turkeys are produced in the United States. Nearly 300 million turkeys are raised each year.

SWINE

The swine industry has increased steadily over the last decade. Though the number of breeding animals decreased in the last 10 years, the number of market animals has increased. Market hogs are young hogs grown for slaughter. They typically weigh about 220 pounds. About 25 billion pounds of pork are annually marketed in the United States.

Iowa leads in hog production with nearly 26 million head. Illinois, North Carolina, Minnesota, and Nebraska follow Iowa.

SHEEP

The 75,000 sheep producers in the United States produced 7.6 million sheep and lambs in 1998. The wool produced was valued at $40 million.

Colorado is the top state in sheep production. Other states in the top five included Texas, California, Wyoming, and South Dakota.

ENTREPRENEURSHIP IN LIVESTOCK AND POULTRY PRODUCTION

An *entrepreneur* is a person who takes a risk by trying to meet the demands of people for a product. Creativity is needed. To be successful, the animal or product must meet a unique demand. The entrepreneur typically owns and operates a business, including farms and ranches. A key part of being an entrepreneur is risk.

Risk is the possibility of losing what has been invested. People must have money—often a large amount—to start some kinds of animal entrepreneurship. If the business fails, they will lose their money. Often, the money is borrowed from banks and other sources.

Entrepreneurs are thought of as their own bosses. They have a big responsibility. The success or failure of their business depends on how they perform. Being an entrepreneur offers many advantages. You can set your own work schedule, have the opportunity to be creative, reap the rewards of your success, and get a great deal of self satisfaction. There are, however, no guarantees of success.

There are many opportunities in animal production for entrepreneurs. These relate to raising animals and meeting the needs of producers for supplies, services, and marketing.

ANIMAL PRODUCER

An *animal producer* is an entrepreneur who raises livestock, poultry, companion animals, or others. Animal producers may be known as farmers or ranchers. Often, they are known by the kind of animal they raise, such as a trout farmer or horse rancher. Producers of companion animals may be known as kennel operators if they produce dogs, or aviary operators if they produce birds.

Livestock and poultry farmers engage in the breeding, feeding, and general business of raising commercial livestock and poultry. They can be classified as beef cattle farmers, dairy farmers, sheep farmers, swine farmers, small-animal farmers, and fish and aquatic-animal producers.

 23-5. Ranchers assume risk in cattle production.

Entering the livestock and poultry production industry as a farmer requires a great deal of capital. Most types of farming require specialized facilities, equipment, and land. Many young or new entrepreneurs begin by working at home or developing partnerships with established farmers. Some start as tenant farmers and eventually are able to save the start-up capital required.

It is best to select an area of interest and concentrate on one or two species of livestock. For example, an entrepreneur could choose to specialize in dairy production. Raising dairy cattle takes much time and effort. The producer must be familiar with approved practices for production and make good business decisions. Although it is possible to have a diversified operation with many types of livestock, a farmer that concentrates on one species can operate more efficiently.

A drawback to focusing on one species is that the producer is tied to one product in the market. When the market is favorable, profits are good. If the market falls, the profits also fall.

If you are considering becoming an entrepreneur, collect as much information about the area as possible. You should know what kind of facility is required, how much it will cost to get started, what the market for your product is like, and have the knowledge to successfully produce your livestock.

PART-TIME ANIMAL PRODUCER

Another option for entrepreneurship lies in becoming a part-time animal producer. A ***part-time animal producer*** is one who raises animals only part of the time. The individual may do other work on a full-time basis.

Many farms in America today are operated on a part-time basis. Some people like to be around nature and animals. They like the thought of making a little extra money from their animals. Some families that have traditionally lived in the cities or suburbs are also starting to enter the part-time farm business. These can provide some income for the families involved, but the major source of income usually comes from the full-time job.

Part-time farms have marked differences from full-time operations. The part-time farms are usually smaller, with fewer acres and fewer numbers of livestock. These operations are commonly more diversified than full-time operations. Several species of livestock, fruit and vegetable production, and some crop land are the norm.

These operations are less expensive to start and can provide some additional income. Many small operations attempt to fill a marketing niche, like

Angora goats or Siamese pot-bellied pigs. If you decide to enter such an enterprise, be very cautious. Be certain that there is a market for your product and that you can raise the animals. Many entrepreneurs have lost money due to bad decisions.

Nonproduction animal entrepreneurship

Producers rely on many supplies, services, and marketing activities to get products to consumers. People are needed to start and operate these businesses. Without these businesses, today's animal producers would not have a high level of production.

Animal Supplies

Providing supplies to animal producers can result in a good opportunity for some individuals. Animal supplies entrepreneurs setup businesses to meet the needs of animal producers for supplies. Many kinds of animal supplies businesses are possible. The kinds needed depend on the animals and products produced. The following are few examples of animal supplies businesses.

- Feed mill—manufactures feed for animal producers
- Feed store—sells feed to animal producers
- Animal medicine store—sells animal medicines and other supplies
- Animal equipment store—sells equipment used in animal production, such as feeders, waterers, milkers, gates, fencing, chutes, and pens

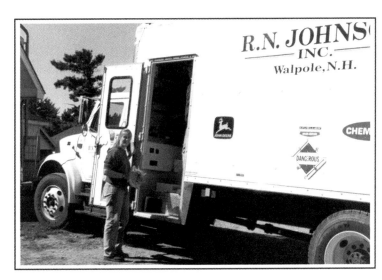

23-6. On-farm delivery of supplies for dairying and other enterprises fills a need.

■ Pet supplies store–sells animals, feed, medications, equipment, and other items needed with small animals

Animal Services

Animal producers often need specialized services to help with animal production. Opportunities often exist for people to be entrepreneurs in these areas. The kind of service depends on the animals and products produced. A few examples of animal services follow.

■ Groomer—trims, bathes, and otherwise cares for companion animals
■ Animal sitter—cares for animals when the owner is away, such as for a dog when its owner is traveling
■ Veterinarian—cares for the health of animals
■ Farrier—puts shoes on the feet of horses
■ Custom shearer—shears wool from sheep
■ Hauler—has specialized equipment to haul live animals, such as cattle trucks and trailers, or their products, such as refrigerated milk tanks on trucks
■ Construction contractor—constructs buildings, fences, and other facilities on farms and ranches for animals

Connection

ADVERTISING WITH A BIG COW

Successful animal businesses need customers. They need people to buy and use their products. This includes all areas ranging from pet supplies stores to ranches with fine breeding cattle and processors of food products.

Advertising is promoting the ideas, goods, or services that are available. Newspapers, radio, television, billboards, and other means are used. The goal is to get the attention of potential customers.

Big cows get attention! This photo shows a big cow being pulled through city streets promoting a brand of milk and ice cream products. The big cow just barely goes under traffic lights and low overpasses. The cow will be stopped in parking lots of stores that sell dairy products. It gets attention!

23-7. Skills are needed to operate a pet grooming facility.

Animal Marketing

Entrepreneurs in animal marketing may fill a wide range of roles. They link the animal producer and the consumer of the animal product. They must have good skills in working with producers and be informed on market information. The following are a few examples of animal marketers.

- Cattle buyer—buys cattle from a producer for a processor or other user
- Sale barn owner—operates a selling facility where producers bring animals to sell to buyers
- Packing house operator—operates a facility that packs and/or processes animal products

23-8. Auctions have been used in marketing livestock. (Courtesy, Mississippi State University)

■ Food processor—takes raw animal products and prepares other products in forms that consumers want, such as making cheese from milk

THE ROLE OF MANAGEMENT

Working hard and producing good animals is important but the success of an animal business requires more—good management. *Management* is using resources to achieve the objectives or goals of the business. The resources are used efficiently and directed toward the objectives.

BUSINESS PLAN

A business plan is essential. A *business plan* is a written plan on how the animal enterprise will be operated so it is successful. Many people try to operate without written plans. With good effort, they may be quite successful on a small scale. Banks and other lending agencies often require a business plan in order to get a loan. Business plans are usually revised each year.

Business plans help in decision-making, which is an important part of management. *Decision-making* is choosing among alternatives that are available. This includes the kind of animal, the facilities to use, the feed and care provided, and when to market the animal. People need good information for decision-making. Bad information may result in bad decisions. And, no one wants to make bad decisions!

23-9. The manager of a pet supplies store arranges merchandise for customer appeal.

MANAGEMENT FUNCTIONS

Management involves many important duties. These are often listed as the five functions of management:

■ Planning—Planning is deciding on a course of action. It is important in preparing a business plan as well as the daily operation of an animal enterprise.

■ Organizing—Organizing is identifying and arranging important activities and resources. This is needed to assure good use of time and effort in meeting objectives.

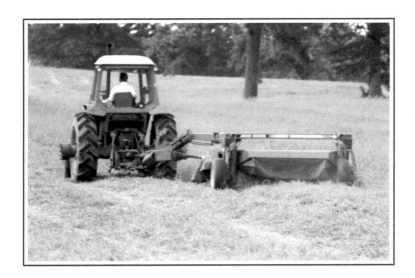

23-10. Being an entrepreneur in livestock production may require the operation of haying equipment.

■ Staffing—Staffing is getting people to help operate an animal enterprise. In some cases, one person can do the work. Larger animal enterprises require more people.

■ Leading—Leading involves getting people to do what needs to be done. It includes the influence a manager has over other people.

■ Controlling—Controlling is assessing how well the enterprise has done. It involves keeping records and following appropriate financial procedures. The information gained is used in revising the goals and objectives.

REVIEWING

*M*AIN IDEAS

The animal industry is large and diverse. Annual receipts from livestock and poultry account for nearly one-half of the income generated from agriculture. Companion and other animals add greatly to the income.

Consumers are people who use animal products. They have preferences about what they will buy. Producers must produce animals that meet the needs of consumers. If not, the consumer will not buy the products and the price the producer receives will be low when the animals are sold.

The needs of consumers are not always met. This gives the opportunity for entrepreneurs to provide new products or services. Careful study and preparation are needed to be successful as an entrepreneur.

There are many opportunities for entrepreneurs in the animal production industry. Most opportunities are in animal farming, either in a full- or part-time operation. Good management is essential for success.

QUESTIONS

Answer the following questions using correct spelling and complete sentences.

1. What is a consumer? What is the role of consumption?
2. Define entrepreneur. Discuss opportunities in the animal industry for entrepreneurship.
3. Select one area of animal production and summarize its scope in the United States.
4. What is the role of an animal producer?
5. What is the importance of management in the animal industry?

EVALUATING

Match the term with the correct definition. Write the letter by the term in the blank provided.

a. animal producer   c. per capita consumption   e. management
b. consumer   d. entrepreneur   f. risk

_____ 1. The possibility of losing what has been invested.
_____ 2. People who eat, wear, and otherwise use goods and services.
_____ 3. The average amount consumed per year by a person.
_____ 4. Using resources to achieve the goals and objectives of an animal enterprise.
_____ 5. One who assumes risk in a business activity.
_____ 6. An individual who raises animals.

EXPLORING

1. Prepare a bar graph that shows amounts of consumption for animal products. Include beef, pork, poultry, eggs, and sheep and mutton. A source of information is: www.usda.gov/

2. Visit an animal producer. Determine the kinds of animals produced and the general nature of the work. Discuss the job and lifestyle of being an entrepreneur in animal agriculture.

3. Investigate the management of an animal enterprise. Select an enterprise, determine the role of management, and develop a list of areas where good decision-making is essential. Prepare a report on your findings.

24

CAREERS IN THE ANIMAL INDUSTRY

You do not have to own animals to work with them! You can work for another person who owns animals.

People who work for others do not have the same risk as entrepreneurs. Many careers are available for those who want to work with animals. Other careers are in animal supplies and services and marketing. These do not always involve contact with animals.

Many off-farm careers in livestock and poultry production support animal production. Less than 2 percent of the United States work force is in farming. However, nearly 20 percent of all workers are in careers related to agriculture. Although some careers have limited contacts with animals, about 535,000 people work in meat and dairy production. In addition, thousands of people are in companion animal, poultry, aquaculture, and equine jobs. Others work in the exotic and service animal fields.

24-1. Animal care occupations involve many areas. This shows an owl being examined. (Courtesy, College of Veterinary Medicine, Texas A&M University)

OBJECTIVES

This chapter introduces careers in the animal industry. It has the following objectives:

1   Explain why choosing a career is important and relate the decision to work ethic and career goals

2   Distinguish between career, occupation, and job

3   Describe career areas in food animal care and production

4   Describe career areas in pleasure and service animals

5   Describe science-oriented careers in animal care and production

6   Describe career areas in animal services

7   Describe career areas in education, extension, and communication

8   Demonstrate safe working habits with animals

9   List and explain important personal skills in career success

10   List and explain important employer expectations

TERMS

career	job
employment	occupation
entry-level job	safety
gainful employment	work ethic
goal	zoonotic disease
goal setting	

SELECTING A CAREER

People want to do useful things. They want to feel that they are contributing to society and have a good income. Choosing a career is an important decision. People work thousands of hours during the 30 years of their employment. They need good information when making choices.

*E*MPLOYMENT

People use their time in different ways. **Employment** is the primary way people use time; employment is often called work. Sometimes, people serve as volunteers and are not paid. Other times, they work for what they gain to help cover the costs of living.

Gainful employment is having a job that provides benefits. People are usually paid a salary or wages for their work. They may also have certain benefits, such as a ranch manager being provided free housing on the ranch. These benefits are a part of the total returns that a person gets for working.

Entrepreneurs have jobs. They are gainfully employed, though they may not be paid a salary. They receive benefits if their animal production enterprise is a success.

*W*ORK ETHIC

People view work differently. A **work ethic** is how we feel about work. Do we think it is essential? Do we think it is the most important thing in our lives?

24-2. A dog is receiving medication. (Courtesy, College of Veterinary Medicine, Texas A&M University)

24-3. Counselors have computer-based materials to assist with goal setting and career information.

How we go about choosing a career and the one we choose is a part of our work ethic. People begin developing attitudes about work at a very young age. They learn from the people around them. Most people learn that work is a natural and integral part of life. It is how people go about having what they need and want.

The way people go about work is also a part of the work ethic. People are expected to be good, dependable, honest workers. They are expected to work hard and do a job well. These attributes help people to be successful.

24-4. Gaining leadership skills through the FFA helps with success in careers.

GOALS

People often think about what they want to do in their lives. They identify things that they want to have and do. Much planning and work are needed to reach goals.

A **goal** is a level of achievement that people set for themselves. Goals have a powerful influence in our lives. We all want to achieve what we set out to do. No one wants to fail. People need to be careful in setting goals to be sure that they reflect what is important to them.

Goal setting is describing what we want to accomplish. It includes how we will achieve the accomplishments. People do it all of the time. Sometimes, our goals may not be very important in the long run. Goals people set about their careers are definitely important! Begin by understanding careers and the needed education. Match this information with what you want.

CAREERS, OCCUPATIONS, AND JOBS

A *career* is the general direction that a person takes with work. Most careers involve a sequence. People begin at a lower level and advance to higher levels. Moving up is climbing the career ladder or advancement.

The lowest level job is the entry level. An *entry-level job* is one that requires little or no previous experience. It is the first job for many people. Education and following instructions will help people advance. Careers may have a series of occupations and jobs.

An *occupation* is an important part of a career. Occupations involve specific work, known as duties. The same occupation has similar duties in different locations. For example, the occupation of a large animal veterinarian has similar duties regardless of where the veterinarian works.

A *job* is specific work. Jobs are carried out at certain sites and with employers. People who have jobs are employed. People can have occupations and not have jobs. For example, if the large animal veterinarian had no job, the individual still has an occupation though unemployed.

Sorting through the meaning of career, occupation, and job helps people make good decisions. The next time you hear about employment, check to see if it is an occupation or a job and how it is a part of a career.

24-5. Hands-on experience develops career skills.

CAREERS IN FOOD ANIMAL AREAS

Owning an animal production business is one way to be involved in the production of livestock. Not everyone can own animals. Many people want to be involved with animals, but not as entrepreneurs. They want employment with other people.

*F*ARM OR RANCH WORKER

Many people who eventually want to own and operate their own farm start out as a farm or ranch worker. Opportunities for this type of employment will vary with the type of agriculture in your community.

Farm and ranch work varies based on the knowledge and ability of the person employed. Duties range from general farm labor in doing routine jobs, such as feeding, sanitation, and moving animals, to more complicated tasks, such as treating sick animals. Small farms may employ a person to do a wide range of tasks. Workers in large operations may specialize in a particular area, such as milking cows.

Many entry-level positions are as a farm or ranch worker. New employees are expected to have a general knowledge of livestock and work competently with animals. They need to quickly learn their job. High school agricultural education is helpful.

24-6. Ranch work may involve using horses to check on cattle. (Courtesy, Paso Fino Horse Association)

*F*ARM MANAGER

The farm manager carries out the everyday operation of the farm. It may also include long-range planning and setting production goals. Farm managers are usually hired by corporate farms, institutional farms, or by individuals who own a farm but do not want to be involved in its daily operation.

Becoming a farm manager requires extensive experience or a combination of experience and education. Employers are often looking for people with education beyond high school.

Tasks a farm manager might do include selecting and marketing animals, formulating feed, maintaining herd health, and supervising employees. Helpful preparation includes high school agricultural education, being a farm worker, and postsecondary education in agriculture.

CAREERS IN PLEASURE AND SERVICE ANIMAL AREAS

Careers in pleasure and service animal areas are numerous. You could have a career raising animals for research or be a pet store owner. Horses, game and fish, service animals, and laboratory animals are a few areas.

HORSE INDUSTRY WORKER

Many people enjoy riding and racing horses. Stables provide a place to keep and ride a horse. In addition, there are plenty of places to enjoy horseback riding.

Increased interest in light horses for pleasure and racing has opened new jobs. These range from stable attendant to race horse trainer. As with any other career, you begin doing hand labor.

Stable hands spend most of their time doing essential work, such as cleaning the barn and feeding the horses. As you become more experienced, you may become a horse trainer, riding instructor, stable manager, or barn boss. Jobs in the race industry differ from job titles in the pleasure horse industry.

24-7. Some people have occupations training animals, such as this mule on a Colorado ranch. (Courtesy, Lucky Three Ranch, Inc.)

GAME AND FISHERIES WORKERS

Do you like wildlife? Do you like being out of doors? Are you concerned about fish and wildlife? Then you may be interested in a career in game and fisheries.

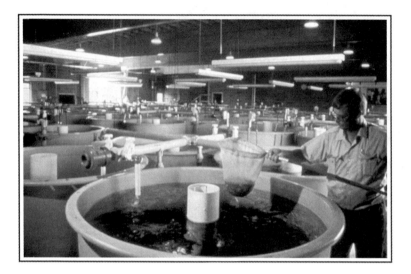

24-8. A fish hatchery technician tends tanks of fish. (Courtesy, Ed Grateau, U.S. Fish and Wildlife Service)

A wildlife conservation officer works to educate the public about fish and wildlife and enforces game and fish laws. The duties may include conducting workshops on wildlife and the environment. Conservation officers are usually hired by state and federal agencies.

Game farm workers raise and manage different species of wildlife. Some are returned to the wild. Others are kept on game farms. Hunting clubs, conservation groups, and government agencies need wildlife to enhance populations. Although no specific qualifications exist, a high school diploma and an interest in working with game birds and animals are desirable.

Fish hatchery technicians raise and manage fish. The fish may be for stocking in public and private ponds, lakes, and waterways. People employed in this field care for brood fish, tend eggs, maintain water quality, feed fish, provide a healthy environment for the fish, and manage the facilities. Hatchery workers are generally under the direction of a fisheries biologist. High school courses in mathematics, science, chemistry, biology, and aquaculture are helpful.

People in game and fish positions should have an interest in wildlife, fisheries, and conservation. Most wildlife conservation officers must pass a test and have needed training. Experience and a college degree in a wildlife area are beneficial.

SERVICE ANIMAL WORKERS

Raising animals that serve humans is important. Service animals, such as leader dogs for the blind, companion animals, and draft animals, fill many roles.

24-9. Grooming a steer for show develops skills in working with animals.

Draft animals are used for work, primarily the power to pull machinery. Most machines today are powered by engines. Some draft animals are used as hobbies and, in a few cases, to do specialized work. In a few countries, animal power is used to grow crops.

Although there are large producers of service animals, many of them are raised by small producers or individuals.

L ABORATORY ANIMAL WORKERS

A laboratory animal is used in research to improve human and animal life. Laboratory animals are raised and kept under the best possible conditions. Without these animals, progress in human medicine and related areas would nearly stop.

Many laboratory animals are delicate animals. They are raised to meet specific qualifications for research. For example, animals are bred with immune systems that are almost identical to human immune systems for AIDS research. Laboratory animal production is a complex field. It requires an in-depth knowledge of science, chemistry, and biology, as well as an interest in animal research.

CAREERS IN
SCIENCE-ORIENTED AREAS

The science of animals is expanding. Animal scientists study animals and use laboratory equipment to solve problems related to animals.

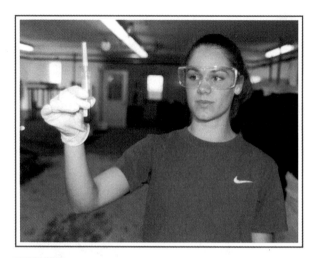

24-10. A blood sample has been collected from a dairy cow for laboratory analysis.

ANIMAL SCIENTIST

Animal scientists study animals and develop improved methods of production. They provide services to the livestock, poultry, and companion animal industry. Positions related to animal science are animal nutrition specialists, animal reproduction specialists, geneticists, chemists, and biotechnologists.

An animal research scientist concentrates in a specific area of interest in animal production. Research scientists work in laboratories with experimental animals and animal production inputs and other scientists. Most people who choose this career have an advanced college degree in the area. Personal needs are a high interest in scientific inquiry and a knowledge of biology, chemistry, and microbiology.

LABORATORY ASSISTANT

Assisting with laboratory animal production or experimentation can be a rewarding career. Animal laboratory assistants help with experiments, care for laboratory animals, and raise laboratory animals. Attention to detail is required. A high school diploma with an emphasis in animal agriculture is a minimum. Most laboratory assistants have some college education. Some have a master's degree in biology, chemistry, animal science, or a related discipline.

AGRICULTURAL ENGINEER

Agricultural engineers often work in the animal industry. They work with animal facilities, waste management, environmental impact, and improving animal products. Engineers must have a bachelor's degree and the ability to assess a problem and prepare a solution to it. Knowledge of agriculture and the animal industry is helpful.

WILDLIFE BIOLOGIST

A wildlife biologist does many of the tasks of a general biologist. They focus on wildlife animals and habitat. A wildlife biologist may study an endangered animal species or concentrate on improving the environment for many species. Their work can include genetics, studying the environmental impact of industry, or repopulating an animal where it no longer exists. Wildlife biologists help with the living conditions of wild animals and fish.

A wildlife biologist must be committed to studying animals and science. A college education is required for entry and higher degrees are needed for advancement. If you want a career in wildlife biology, you need a high interest in biology, chemistry, animal science, mathematics, and research.

SERVICE CAREER AREAS

People who provide for the needs of animals and help with marketing are important in the animal industry. Producers often solicit help from animal service professionals.

Connection

PARTICIPATION DEVELOPS SKILLS

People often have many opportunities to develop job skills. Some people are reluctant to participate. Not participating can later be a handicap in getting and succeeding on a job.

Students in agriculture classes are in the FFA. Many FFA activities develop skills for job success. There is one catch. You must be active. Get involved. Participate.

Leadership, personal, and career skills are developed through the FFA. Those who are most active, gain the most from participation. The Montana student in this photo is developing valuable skills in relating to other people. She is presenting a project she has developed to an adult leader.

VETERINARIAN

Veterinarians promote animal health. They provide care for sick and injured animals as well as prevent illness. A veterinarian may specialize with specific livestock or generalize in all animals. A few now specialize in fish health.

Becoming a veterinarian takes time and dedication. Preveterinary students often earn a bachelor's degree in animal science or biology. After completing the courses, they apply for admission to a school of veterinary medicine. If they are accepted, they complete several more years of education and training. Although much is required to become a veterinarian, it is a very rewarding career.

SALES AND SERVICE

The animal industry has many professionals in supplies sales and service. These people sell feed, ensure quality products, sell retail products, service equipment, operate stock yards, and perform many other services.

Many sales positions related to livestock and poultry production require a college education. The education should be in animal science or related area. Examples include machinery service, stockyard worker, veterinary assistant, and feedmill workers. Although not a requirement, some postsecondary education is desirable. Agricultural education in high school and experience in the animal industry are beneficial.

24-11. Pet grooming is a service occupation with good opportunities in some cities.

LIVESTOCK BUYER

Knowing what the industry and consumer demand in a meat animal leads to success in livestock buying and selling. A good "eye" for livestock quality on the hoof is essential. Buyers and sellers may work for themselves or have clients in the production or packing industry.

Training in animal selection is a requirement for this career. High school education in agriculture, biology, and mathematics is helpful. Additional training in animal science, business management, and public relations at the postsecondary level will be beneficial.

EDUCATION, EXTENSION, AND COMMUNICATION AREAS

Careers available in education, extension, and communication emphasize the "people" side of the animal industry. All levels are involved. Teaching and communicating information are necessary.

AGRISCIENCE TEACHER

Agriscience teachers work in the public schools. They teach basic agriculture and specialized classes. They work with students of all ages—kindergarten through adult. They are agriculture experts in their local communities. A bachelor's degree in agriculture or related area and certification to teach agriculture are needed.

Teachers involve students in the FFA. The FFA is a student organization that enhances agriculture instruction. It also provides personal and leadership skill development. Teachers also involve students in supervised experience programs that include areas of the animal industry.

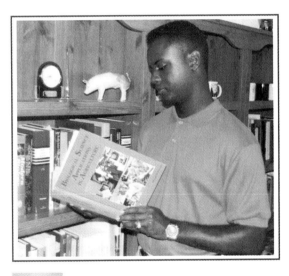

24-12. Good education is needed in all areas of animal science.

AGRICULTURAL EXTENSION AGENT

Agricultural extension agents work with local producers to help them improve production or solve livestock and other agricultural problems. Extension agents that specialize in livestock provide help to producers.

A bachelor's degree in agriculture or related area is needed to be an extension agent. Many states require a master's degree within a few years. Some specialize in animals. This requires a strong background in animal production.

AGRICULTURAL COMMUNICATION SPECIALIST

Agricultural communication specialists share information about animals. They may work with newspapers, in public relations, and with broadcast media (radio and television). An agricultural communicator may host a television show, write for a livestock magazine, or produce videos and computer-based materials.

A career in agricultural communication will require a university degree in agriculture or communication. Sometimes, people combine the study of agriculture and communication. People employed in this field need to have a broad knowledge of animals and have a people-oriented personality.

24-13. Agricommunication specialists are important in keeping the public informed. This shows an agriculture television editor.

SAFE PRACTICES IN WORKING WITH ANIMALS

Working with animals can be dangerous. Some animals are large and others have sharp teeth and claws. Animals are often frightened when people work with them. People can be injured when they try to protect themselves. *Safety* is preventing injury or loss. It is important to prevent injury to yourself, other people, and animals. Safety often involves equipment and how it is used.

Here are several practices that promote safety with animals:

■ Avoid Zoonotic Disease—A *zoonotic disease* is an illness that is passed from animals to humans. A few zoonotic diseases are ringworm, cat scratch fever, salmonella, rabies, brucellosis, and tuberculosis. The best way to prevent

24-14. Knowing animals and how to approach them improves safety. This shows a farrier repairing a horse's shoe. (Courtesy, American Paint Horse Association)

zoonoses from spreading is to keep healthy animals! Make sure the animals have clean quarters and are vaccinated against diseases. Avoid contact with infected animals and take precautions to avoid bites and scratches. Always wash your hands after handling animals.

■ Never Work Alone with an Unfamiliar Animal—Learn how to handle animals before attempting to work with them. You may not be familiar with specific behaviors or dangers. Never work alone with an animal without someone knowing where you are and what you are doing. Never work with an animal without permission from your instructor, supervisor, or its owner.

■ Work in Teams or Groups—When possible, you should have a team or group to work with. At the very least, you should have a partner when working with animals in a laboratory or in a production facility. Accidents happen in the best facilities with the best workers. Always have another person with you just in case!

■ Close Doors You Open—Close all doors and gates that you open. Make sure entrances and exits are secured. Having an animal escape can create big problems, especially if it is a large animal. There are many stories of cattle running through the streets of a town! Take precautions to keep animals restrained.

■ Handle One Kind of Animal at a Time—Work with one type of animal at a time. This keeps you focused on your work and you will be better able to handle the animal. Remember that animals have good senses and can detect the smell of other species easily. Working with mice and then with snakes could end in a bite to your hand!

■ Remain Calm—Animals can sense if you are nervous or excited. Try to remain calm at all times. Do not get frustrated or angry when the animals are not doing what you want. By remaining calm, you will help the animals stay

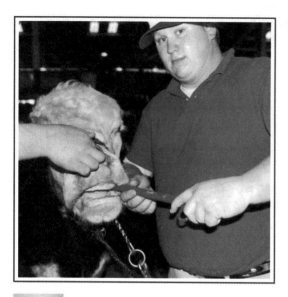

24-15. Animal keepers often do a variety of tasks, such as giving medicine.

at ease. This will make them easier to work with and reduce the possibilities of injuring the animal and you.

■ Be Gentle With Animals—Avoid using excessive force to get animals to cooperate. Most of the time, gentle nudging, encouragement, and patience will far outperform brute force. Using force increases the possibility of injuring the animal and yourself. Do not abuse animals.

■ Get Away from Aggressive Animals—Get away from aggressive animals as soon as possible. Properly report the behavior. There is a reason for animals to become aggressive. Work with someone to find the reason and correct the behavior.

■ Properly Restrain Animals—Make sure that the animals you are treating with medicine or grooming are properly restrained. If you are working with large animals, they should not be able to injure themselves or others by sudden movements. Small animals should also be handled correctly and restrained when given medical treatment or groomed.

PERSONAL SKILLS FOR CAREER SUCCESS

Personal skills help people to be successful on the job. These are the skills in relating to other people. Seven key personal attributes are included here.

■ Have a Positive Life Style—Live a life that speaks well of you. Get plenty of rest. Avoid substances that impair your ability to work. Obey laws and be courteous to others.

■ Get Along with Co-Workers—No one likes to work with troublesome people. They are unpleasant to be around. Make it a point to get along with others. Avoid creating trouble between people.

■ Have a Sense of Humor—Have a sense of humor! Especially if it helps you laugh at yourself when you make mistakes. Being able to see the humor in the moment will make you a more effective employee. Humor is a great personal trait that will add to your value in most careers.

24-16. People often work together as a team, such as this team branding calves.

■ Be Honest—Be honest in all you do every day. Honesty is a virtue that will carry you far in all walks of life. It is especially important in your career. Honesty is measured in what you say, in what you do, and in the promises you keep. It takes much time to gain the trust of others, but you can lose trust in a matter of seconds through dishonest acts.

■ Be Willing to Learn—One of the most important personal traits is the willingness to learn new things. Knowledge keeps our mind alive and keeps our curiosity healthy. Being unwilling to learn new things leads to closed-mindedness, stale ideas, stubbornness, and a decline in knowledge and skill.

■ Eliminate Biased Opinions—Eliminating bias of others is a positive step toward including all people in the decisions we make. Include all people in your decision-making. The more minds working on a problem, the better the solution!

■ Love Your Work—You can often hear people complaining about their work. Work is easy for people who love their job. They look forward to going to work. They like the people and customers they work with and they have a purpose for doing their work. People who love their work are in great demand. They are the best employees. Find a career that you can love.

EMPLOYER EXPECTATIONS OF WORKERS

Employers expect many things from their employees. Simple tasks like coming to work on time or working for a full day are easily identified.

Six employer expectations that can make you more successful at work are listed here.

- Communicate with Others—When employers are asked to list expectations of employees, they often identify communication skills as a top priority. The ability to speak and write, to clearly present your ideas, is necessary in most jobs today. Practice writing and speaking at every opportunity. Become a good communicator in preparation for entering the work force.

- Look Professional—Looking the part is important in all careers. If you went to the dentist's office and were met by a woman who claimed to be a dentist but was wearing cutoff shorts, a tank top, and was missing two front teeth, you probably would want to find another dentist! The woman may be a fine dentist, but your first impression would make you doubt that fact. If you are a veterinarian, a salesperson, or a laboratory technician, you should be committed to looking the part for your job. Employers also expect good personal hygiene habits with your appearance, clothes, and manners.

- Knowledge and Skill—Employers would like to hire people who have the skills they need to do the job. In reality, employers will not be able to find individuals who possess all of the skills they need. Employers hire employees with the expectation that they will learn the knowledge and skills to perform the job. Modern animal production is changing at a rapid pace. Even if employers could hire people with all the skills needed today, next year those skills will be outdated. Prepare yourself with current skills, but be willing and able to learn new skills to keep you on top of your profession.

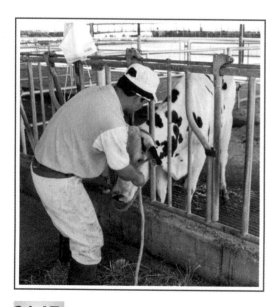

24-17. Employers expect employees to care for the well-being of animals by treating those that are diseased.

- Be a Team Player—Employers expect their workers to be able to get along with each other and be able to operate as a team. Being a team player means that you are able to work with others to accomplish a common goal. You are not a team player if you are concerned only with your personal success and recognition. Team players are happy to be a part of the group and work to support the group's efforts and goals.

- Be Responsible—Being responsible for your own actions is an expectation of most employers. If you make a mistake, admit it. There are few things that will make an employer angrier than an employee who cannot accept responsibility for his or her actions.

- Practice the Work Ethic—Employers expect employees to have and practice a strong work ethic. A strong work ethic is shown by having pride in work, working a full day for a day's pay, and doing the best work you are capable of doing. Never miss work unless it is an emergency or scheduled vacation. Always be at work early and stay on the job. Do not leave work early. A strong work ethic will make the employer happy and will serve customers well. A strong work ethic makes you a valuable employee.

REVIEWING

MAIN IDEAS

Exciting careers are available in the animal industry. There are opportunities in many areas. The meat animal industry provides wholesome food. The pleasure and service animal areas meet important needs. Careers are also found in science-based fields, in animal management and service, and in agricultural communication.

Careers in the meat animal industry include becoming an entrepreneur in animal or poultry production, working for someone on a farm or ranch, or managing a farm or ranch. The pleasure and service animal industries also offer many opportunities that include working with horses, game and fisheries, draft animals, and laboratory animals.

Science-related careers include working in animal laboratories, conducting animal research, discovering new medical treatments, and designing animal facilities. Animal management and service careers include veterinary science, providing inputs for animal production, and marketing animal products. Many people are involved in the people side of the animal industry through education, extension, and agricultural communications.

Safe handling is important when working with animals. You should respect animals and remain calm when working with them. Avoid zoonotic diseases by maintaining animal health and using preventive measures.

There are several employer expectations and personal skills you should develop in yourself to make you a more valuable employee. A good work ethic will carry you toward success.

QUESTIONS

Answer the following questions using correct spelling and complete sentences.

1. What is employment? Gainful employment?
2. What attributes of a person indicate a good work ethic?
3. How is goal setting a part of career planning?
4. What are the responsibilities of a farm manager?
5. Describe one career area in pleasure and service animals.

6. What is the nature of the work of an animal scientist?

7. Why is safety important in working with animals?

8. What safety practices should be considered in working with animals?

9. Name seven personal skills important in job success. Select one of the skills and describe it in detail.

10. What are the expectations employers have of workers? Name six and describe one in detail.

EVALUATING

Match the term with the correct definition. Write the letter by the term in the blank provided.

a. work ethic   c. career   e. entry-level job
b. goal   d. job   f. safety

_____ 1. Preventing injury or loss.

_____ 2. How a person feels about work.

_____ 3. Specific work carried out at a site.

_____ 4. The level of achievement people set for themselves.

_____ 5. The general direction that a person takes with work.

_____ 6. The first employment a person takes.

EXPLORING

1. Interview professionals in the animal industry. Discuss the job description, daily activities, and career preparation. Also, find out what the salary for beginning professionals would be. Prepare a written report that summarizes what you learned.

2. Arrange to discuss post-high school education for career opportunities with your guidance counselor and your agriculture teacher. Take the *AgriScience Interest Inventory* to help assess your interests in a career in animal production. (The Inventory is available from Interstate Publishers, Inc., P. O. Box 50, Danville, IL 61834.)

3. Construct a poster or computer-based presentation of a career that interests you in the animal industry. Share it with the class.

4. Make a wall chart that lists rules for safely handling animals. Post the chart in your school lab.

APPENDIXES

APPENDIX A

Mineral Needs, Functions, and Deficiency Symptoms

Macro-minerals	Function	Deficiency Symptoms
Calcium (Ca)	structural component of the skeleton; controls the excitability of nerves and muscles; required for coagulation of the blood	rickets or osteomalacia develop—the bones become soft and deformed; milk fever occurs in cows; blood clotting time increases
Phosphorus (P)	component of the skeleton providing structural support; aids in lipid transport and metabolism; component of AMP, ADP, and ATP	rickets; depressed appetite; may chew on wood or other objects
Magnesium (Mg)	needed for normal skeletal growth; needed in chemical reactions in muscles; activates enzymes during the Kreb cycle	anorexia; reduced weight gain; reduced magnesium in the blood; hyperemia of the ears and other extremities
Potassium (K)	required to move sodium; helps to maintain acid-base balance in the body; helps to uptake glucose and carbohydrates	slowed growth; unsteady walk; overall muscle weakness; Mg deficiency causes K deficiency
Sodium (Na)	maintains osmotic pressure; maintains acid-base balance; required in transmission of nerve impulses	reduced rate of growth; reduced feed efficiency; decreased milk production; weight loss in adults
Chlorine (Cl)	regulates extracellular osmotic pressure; maintains acid-base balance in the body	slowed growth; fall forward with legs extended backward when startled by a sudden noise
Sulfur (S)	required for protein synthesis; used in cartilage; in birds S is used in feathers, the lining of the gizzard, and in muscle	reduced growth; reduced weight gain
Cobalt (Co)	makes up vitamin B_{12}	loss of appetite; reduced growth; anemia; if untreated, death will result

(Continued)

Appendix A (Continued)

Macro-minerals	Function	Deficiency Symptoms
Iodine (I)	component of thyroxin—controls the oxidation of cells	Goiter—enlargement of the thyroid gland; dry skin; brittle hair; young born without hair; reproductive problems
Zinc (Zn)	activates enzymes; used in DNA and RNA	slowed growth; anorexia; scaling and cracking of paws; poor feathering; rough hair
Iron (Fe)	found in hemoglobin in the blood; found in myoglobin in muscle	anemia—common problem among newborns; pale color; shallow breathing; rough hair; slow growth; iron deficiency affects about half of the world's human population
Copper (Cu)	enzyme activity to uptake iron; needed to maintain the central nervous system	anemia; incoordination
Manganese (Mn)	needed in bone structure	skeletal deformation—enlarged joints, lameness, shortening of legs, and bowing of the legs
Selenium (Se)	component of numerous enzymes	nutritional muscular dystrophy

APPENDIX B

Functions and Deficiency Symptoms of Vitamins

Fat Soluble Vitamins	Function	Deficiency Signs
Vitamin A	required in retinol for night vision; needed in epithelial cells, which cover body surfaces; needed for bone growth	night blindness; dry and irritated eyes; respiratory infection; reproductive problems
Vitamin D	enhanced CA and P levels allowing bone mineralization; prevents tetany	abnormal skeletal development—lameness, bowed, and crooked legs; slowed growth
Vitamin E	promotes health	failure of the reproductive system; changed cell permeability; muscular lesions
Vitamin K	required for blood clotting	long blood clot time; hemorrhages; in severe cases, death
Thiamin (B1)	promotes health	anorexia, beriberi in humans—numbness, weakness, and stiffness in thighs, unsteady walk, edema in feet and legs, and pain along the spine
Riboflavin (B2)	functions in coenzymes	reduced growth rate; skin lesions; hair loss
Niacin	used by cells in energy metabolism	retarded growth; decreased appetite; diarrhea; vomiting; dermatitis
Pantothenic Acid	needed in energy metabolism	slowed growth; dermatitis; graying of the hair; fetal death; skin lesions

(Continued)

Fat Soluble Vitamins	Function	Deficiency Signs
Vitamin B6	help with protein and nitrogen metabolism; involved in formation of red blood cells	convulsions; lesions around feet, face, and ears
Vitamin B12	needed in several enzyme systems	anemia*; retardation; skin pigmentation
Folacin	used in a variety of metabolic reactions	slow growth rate; anemia
Biotin	needed for several enzyme systems	scaly skin; abnormalities of the circulatory system
Choline	aids in transmission of nerve impulses	fatty liver; hemorrhaging kidney
Ascorbic Acid (Vitamin C)	prevents scurvy; causes several metabolic reactions to occur	scurvy—edema, weight loss, and diarrhea

*This can occur in monogastrics fed entirely from plant material. Babies nursing from vegetarian mothers may also develop a vitamin B12 deficiency.

APPENDIX C
Scientific Classifications of Agricultural Animals

DAIRY AND BEEF

a. Kingdom Animalia—animals collectively; the animal kingdom

b. Phylum Chordata—one of the approximately 21 phyla of the animal kingdom, in which there is either a backbone (in the vertebrates) or the rudiment of a backbone, the chorda

c. Class Mammalia—mammals or warm-blooded, hairy animals that produce their young alive and suckle them for a variable period on a secretion from the mammary glands

d. Order Artiodactyla—even-toed, hoofed mammals

e. Family Bovidae—ruminants having polycotyledonary placenta; hollow, nondeciduous, up-branched horns; and nearly universal presence of a gall bladder

f. Genus *Bos*—ruminants quadruped, including wild and domestic cattle, distinguished by a stout body and hollow, curved horns standing out laterally from the skull

g. Species *Bos taurus* and *Bos indicus*—*Bos taurus* includes ancestors of European cattle and of the majority of cattle found in the United States; *Bos indicus* is represented by the humped cattle (Zebu) of India and Africa and the Brahman breed of America

SHEEP

a. Kingdom Animalia

b. Phylum Chordata

c. Class Mammalia

d. Order Artiodactyla

e. Family Bovidae

f. Genus *Ovis*—the genus consisting of the domestic sheep and the majority of wild sheep; horns form a lateral spiral

g. Species *Ovis aries*—domesticated sheep

Swine

a. Kingdom Animalia

b. Phylum Chordata

c. Class Mammalia

d. Order Artiodactyla

e. Family Suidae—the family of nonruminant, ariodactyl ungulates, consisting of wild and domestic swine

f. Genus *Sus*—the typical genus of swine, restricted to the European wild boar and its allies, with the domestic breed derived from them

g. Species *Sus scrofa* and *Sus vittatus*—*Sus scrofa* is a wild hog of continental Europe from which most domestic swine have been derived; *Sus vittatus* was the chief, if not the only, race or species of East Indian pig that contributed to present day domestic swine

Horse

a. Kingdom Animalia

b. Phylum Chordata

c. Class Mammalia

d. Order Perissodactyla—nonruminant, hoofed mammals, usually with an odd number of toes, the third digit the largest, and in line with the axis of the limb; includes the horse, tapir, and rhinoceros

e. Family Equidae—members of the horse family may be distinguished from the other existing perissodactyla (rhinoceros and tapir) by their comparatively more slender and agile build

f. Genus *Equus*—includes horses, asses, and zebras

g. Species *Equus caballus*—the horse is distinguished from asses and zebras by the longer hair on the mane and tail; the presence of the "chestnut" on the inside of the hind leg, and by other less-constant characteristics, such as larger size, larger hooves, more arched neck, smaller head, and shorter ears

Poultry

a. Kingdom Animalia

b. Phylum Chordata

c. Class Aves—those animals having feathers

d. Subclass Neornithes—modern birds (as opposed to fossil birds)

e. Suborder Carinatae—those birds with keel-like breast bones; this superorder has 2,810 genera and 8,616 species; in addition, there are over 25,000 subspecies; the genus-species of domesticated birds are:
    1. chicken—*Gallus domesticus*
    2. duck—*Anas domestica*
    3. goose—*Anser domestica*
    4. guinea fowl—*Numida meleagris*
    5. pigeon—*Columba domestica*
    6. turkey—*Mealeagris gallapavo*

GLOSSARY

Abomasum—the compartment in a ruminant stomach that secretes digestive juices, which kill microbes that have acted on food in preparation for digestion

Accelerated lambing—improving reproductive efficiency with sheep by using out-of-season lambing techniques

Additive—a substance added to feed for a particular purpose, such as to increase vitamin content of the feed

Aeration—physical process of bubbling air into water or splashing the water

Aging—collecting water in an open container and allowing it to stand for volatilization of substances in the water that are harmful to fish

Albumen—the white part of a poultry egg that surrounds the yolk

Allele—different forms of genes

Alveoli—tiny structures in a mammary gland that remove nutrients from the blood and convert the nutrients into milk

Amino acids—the building blocks of protein

Amphibian—animal species that spend part of their life in water and part on land

Anaplasmosis—a parasitic disease, primarily of cattle, caused by a protozoan that attacks the red blood corpuscles

Anatomy—study of the form, shape, and appearance of animals

Anestrous—when a female does not cycle through estrous; nonbreeding season

Animal biotechnology—the application of biotechnology methods to improve animals

Animal by-product tissues—tissues taken from animals when they are used for other purposes, such as from slaughtered cattle

Animal domestication—taking animals from nature and raising them in a controlled environment

Animal health—condition in which an animal is free of disease

Animal industry—all of the activities in raising animals and meeting the needs of people for animal products

Animal marketing—movement of animals from the farm or ranch to the consumer; includes preparing the products for consumption

Animal model—the genetic evaluation for dairy cattle production using pedigree data and production related factors

Animal processing—preparing animals or their products for the consumer; includes slaughtering, grading, fabricating, and other activities

Animal producer—a person who raises livestock, poultry, or other animals

Animal refuge—protected place where animals are cared for to assure their well-being

Animal selection—choosing animals to achieve desired goals, such as improving herd quality

Animal services—assistance provided in animal production from off-farm sources

Animal supplies—materials needed in animal production that are from off-farm sources

Animal tissue cultures—tissues taken from living animals for laboratory study

Animal well-being—caring for animals so their needs are met and they do not suffer

Animal wildlife—any animal that has not been domesticated

Anthelmintic—chemical compounds used for deworming cattle

Anthrax—an acute infectious disease that can affect most endothermic animals

Antibiotic—a substance produced by one organism that will inhibit or kill another organism when used in medicine

Antibody—immune substance produced in the body of an animal

Applied research—research carried out to achieve a specific purpose

Aquacrop—commercially produced aquatic plants, animals, and other species

Aquaculture—the production of aquatic plants, animals, and other species

Aquarium maintenance schedule—a listing of important activities in keeping a good aquarium

Aquarium—container used to hold water for ornamental fish

Arthropoda phylum—animals with exoskeletons, segmented bodies, and jointed legs

Artificial insemination—when semen collected from a male animal is placed in the reproductive tract of a female of the same species

Atom—smallest unit of an element

Automated feeding—mechanical animal feeding

Aviary—a house-like enclosure for birdkeeping

Backgrounding—production system that takes calf from the time of weaning to the feedlot phase; weight is added before going to a feedlot

Bacteria—one-celled organisms that may cause disease; disease examples include tuberculosis, brucellosis, and mastitis

Balanced ration—a ration that contains all the nutrients that the animal needs in the correct proportions

Balling gun—a tool or instrument used for orally administering boluses or capsules to large animals

Barrow—a male hog castrated before sexual maturity

Basic research—research carried out to increase the knowledge and understanding of an area

Beef—the meat of cattle

Behavior—reaction of an organism to a stimulus; how an organism responds to its environment

Billy—male goat; also known as buck

Biodiversity—having a wide variety of members in a population, especially a wildlife population

Biological—a type of medicine used to prevent disease

Biological filtration—using bacteria and other organisms to convert harmful materials in aquaculture water into forms that are less harmful

Biological oxygenation—using phytoplankton (tiny plants) to replenish oxygen in water

Biologist—a person with specialized training in living organisms

Biotechnology—the management of biological systems for the benefit of humans; biology is used to produce new products

Bird cage—a box-like enclosure for birdkeeping

Birding—watching and studying birds, especially song birds

Birdkeeping—raising birds in captivity, especially ornamental birds

Bitch—female dog

Bivalve mollusk—any mollusk species with a hinged shell

Blackleg—an acute, highly infectious disease that primarily affects young cattle; usually results in death

Boar—sexually mature male hog

Bos indicus—cattle of Indian origin; descendants of the zebu and have humps on their necks, large droopy ears, and loose skin; include cattle with Brahman blood

Bos taurus—cattle of European origin; includes common breeds, such as Hereford and Angus

Botany—the study of plants

Brackish water—water that is a mixture of saltwater and freshwater

Brand—a permanent method of cattle identification involving burning a scar on the hide of cattle

Breed—a group of animals of the same species that share common traits

Breeding period—the time of the year when the female has estrus periods and estrous cycle

Breeding—helping animals reproduce and grow so the desired offspring result

Broiler—young meat-type chicken between 5 and 12 weeks of age

Broodfish—sexually mature fish kept for reproduction

Browse—woody and broadleaf plants; forage for goats, deer, and certain other animals

Brucellosis—a disease of the reproductive tract of cattle, sheep, goats, and hogs that causes abortion; also known as Bang's disease

Buck—male goat; also known as billy

Business plan—a written plan on how an enterprise will be operated to help assure success

By-product—the products made from parts of animals that are not used for food

By-product feeds—feeds made from by-products of manufacturing foods, such as pulp from citrus or beets

Cage—a structure for confining aquacrops in water; cages float on the water

Calorie—a measure in nutrition of the amount of heat required to raise the temperature of water 1 degree C

Candling—shining light through an egg to see its features

Capon—castrated male chicken

Carbohydrate—food component that provides energy; includes sugars, starches, and cellulose

Career—general direction that a person takes with work

Carnivore—flesh-eating animal

Carrying capacity—with wildlife, the number of animals a given habitat area can support

Castration—removing or destroying the testicles of a male so it does not breed

Cavy—a guinea pig

Cell—basic unit of life; building blocks for organisms

Cell division—the process of a cell splitting into two cells

Cell specialization—differences in cells that allow them to perform unique activities

Chammy—soft, pliable leather made from the skin of sheep and goats

Chemical filtration—using chemical processes to filter water; activated charcoal is often used for chemical filtration in an aquarium

Chromosome—thread-like parts of a cell nucleus containing the genetic material

Circulatory system—body system that moves nutrients, oxygen, and metabolic wastes

Circus—variety show with animal and human performers

Classification system—a scientific approach used to distinguish between animals of different species

Clinical research—research that takes place in a medical clinic or other laboratory

Cloning—the production of one or more exact genetic copies of an animal

Coccidiosis—a parasitic disease affecting poultry; prevented and treated with anticoccidials in feed and water

Cock—male chicken over one year of age

Cockerel—male chicken less than one year old

Cold housing—using unheated buildings to house animals

Colostrum—the first milk given by an animal after parturition; baby animals need colostrum to help develop disease immunity

Common name—the name used for animals of a species in everyday conversation

Companion animal—a domesticated animal kept by humans for enjoyment in a long-term relationship

Companion fish—ornamental fish kept for human companionship or as pets

Concentrate—a feed high in energy, such as corn, oats, and wheat

Conception—the time when an egg is fertilized to create new life

Confinement method—raising sheep, hogs, and other animals in confined areas; complete environmental control may be used

Conformation—the type, shape, and form of an animal

Connective tissue—tissue that holds and supports body parts

Consumer—people who eat, wear, and otherwise use goods and services

Consumption—the use of products and services that have value

Contagious disease—a disease spread by direct or indirect contact

Contract production—when an agreement is made between a producer and buyer before the hogs or other animals are raised

Copulation—sexual union of a male and female animal

Cow-calf system—a cattle production system that involves raising calves for sale to other growers

Creep feeding—feeding calves supplementally so they grow faster while still nursing

Cria—young llama

Crossbreeding—when animals of the same species, but of a different breed, are mated

Crustacean—animal with an exoskeleton, such as shrimp and lobsters

Culling—removing dairy cows from the herd based on established criteria

Curry comb—oval shaped plastic or metal device use to loosen sweat, manure, and other foreign materials in animal hair

Cutability—the amount of saleable retail cuts that may be obtained from a carcass

Dairy cattle—cattle raised and kept for milk production

Dairy product—milk and any products in which milk is the major ingredient

Dairy Herd Improvement Program—a national dairy production testing and record-keeping program

Debeaking—removing the tips of beaks so birds cannot peck each other

Decision-making—choosing available alternatives, especially as related to a business enterprise

Declawing—cutting the claws from the feet of a cat

Dehorning—removing the horns from horned breeds of cattle

Deoxyribonucleic acid (DNA)—protein-like nucleic acid on genes that controls inheritance

Diet—the type and amount of feed an animal receives in its ration

Digestive system—system that breaks food into smaller parts for use by the body

Disease—disturbance of body functions or structures

Disinfectant—a cleaning substance that destroys the microbial causes of disease

Disposal pit—a pit in the earth for disposing of dead birds; also known as a compost pit; microorganisms decompose the dead birds

Dissolved oxygen—gaseous oxygen that is suspended in water; used by aquatic organisms; measured as DO

Diurnal—animals that sleep during the night and are awake during the day

DNA sequencing—determining the order of nucleotides in a DNA fragment

Docking—cutting of all or part of the tail; with sheep, docking keeps manure from soiling the wool

Doe—female goat; also known as a nanny

Dose syringe—a tool or instrument for orally administering liquid medicine to animals

Dominant trait—trait that covers up or masks the alleles for recessive traits

Down—soft, feathery covering on young poultry

Draft—pulling of an animal to move an object

Draft animal—an animal that has been trained and is used for pulling heavy loads

Draft horse—a horse standing 14.2 to 17.2 hands high and weighing over 1,400 pounds (635 kg)

Drake—mature male duck

Drenching—orally giving liquid medications

Driving horse—a group of horses used for showing and entertaining

Dry cow—a cow that has stopped producing milk; most cows in a herd produce milk 305 days for each lactation

Dual-purpose breed—breed of cattle raised for both meat and milk

Dubbing—removing the comb and wattles on day-old chicks

Duckling—young duck that has not grown feathers

Earthen pond—water impoundments dug into the earth; small ponds are used for ornamental fish

Ectotherm—an animal that adjusts its body temperature to its environment

Egg injection—a method of vaccinating chicken embryos in the egg on the eighteenth day of incubation

Egg—sex cell of the female animal

Egg-laying fish—fish that reproduce by laying eggs

Elastrator—a process of castration and docking used primarily in beef cattle and sheep; it involves placing a rubber band-type device around the scrotum or tail, which cuts off circulation so the scrotum or tail dries up and falls off; a bloodless method of castration and docking

Embryo—a young animal developing in the reproductive tract of a female; from fertilization to the completion of differentiation

Embryo transfer—moving an embryo from the reproductive tract of one female animal to that of another female animal of the same species

Employment—the primary way people use time and earn a living

Endangered species—animals that are close to becoming extinct

Endocrine system—body system that secretes hormones that regulate metabolism, growth, and reproduction

Endotherm—an animal that maintains a constant body temperature; sometimes called warm-blooded

Entrepreneur—person who takes risk by trying to meet the demands of people for a product

Entry-level job—a job that requires little or no previous experience

Environment—the surrounds of an organism

Epithelial tissue—tissue that covers body surface

Equine sleeping sickness—a viral disease of horses, mules, and wild rodents; transmitted by biting insects

Equitation—riding and managing horses

Estrous cycle—time between periods of estrus

Estrous synchronization—bringing a group of female animals into heat at the same time

Estrus—time when the female is receptive to the male and will stand for mating; also known as heat

Ewe—female sheep

Euthanasia—providing easy and painless death; relieving an animal of suffering

Excretory system—system that rids the body of wastes

Exotic animal—an animal that is not native or is being given a new use; exotic animals often are not domesticated

Exotic animal dealer—buyer and seller of exotic animals

Exotic animal investment species—exotic animal that is raised in hopes of making much money; giving a good return on investment

Expressing—manually squeezing the eggs from a female fish; sperm may be squeezed from a male using a similar procedure

External parasite—a parasite that lives on the external parts of its host

Factory farm—a large-scale animal production facility that efficiently produces animals in a controlled environment

Farm flock method—a flock of sheep kept on a farm to produce market lambs

Farrier—a person who cares for the feet of horses, especially the design and fitting of shoes

Farrowing—the process of a sow giving birth to pigs

Fat-soluble vitamin—kinds of vitamins that are stored in body fat and released by the fat as the body needs them

Fats—food component that provides energy; contains 2.25 times more energy than carbohydrates; form in which excess energy is stored by an animal

Feather—long hair on legs of horses

Feed—what an animal eats to get nutrients

Feed analysis—the process of determining the nutrients in a feedstuff or manufactured feed

Feeder pig—a pig that weighs 30 to 60 pounds and is sold to another farm to feed to market weight

Feedstuff—ingredients used in the feeds for animals

Feral animal—an animal that was once domesticated but now lives in the wild

Fertilization—union of sperm and egg

Fetus—a young animal developing in the reproductive tract of a female after the organs have begun to form; the fetus matures for birth

Fiber—the material left after food has been digested

Filly—female horse under three years of age, except for Thoroughbreds which are under four years of age

Filtration—process of removing solid materials from water

Fin—structures on a fish that help with locomotion and balance in the water

Fingerling—an immature fish larger than a fry

Finishing system—beef cattle production system that feeds growing calves for slaughter

Fish—vertebrate aquatic animals

Fledgling—a young bird that has left the nest but cannot feed on its own

Floating—using a file to remove the sharp edges on the teeth of horses

Foal—young horse of either sex that has not been weaned

Foaling—process of a female horse giving birth

Food fish production—raising fish for human use as food

Foot and mouth disease—a viral disease affecting animals with cloven hoofs; no known treatment

Forage—feeds made mostly of leaves and tender stems of plants

Forager—cattle that eat grasses and plants and make good use of these foods in meeting their nutritional needs

Free access—when animals eat all the feed they want; feed is always available

Freshwater—water with little or no salt content

Freshwater ornamental fish—ornamental fish that live in freshwater

Frog—the V-shaped elastic pad in the middle of the sole of the hoof of a horse

Fry—young, newly hatched fish

Functional type—the traits of a cow that are good enough to allow her to complete a useful life in a dairy herd

Fungi—one of five kingdoms of living organisms; fungi are unicellular (one-celled) organisms; some fungi may cause disease

Gaggle—a group of geese

Gainful employment—employment that provides benefits, such as salary or wages

Gait—the way a horse moves (walks, runs, etc.)

Game—animals hunted for sport and food

Gamete—sex cell that can unite to form other cells

Gander—mature male goose

Gelding—male horse castrated before reaching sexual maturity

Gene—segments of chromosomes that contain hereditary traits of organisms

Gene transfer—moving a gene from one organism to another

Genetic code—sequence of nitrogen bases in a DNA molecule

Genetic engineering—a form of biotechnology in which genetic information is changed to make a new product

Genetics—laws and processes of inheritance by offspring from parents

Genome—the genetic material for an organism

Genotype—the genetic makeup of an organism

Gestation—the period of pregnancy

Gilt—a young female hog that has not farrowed

Gizzard—muscular organ inside a bird that grinds food

Glucose—a simple sugar in food that is the source of energy for most cells

Goal—level of achievement that people set for themselves

Goal setting—establishing goals to be achieved and ways and means of reaching them

Gosling—baby goose of either sex

Greenfood—nondried vegetables or plant materials with sap still present, example are cabbage and green peas

Grade cattle—cattle that are not registered; may be purebred or of mixed breed.

Gregarious behavior—instinct of animals to form groups, such as flocks or herds

Grit—finely ground abrasive material fed to birds to promote grinding in their gizzard

Grooming—washing, combing, brushing, trimming, and otherwise caring for a dog or other animal

Growth—increase in size of muscle, bones, and organs of the body of animals

Grubs—internal parasites caused by heel flies; move through the body forming bumps under the skin on the backs of cattle; also known as warbles

Guard dog—a dog trained to protect property

Habitat—place where animal and plant wildlife live

Hairball—wad of hair that collects in the digestive tracts of cats; hair is from licking their coats, known as grooming

Halter broken—an animal that is comfortable being led with a halter and responding to commands of its handler

Hand—a measurement for horses equal to 4 inches (10.2 cm)

Handler—a person who trains and works with K-9 dogs or other animals

Harness—attachments placed on an animal to provide control and allow it to pull

Harvesting—capturing an aquacrop so it can be marketed

Hatchery—place where eggs are incubated for hatching

Health—the condition of an animal; how well the functions of life are performed

Herbivore—an animal that eats foods from plant sources

Herding dog—dog trained to assist in herding sheep and other animals

Heredity—the passing of traits from parents to offspring

Heterozygous—an organism having different alleles for a particular trait

Hog cholera—a viral disease of swine; nearly eradicated in the United States

Homeostasis—a characteristic of animals with organ systems in which a relatively constant internal environment is maintained by the animal

Homogenization—breaking fat droplets in milk into very small particles so they stay in suspension

Homozygous—an organism having similar alleles for a trait

Horsemanship—equitation; riding and managing horses

Hound—dog that tracks; follow scent left by animals or people

Hunting—catching or killing wild animals for sport or food; guns or other weapons are often used

Hunting and jumping horse—a horse specially bred and trained for use in fox hunting

Hutch—a cage or similar structure specially designed for keeping rabbits

Immunity—resistance of an animal to disease

Immunoglobulin—antibodies passed from a cow to her calf in colostrum; provides passive immunity to disease

Implant—small pellet-type material placed underneath the skin for the slow release of medicine, growth regulators, or other substances

Incinerator—a device for burning dead chickens or other animals

Incubation—the time required for the embryo to develop in a fertile egg; with fish it is the time between spawning and hatching

Injection—shot; administering medications directly into the bloodstream, tissues, muscles, or body cavity; normally performed with a hypodermic needle and syringe

Insemination—placing sperm in the reproductive tract of a female

Integumentary system—the skin and other outer covering that protects an animal and helps regulate temperature and body processes

Internal parasite—a parasite that lives inside a host

Invertebrate—animal that does not have a backbone

Isolation—separating diseased and non-diseased animals

Jack—a male donkey of breeding age

Jenny—female donkey

Job—specific work; work carried out at a site

Jog—a two-beat horse gait

K-9—trained police dog

Kid—young goats under a year of age

Kidding—process of a doe (nanny) giving birth

Kingdom Animalia—the major division of living organisms that contains all animals

Kitten—baby cat

Kittening—a female cat giving birth

Laboratory animal—an animal raised and/or used in research laboratories

Lactation—secretion of milk by the mammary glands; the production of milk by a female mammal

Lactation ration—a ration high in nutrients needed by lactating animals

Lamb—a young sheep less than one year old; the meat from a young sheep

Lamb feeding—specialized lamb production where the lambs go to feedlots after weaning

Lambing—the process of a ewe giving birth

Layer—a mature female chicken that is producing eggs; a chicken kept for egg production

Leptospirosis—a bacterial disease of cattle, sheep, and other animals

Lice—species of external biting and sucking parasites of cattle, hogs, and other species

Life science—science dealing with living things

Light horse—a horse standing 14.2 to 17 hands and weighing 900–1,400 pounds (408–635 kg)

Limiting factor—the missing component of a wildlife habitat so wildlife cannot survive

Linear evaluation—coding 15 primary traits of dairy cattle for use in corrective mating

Lipid—a nutrient that can be dissolved with ether (ether is a solvent used in nutrition research)

Litter—the floor covering in a poultry house; also applies to floor coverings for other animals; a group of young animals born at a single birth, such as pigs or kittens

Litter pan—container with absorbent material for pets to use in urinating and defecating

Livebearing fish—fish that reproduce by giving birth to live young

Livestock—animals produced on farms and ranches for food and other products

Living research animal—living animals used in research

Long-haired cat—a class of cat breeds having long hair

Lope—three-beat horse gait

Lymphatic system—body system that circulates a fluid known as lymph; lymph protects the body from disease

Maintenance—no loss or gain in weight by the body

Mammal—a class of animals whose offspring are fed with milk secreted by the mammary glands of the female

Management—using resources to achieve the goals or objectives of an enterprise

Management intensive grazing—a pasture system that concentrates large numbers of dairy cows on small pasture areas to harvest forage by grazing plant leaves and stems; the cattle are moved to another pasture after 12 to 48 hours of grazing

Marbled—intramuscular fat that makes beef more palatable to humans

Mare—mature female horse; over four years of age except for Thoroughbreds, which are over five years of age

Mariculture—producing aquacrops in saltwater

Market hog—young hog raised for meat and weighing about 220 pounds

Mastitis—a bacterial disease of the mammary glands; the udder of cattle may be warm and hard to the touch; causes lost dairy production

Maternity box—a box prepared for giving birth, typically for cat parturition

Mathematical and computer model—carrying out research related to animals without using animals; the models mimic animal behavior

Meat animal—animal raised for meat

Meat animal by-product—products made from the parts of animals not used as food

Meat-type hog—a hog whose carcass gives the greatest amount of lean meat in the areas of high-value cuts

Mechanical filtration—using devices to remove particles from the water and keep the water clear

Mechanical oxygenation—using devices that bubble or splash water to add dissolved oxygen

Medication—a substance used to prevent, control, or treat disease

Meiosis—cell division for sexual reproduction

Metabolic disorder—disease related to nutrition

Microinjection—process of injecting DNA into a cell using a fine-diameter glass needle and microscope; used to produce transgenic animals

Milk fever—a metabolic disorder characterized by low blood calcium and paralysis of cows, primarily in dairy cows

Milking parlor—a concrete platform raised above the floor to make it easier for people to attach milkers to dairy cows and perform other functions on the udder

Mineral—inorganic elements needed for a healthy body

Mitosis—cell division for growth and repair

Mixed-breed dog—dog of unknown ancestry and likely with parents of different breeds

Models for animal research—controls over the use of laboratory animals to assure that the use will contribute to knowledge and not abuse the animal

Mohair—a product from Angora goats used in making clothing

Molecular biotechnology—changing the structure and parts of cells to change an organism

Molecule—smallest unit of a substance

Mollusca phylum—animals with hard shells

Mollusk—aquatic animal with thick, hard shells

Molting—process of poultry shedding and renewing feathers

Mule—hybrid offspring of a male ass (jack) and female horse (mare)

Multi-purpose dog—K-9 dog used for more than one activity

Muscular system—body system that creates bodily movements; acquires materials and energy

Muscular tissue—tissue that creates movement of body parts

Mutation—changes that naturally occur in the genetic material of an organism

Mutton—meat from a sheep more than one year old

Nail bed—the growing point of a nail, such as the nail on a dog's foot

Nanny—female goat; also known as doe

Natural insemination—when a male animal deposits semen in the reproductive tract of a female during copulation

Natural selection—when animals breed without control by humans

Needle teeth—eight sharp teeth in pigs at birth; usually clipped off to prevent injury to the sow

Nervous system—a body system that coordinates body functions and activities

Nervous tissue—tissue that responds to stimuli and transmits nerve impulses

Neutering—the process of sexually altering animals

Niche—the special place in a habitat where every animal fits

Nitrogen cycle—process of converting animal wastes into ammonia, ammonia into nitrites, and nitrites into nitrates; naturally occurs in water

Nocturnal—animals that sleep during the day and are awake during the night

Non-game animal—animal that does not provide products, such as meat and fur

Non-human primate—primates closely resembling humans; often used in research; examples are monkeys and chimpanzees

Non-living research system—using mechanical models that resemble animal activity

Non-sporting dog—dog developed for a specific purpose; primarily used as pets

Noncontagious disease—a disease that is not spread by animal contact; includes nutritional, physiological, and morphological diseases

Nutrient—substance that is necessary for an organism to live and grow

Nutrient dense—a food or feed with large amounts of nutrients relative to calories

Nutrition—food needs and how food is used by an animal; process by which animals eat and use food

Occupation—specific work that can be described and has similar duties in different locations

Omasum—the third compartment in the stomach of a ruminant; filters food materials as they move into the next compartment

Omnivore—an animal that eats foods from both plant and animal sources

Operculum—structure on a fish that covers the gills

Oral medication—medications given through the mouth

Organ—a group of tissues with a similar function

Organ system—a system formed when two or more organs work together to perform an activity

Organism—any living thing

Organismic biotechnology—improving intact or complete organisms without artificially changing the genetic makeup

Ornamental fish—fish kept for their appearance and personal appeal

Orphaned lamb—a lamb whose mother has died

Ovary—primary reproductive organ of the female

Ovulation—when a mature ovum is released by the ovary

Ox—a bovine draft animal

Oxygenation—process of keeping adequate dissolved oxygen in water

Palatability—how well an animal likes food; how food feels in the mouth and on the tongue

Parasite—multicellular animal organisms that live in or on another animal

Particle injection—using a microprojectile unit to shoot tiny particles coated with DNA into cells; a genetic engineering technique

Part-time animal producer—a person who raises animals on a part-time basis; the person may have income from other sources

Parturition—process of giving birth

Pasteurization—heating milk to destroy bacteria

Peacock—male peafowl

Peahen—a female peafowl

Pearson Square Method—a simple way to hand calculate a ration for an animal

Pedigree—record of an individual's heredity

Pelt—the whole hide of an animal with the fur attached

Pen—a structure that confines aquacrops in water; pens are attached to the earth at the bottom of the water

Per capita consumption—the average amount consumed in a year by a person

Performing animal—an animal taught to go through a routine of unusual activity

Permanent pasture—a pasture comprised of perennial grasses and other plants; pasture grows from year to year without planting

Pet carrier—a carrying case to safely and securely carry a pet

Pharmaceutical—medicine used to treat disease

Phenotype—an organism's physical or outward appearance

Physiology—study of the functions of cells, tissues, organs, and systems of an organism

Piglet—baby pigs

Plankton—tiny plants and animals that float in water; serve as food to some aquatic animals

Pleasure animal—animals kept as pets or to use in other ways for the enjoyment of people

Plug—a horse with poor conformation and common breeding

Polled—cattle naturally without horns

Polo mount—a horse especially bred for use in a polo match

Pond—water impoundment made by building earthen dams or levees

Pony—horse standing under 14.2 hands high and 500 to 900 pounds (227–408 kg); small horse

Porcine somatotropin (pST)—a growth hormone for swine

Porcine stress syndrome—a nonpathological disorder in heavily muscled hogs that results in sudden death

Poult—young turkey not grown to the point where its sex is readily identifiable

Poultry—fowl raised on farms for use as food

Poultry science—study and use of science in raising poultry

Power—the combination of pulling capacity and speed; measured in horsepower

Preconditioning—preparing animals for stress associated with hauling

Predator—animal that lives off of other animals; kills, mauls, or preys on them

Predicted transmitting ability—estimate of the traits an animal will transmit to its offspring

Pregnant—condition of a female carrying unborn young

Primate—animal with an opposable thumb used for grasping things

Probability—the likelihood or chance that a trait will occur

Probe—a tool for measuring backfat on a hog

Production cycle—the complete cycle in the production of a crop or growth of an animal

Production intensity—the number or weight of aquatic species in a volume of water; biomass

Progeny—offspring of animals

Prolific—producing large numbers of young or offspring

Protein—food substance important for animal growth, maintenance, reproduction, and other functions

Protoplasm—the mixture of water and proteins in cells that carry out life processes

Protozoa—unicellular organisms; simplest form of animal life; some protozoa may cause disease

Puberty—time when sexual maturity is reached; the animal is capable of reproduction

Pullet—young female chicken

Punnett square—in genetics, a technique for predicting phenotype and genotype

Purebred—animal eligible for registry in a breed association; parents must meet standards of the association

Purebred flock—flock of sheep of purebred breeding

Queen—female cat; also known as pussy

Quill—a modified hollow hair that is stiff and has a point; found on hedgehogs and porcupines

Rabies—a virus disease in endothermic animals accompanied by frenzied behavior, fever, vomiting, diarrhea, drawn lips, and often foaming at mouth; autopsy of brain needed for accurate diagnosis

Racehorse—running or harness horses used in racing

Raceway—water impoundment that uses flowing water

Ram—male sheep kept for breeding purposes

Range band method—production system used with sheep where a herder moves the flock over a large area of land

Rare breed—a formerly popular breed of domesticated animal that is now disappearing

Ration—the food an animal consumes each day; the total amount of feed an animal is given in a 24-hour period

Ratite—a group of large, flightless birds, such as the ostrich and emu

Recessive trait—genetic traits masked by dominant traits

Recombinant DNA—gene splicing; genes are cut and moved to a cell to be altered

Regurgitate—when ruminants return eaten but unchewed food to their mouths for chewing

Reproduction—process by which offspring are produced

Reproduction ration—a ration high in nutrients needed by breeding animals

Reproductive efficiency—timely and prolific replacement of a species

Reproductive system—system by which new individuals of the same species are created

Reptile—an ectothermic animal with dry scaly skin and lungs for breathing; examples are lizards and snakes

Research—a systematic effort to answer questions; carefully planned and designed procedures are followed

Respiratory system—body system that regulates gas exchange of an organism

Reticulum—the second compartment of ruminant stomach; its many layers trap nonfood materials

Riding horse—a horse that is ridden for pleasure

Risk—the possibility of losing; in entrepreneurship, the possibility of losing financial investment

Roan—mixture of white and colored hairs on cattle and horses

Roaster—young chicken somewhat older and larger than a broiler

Rodent—an animal with two large, upper, front teeth for gnawing

Roughage—feedstuffs of stems and leaves; high fiber content

Rumen—the first and largest section of the stomach of a ruminant; receives ingested food materials

Ruminate—when a ruminant chews their cud

Sac fry—newly hatched fish with the yolk sac still attached

Safety—preventing injury or loss by following practices that are wise and prudent

Safety animal—an animal that helps protect people and property

Salinity—the amount of salt in water; measured as ppt (parts per thousand)

Saltwater—water that has a high salt content; typically 33 to 37 ppt salt

Saltwater ornamental fish—ornamental fish that live and grow in saltwater

Sanitation—using practices that prevent development of disease-causing organisms

Scheduled feeding—providing feed at certain times of the day

Science—knowledge about the world we live in

Scientific name—the name of an organism based on its taxonomy; usually two parts to name though some have three parts; written in italics or underlined

Scratching post—a device for cats to use in scratching; exercise post for cats

Scrotum—two-lobed sac in mammals that holds the testicles and helps regulate testicle temperature

Selection—choosing or picking the animals for a herd, especially with dairy cattle

Self-feeding—when animals have feed available all of the time

Service animal—an animal that assists people in living and work

Sexed semen—semen that has been prepared in a lab to produce all male or all female offspring

Sexual reproduction—union of sperm and egg to form a new individual

Shipping fever—an environmental disease of cattle and sheep caused by stress; more likely to affect underfed animals

Shoal—a group of five or more fish living together

Short-haired cat—a class of cat breeds having short hair

Single-purpose dog—K-9 dog used for just one activity

Skeletal system—bony structure that gives a framework for the body

Sleeping sickness—a viral disease of horses transmitted by insect and rodent bites; animals appear sleepy, cannot swallow, and die within two to four days

Smolt—immature salmon larger than a fry

Social ranking—the order that an animal falls within a group of animals of the same species

Solitary—living along; not needing companions

Sow—sexually mature female hog

Spawning—release of eggs by a female fish and subsequent fertilization by a male of the same species

Spawning nest—a place where fish spawn; artificial spawning nests are used in aquaculture

Spaying—surgical procedure that removes the ovaries and uterus of a bitch to prevent breeding; also used with other animals

Specific pathogen free—animals free of disease at birth and raised in an aseptic environment; most often associated with hogs but used with other species, including shrimp

Spent hen—hen no longer used for egg production; processed into cooked poultry products

Sperm—sex cell of the male animal

Sporting dog—dog trained to help hunters find and retrieve game

Spring water—water from natural openings in the earth; similar to well water

Stallion—mature male horse; over four years of age except for Thoroughbreds, which are over five years of age

Steer—a male bovine castrated while young and before secondary sexual characteristics had developed

Stock horse—popular mixed-breed horses often used with cattle

Stomach—muscular organ that stores and prepares food for digestion in the small intestine

Stud dog—a male dog used for breeding

Stud horse—male horse kept specifically for breeding

Superovulation—getting a female to release more than the usual number of eggs during a single estrous cycle

Supplement—a feed material high in specific nutrients; used to assure an animal has sufficient nutrients

Surface runoff—excess water from precipitation

Swim bladder—a small, balloon-like organ that helps a fish maintain buoyancy in the water

Synthetic biology—using chemical substances to create systems with some of the characteristics of living organisms

Synthetic seawater—made for saltwater aquaria by using a commercially available mix

Tail docking—clipping the tail from baby pigs to prevent tail biting

Tapwater—the water from faucets in homes and offices; usually must be treated before using with fish

Taxonomy—science of classifying organisms

Temporary pasture—forages used for grazing one season and replanted

Terrier class—a class of 25 breeds of terriers

Territory—the area an animal defends and protects

Testicle—primary reproductive organ of the male

Thermostat—a control device for maintaining a constant environmental temperature in air or water

Tissue—a cluster of cells that are alike in structure and activity

Tom—male turkey

Tomcat—male cat

Topical medication—a medication placed on the skin or surface of an animal

Total mixed ration—a ration that provides all needed feed ingredients in each mouthful of feed a cow eats

Toy breed—small dogs used for companionship

Training—the process of getting a dog or other animal to have the desired habits or qualities

Transgenic animal—an animal that has stably incorporated a foreign gene into its cells

Trapping—capturing animals for their products using traps that keep the animals alive until detected by the trapper

Tri-purpose animal—an animal used for work, milk, and meat; oxen is an example

Tropical fish—small, brightly-colored fish that are popular in home and office aquaria

Type production index—a system for ranking dairy cows based on overall performance

Udder—mammary glands, teats, and associated structures on female mammals

Ultrasonics—equipment that uses bursts of high-frequency sound in assessing the conditions of animals, including amount of lean meat and reproductive condition

Urbanization—building cities, suburbs, and other developments for the convenience of people; often destroys wildlife habitat

Vat—large tank made of concrete, fiberglass, or similar material for growing fish

Vertebrae—bones that form a segmented spinal column in animals

Vertebrate—animal with a backbone

Vertical integration—when an agribusiness is involved in more than one step in providing agricultural products; chicken industry is a good example

Veterinary medicine—the discipline that deals with the health and well-being of animals

Virus—a tiny particle too small to be seen with an ordinary microscope that often causes disease

Vitamin—organic food substances needed in small amounts for good health; important in body functions

Walk—a four-beat horse gait

Warm housing—heating buildings in which animals are kept in the winter

Water facility—structures in which aquacrops are grown

Water quality—suitability of water for a particular use

Water-soluble vitamins—vitamins dissolved by water and need to be consumed each day

Weaning—withdrawing the need or opportunity for a puppy or other young animal to nurse its mother

Well water—water pumped from aquifers deep in the earth

Wether—male sheep castrated before sexual maturity

Whelp—process of a bitch giving birth

Whelping box—a specially made box for dog birthing

Withdrawal time—period before slaughter during which an animal should not receive a medication or other restricted substance

Wool—soft coat on sheep used as a fiber for human clothing and other products

Work ethic—how people feel about work

Working dog—dog used by people to get work done

Yoke—wooden bar in a harness that holds two animals together so they work together as a team; primarily used with oxen; also a device placed on the necks of animals to keep them in a pasture or pen

Yolk—the center, yellow part of a chicken egg; provides nourishment for a developing embryo

Zoo—zoological garden; park-like area where animals are kept for viewing

Zoo curator—person trained in exotic animal zoology who looks after the animals in a zoo

Zoology—the study of animals

Zoonotic disease—any disease that humans can contract from other animals

Zygote—a fertilized ovum

BIBLIOGRAPHY

Ackefors, Hans, Jay V. Huner, and Mark Konikoff. *Introduction to the General Principles of Aquaculture*. Binghamton, New York: The Hawarth Press, Inc., 1994.

Alderton, David. *You and Your Pet Bird*. New York: Alfred A. Knopf, 1997.

American Kennel Club, The. *American Kennel Club Dog Care and Training*. New York: MacMillan Publishing Company, 1991.

American Kennel Club, The. *The Complete Dog Book*. New York: MacMillan Publishing Company, 1992.

Baker, MeeCee and Robert E. Mikesell. *Animal Science Biology & Technology*. Danville, Illinois: Interstate Publishers, Inc., 1996.

Blasiola, George C. *The New Saltwater Aquarium Handbook*. Hauppauge, New York: Barron's Educational Services, Inc., 1991.

Burton, DeVere. *Ecology of Fish and Wildlife*. Albany, New York: Delmar Publishers, Inc., 1995.

Carlson, Delbert G. and James M. Giffin. *Dog Owner's Home Veterinary Handbook*. New York: Howell Book House, 1992.

Carlson, Delbert G. and James M. Giffin. *Cat Owner's Home Veterinary Handbook*. New York: Howell Book House, 1995.

Cheeke, Peter R. *Impacts of Livestock Production*. Danville, Illinois: Interstate Publishers, Inc., 1993.

Cheeke, Peter R. *Applied Animal Nutrition*, 2nd Edition. Upper Saddle River, New Jersey: Prentice-Hall, Inc., 1999.

Damerow, Gail. *The Chicken Health Handbook*. Pownal, Vermont: School House Books, 1994.

Ensminger, M. E. *Beef Cattle Science*, 7th Edition. Danville, Illinois: Interstate Publishers, Inc., 1997.

Ensminger, M. E. *Horses and Horsemanship*, 7th Edition. Danville, Illinois: Interstate Publishers, Inc., 1999.

Ensminger, M. E. *Animal Science*. Danville, Illinois: Interstate Publishers, Inc., 1991.

Ensminger, M. E. and R. O. Parker. *Swine Science*, 6th Edition. Danville, Illinois: Interstate Publishers, Inc., 1997.

Ensminger, M. E. and R. O. Parker. *Sheep and Goat Science*. Danville, Illinois: Interstate Publishers, Inc., 1986.

Ensminger, M. E. *Poultry Science*. Danville, Illinois: Interstate Publishers, Inc., 1992.

Ensminger, M. E. *Dairy Cattle Science*. Danville, Illinois: Interstate Publishers, Inc., 1993.

Ensminger, M. E., J. E. Oldfield, and W. W. Heinemann. *Feeds & Nutrition Digest*. Clovis, California: Ensminger Publishing Company, 1990.

Gibson, Jerry D. *Equine Science and Management*. Danville, Illinois: Interstate Publishers, Inc., 1999.

Gillespie, James R. *Modern Livestock and Poultry Production*. Albany, New York: Delmar Publishers, Inc., 1992.

Harrington, Rodney B. *Animal Breeding*. Danville, Illinois: Interstate Publishers, Inc., 1995.

Herman, H. A., Jere R. Mitchell, and Gordon A. Doak. *The Artificial Insemination and Embryo Transfer of Dairy and Beef Cattle*. Danville, Illinois: Interstate Publishers, Inc., 1994.

Hieronimus, Harro. *Guppies, Mollies, Platys and other Live-bearers*. Hauppauge, New York: Barron's Educational Series, Inc. 1991.

Holland, Barbara. "An Exotic Creature Makes a Useful Change in Careers," *Smithsonian*. August, 1994.

Lee, Jasper S. and Diana L. Turner. *Introduction to World AgriScience and Technology*, 2nd Edition. Danville, Illinois: Interstate Publishers, Inc., 1997.

Lee, Jasper S. and Michael E. Newman. *Aquaculture: An Introduction*, 2nd Edition. Danville, Illinois: Interstate Publishers, Inc., 1997.

McNitt, James I.; Nephi M. Patton; Peter R. Cheeke; and Steven D. Lukefahr. *Rabbit Production*, 7th Edition. Danville, Illinois: Interstate Publishers, Inc., 1996.

Moody, E. Grant. *Raising Small Animals*. Alexandria Bay, New York: Diamond Farm Enterprises, 1991.

Palika, Liz. *Reptiles and Amphibians*. New York: Alpha Books, 1998.

Prince, Eleanor F. and Gaydell M. Collier. *Basic Horsemanship*. New York: Doubleday, 1993.

Romans, John R., William J. Costello, C. Wendell Carlson, Marion L. Greaser, and Kevin W. Jones. *The Meat We Eat*. Danville, Illinois: Interstate Publishers, Inc., 1994.

Schraer, William D. and Herbert J. Stoltze. *Biology: The Study of Life*. Englewood Cliffs, New Jersey: Prentice Hall, 1995.

Spotte, Stephen. *Marine Aquarium Keeping*. New York: John Wiley & Sons, Inc., 1993.

Stutzenbaker, Charles D.; Brenda J. Scheil; Michael K. Swan; Jasper S. Lee; and Jeri D. Mattics. *Wildlife Management*. Danville, Illinois: Interstate Publishers, Inc., 1999.

Taylor, David. *Small Pet Handbook*. London: Harper Collins Publishers, 1996.

Taylor, Robert E. *Scientific Farm Animal Production*, 5th Edition. Englewood Cliffs, New Jersey: Prentice Hall, 1995.

Warren, Dean M. *Small Animal Care and Management*. Albany, New York: Delmar Publishers, 1995.

Wrobel, Dawn. *The Hedgehog*. New York: Howell Book House, 1997.

____. *Sire Summaries: 1995*. Battleboro, Vermont: Holstein Association, 1995.

INDEX

A

Accelerated lambing, 115
Alligator, 561-562
Amphibians, 475-476
Anatomy, 35
Angelfish, 493
Animal industry, 3
Animal processing, 6
Animal refuge, 568-569
Animal services, 5, 584
Animal supplies, 5, 583-584
Antibiotic, 103
Aquaculture, 335
Aquaria, 494-495, 497-500
Aviary, 464

B

Bag limit, 515
Barbs, 491
Basslets, 493
Beef cattle, 165-202
Biodiversity, 516
Biotechnology, 139-164
Birds, 30, 459-464, 519-520
Bison, 562
Breeding, 114-116
Breeds,
    beef cattle, 171-185
    bird (groups), 459-462
    cats, 444-445
    dairy cattle, 272-277
    dogs, 419-422
    draft horses, 396-398
    goats, 248-251
    horses, 372-375
    sheep, 244-248
    swine, 209-213
Butterfly fish, 493

C

Carbohydrates, 53-54
Careers, 589-608
Carnivore, 60
Castration, 17, 195, 223
Cats, 439-456, 551
    selection, 445-446
Cells, 33-35
Chickens, 303-305
Chocolate, 426
Classification, 27-28
Cloning, 135
Clothing, 8-9
Companion animal, 4, 9-10
Concentrates, 68
Consumer, 577-578
Consumption, animal products, 578-580
Cow-calf system, 187-189
Crustaceans, 344-345

D

Dairy cattle, 7, 265-299
Dairy goats, 248-249
Dairy Herd Improvement Program, 278-279

Declaw, 443
Dehorning, 194-195
Digestive systems, 61-63
Disease, 83-84, 89-98,
    228-229, 258-260, 329
DNA, 148-150
Dogs, 415-438, 551-552
    selection, 423-424
Domestication, 12
Draft animals, 393-414
Ducks, 306-307

E

Egg production, 322-325
Elk, 562-563
Embryo transfer, 134-135,
    156-158
Employer expectations,
    605-606
Employment, 591
Entrepreneurship, 575-588
Environment, 17
Equitation, 387-391
Estrous cycle, 121-124
Exotic animals, 559-574

F

Facilities,
    beef cattle, 196-198
    dairy cattle, 294-298
    dogs, 430431
    draft animals, 408-412
    horses, 385-387
    poultry, 318-322
    sheep, 260-262
    swine, 229-233
Factory farm, 11, 207
Farrowing, 221-222
Fats, 54-55
Feed, 65-70, 379-381
Feeder pigs, 217, 224-225

Ferrets, 479-480
FFA, 599, 601
Fish, 30, 333-366
    game, 518-519
    ornamental, 485-509

G

Game, 513-514, 516
Geese, 307
Genetic engineering, 136,
    152-155
Genetics, 142-147
Gerbils, 465-466
Goats, 235-264
Goldfish, 488
Gouramis, 489
Greenfood, 463
Grooming
    cats, 447-449
    dogs, 431-435
Guinea pigs, 472-473, 549
Guppies, 492

H

Habitat, wildlife, 520-522
Hamsters, 467-469, 549
Hatcheries, fish, 352-354
Health, 16, 81-110,
    189-191, 214-215,
    291-293, 381-383
    ornamental fish, 507-508
Hedgehogs, 481-483
Herbivore, 60
Hoof care, 384-385
Horses, 367-392
    types, 372-375
Horse worker, 595

I

Identification, 191-194
Immunization, 101-108,
    223, 327
Incubation, 128-129, 315
Injections, 103-106
Insemination,
    artificial, 114, 130-134,
    219
    natural, 113-114, 219

J

Job, 593

K

Kidding, 243-244
Kittening, 450-451
Koi, 488

L

Laboratory animals, 543-
    558
Lactation, 64, 127,
    288-289
Lamb, 238
Lambing, 255
Life processes, 32
Linear evaluation, 280
Litter, cat, 454
Livestock, 3
Llama, 561

M

Mammals, 31-32
Management, 586-587
Marketing, 6, 585-586

Maternity box, 451
Meat animal, 7
Medicine, 103-108
Mice, 469-471, 548
Minerals, 57-58
Mollusks, 345-346
Mollies, 492
Molting, 325
Mules, 398-399
Mutation, 151

N

Names, 29
Neutering, 17
Nutrition, 15-16, 49-80,
    225-226, 256-258,
    285-286, 358-360

O

Omnivore, 61
Organ systems, 39-46
Oxen, 395, 401, 404-405

P

Parasites, 90-98, 227-228,
    260, 329-330
Pasteurization, 267-268
Pasture, 66-68
Personal skills, 604-605
Pet carrier, 453
Physiology, 35
Platys, 492
Ponds, aquaculture,
    346-347
Poultry, 4, 301-332
Preconditioning, 101
Predators, 362
Primates, 552-553
Protein, 55-56

Puppies
    feeding, 428
    orphaned, 426-427
    weaning, 427-428

R

Rabbits, 476-479, 550-551
Raceways, 347
Ranch worker, 594
Rations, 51-52, 69-70,
    72-77, 285
Ratites, 309–310
Rats, 469-471, 549
Reproduction, 16-17, 64,
    111-138, 282-284,
    314-315
Reproductive,
    development, 124-127
    systems, 116-121
Reptiles, 474-475
Research, 545-547
Rodents, 464-474

S

Safety, 602-604
Sanitation, 100, 326-327,
    435-436
Science, 13
Seining, 362-365
Sheep, 235-264
Showing, 198-201
Semen, sexed, 133-134
Spawning, 344, 350-352
Spaying, 17
Specific pathogen free,
    214-215
Superovulation, 156-157
Swim bladder, 344
Swine, 203-234
Swordtails, 492
Synthetic biology, 142

T

Tarantula, 482
Taxonomy, 27
Technology, 14-15
Teeth, care
    horses, 383-384
Tetras, 491
Turkeys, 305-306

U

Udder, 271-272

V

Vertebrate, 28
Vertical integration,
    312-313
Vitamins, 58-59

W

Wastes, 20-21
Water
    environments, 339-341,
        501-505
    management, 356-358
Well-being, 19-20
Whelping, 425
Wildlife, 511-542
Withdrawal time, 226-227
Wool, 239
Work ethic, 591-592

Z

Zoo, 566-567
Zoonotic disease, 602-603